The Handbook of Environmental Chemistry

Volume 3 Part C

Edited by O. Hutzinger

Anthropogenic Compounds

With Contributions by
E. Atlas, L. Fishbein, C. S. Giam, J. E. Leonard,
D. C. G. Muir, M. A. Powers, Jr., J. Schoer

With 31 Figures

Springer-Verlag
Berlin Heidelberg GmbH 1984

Professor Dr. Otto Hutzinger
University of Bayreuth
Chair of Ecological Chemistry and Geochemistry
Postfach 3008, D-8580 Bayreuth
Federal Republic of Germany

ISBN 978-3-662-15233-1

Library of Congress Cataloging in Publication Data
Main entry under title: The Handbook of environmental chemistry.
Includes bibliographies and indexes.
Contents: v. 1, pts. A–B. The natural environment and the biogeochemical cycles/with contributions by P. Craig ... [et al.] – v. 2, pts. A–B. Reactions and processes/with contributions by W. A. Bruggeman ... [et al.] – v. 3, pts. A–C. Anthropogenic compounds/with contributions by R. Anliker ... [et al.]
1. Environmental chemistry – Collected works. I. Hutzinger, O.
QD31.H335 [QH545.A1] 574.5′222 80-16607
ISBN 978-3-662-15233-1 ISBN 978-3-540-38819-7 (eBook)
DOI 10.1007/978-3-540-38819-7

Originally published by Springer-Verlag Berlin Heidelberg New York Tokyo in 1984
Softcover reprint of the hardcover 1st edition 1984

2154/3140-543210

Preface

Environmental Chemistry is a relatively young science. Interest in this subject, however, is growing very rapidly and, although no agreement has been reached as yet about the exact content and limits of this interdisciplinary discipline, there appears to be increasing interest in seeing environmental topics which are based on chemistry embodied in this subject. One of the first objectives of Environmental Chemistry must be the study of the environment and of natural chemical processes which occur in the environment. A major purpose of this series on Environmental Chemistry, therefore, is to present a reasonably uniform view of various aspects of the chemistry of the environment and chemical reactions occurring in the environment.

The industrial activities of man have given a new dimension to Environmental Chemistry. We have now synthesized and described over five million chemical compounds and chemical industry produces about hundred and fifty million tons of synthetic chemicals annually. We ship billions of tons of oil per year and through mining operations and other geophysical modifications, large quantities of inorganic and organic materials are released from their natural deposits. Cities and metropolitan areas of up to 15 million inhabitants produce large quantities of waste in relatively small and confined areas. Much of the chemical products and waste products of modern society are released into the environment either during production, storage, transport, use or ultimate disposal. These released materials participate in natural cycles and reactions and frequently lead to interference and disturbance of natural systems.

Environmental Chemistry is concerned with *reactions in the environment*. It is about distribution and equilibria between environmental compartments. It is about reactions, pathways, thermodynamics and kinetics. An important purpose of this Handbook is to aid understanding of the basic distribution and chemical reaction processes which occur in the environment.

Laws regulating toxic substances in various countries are designed to assess and control risk of chemicals to man and his environment. Science can contribute in two areas to this assessment; firstly in the area of toxicology and secondly in the area of chemical exposure. The available concentration ("environmental exposure concentration") depends on the fate of chemical compounds in the environment and thus their distribution and reaction behaviour in the environment. One very important contribution of Environmental Chemistry to the above mentioned toxic substances laws is to develop laboratory test methods, or mathematical correlations and models that predict the environmental fate of new chemical compounds. The third purpose of this Handbook is to help in the basic understanding and development of such test methods and models.

The last explicit purpose of the Handbook is to present, in concise form, the most important properties relating to environmental chemistry and hazard assessment for the most important series of chemical compounds.

At the moment three volumes of the Handbook are planned. Volume 1 deals with the natural environment and the biogeochemical cycles therein, including some background information such as energetics and ecology. Volume 2 is concerned with reactions and processes in the environment and deals with physical factors such as transport and adsorption, and chemical, photochemical and biochemical reactions in the environment, as well as some aspects of pharmacokinetics and metabolism within organisms. Volume 3 deals with anthropogenic compounds, their chemical backgrounds, production methods and information about their use, their environmental behaviour, analytical methodology and some important aspects of their toxic effects. The material for volume 1, 2 and 3 was each more than could easily be fitted into a single volume, and for this reason, as well as for the purpose of rapid publication of available manuscripts, all three volumes were divided in the parts A and B. Publisher and editor hope to keep materials of the volumes one to three up to date and to extend coverage in the subject areas by publishing further parts in the future. Readers are encouraged to offer suggestions and advice as to future editions of "The Handbook of Environmental Chemistry".

Most chapters in the Handbook are written to a fairly advanced level and should be of interest to the graduate student and practising scientist. I also hope that the subject matter treated will be of interest to people outside chemistry and to scientists in industry as well as government and regulatory bodies. It would be very satisfying for me to see the books used as a basis for developing graduate courses on Environmental Chemistry.

Due to the breadth of the subject matter, it was not easy to edit this Handbook. Specialists had to be found in quite different areas of science who were willing to contribute a chapter within the prescribed schedule. It is with great satisfaction that I thank all 52 authors from 8 countries for their understanding and for devoting their time to this effort. Special thanks are due to Dr. F. Boschke of Springer for his advice and discussions throughout all stages of preparation of the Handbook. Mrs. A. Heinrich of Springer has significantly contributed to the technical development of the book through her conscientious and efficient work. Finally I like to thank my family, students and colleagues for being so patient with me during several critical phases of preparation for the Handbook, and to some colleagues and the secretaries for technical help.

I consider it a privilege to see my chosen subject grow. My interest in Environmental Chemistry dates back to my early college days in Vienna. I received significant impulses during my postdoctoral period at the University of California and my interest slowly developed during my time with the

National Research Council of Canada, before I could devote my full time to Environmental Chemistry, here in Amsterdam. I hope this Handbook may help deepen the interest of other scientists in this subject.

O. Hutzinger

Preface to Parts C of the Handbook

Parts C of the three series
- The Natural Environment and the Biogeochemical Cycles (Vol. 1)
- Reactions and Processes (Vol. 2)
- Anthropogenic Compounds (Vol. 3)

are now either available or in press. During their preparation it became obvious that further parts will have to follow to present the respective subject matters in reasonably complete form.

The publisher and editor have further agreed to expand the Handbook by three new series: Air Pollution, Water Pollution and Environmental Trace Analysis.

Again, I thank all authors as well as collaborators at the Springer Publishing House for their cooperation and help. Thanks are also due to many environmental chemists and reviewers in particular for their critical comments and their positive reception of the Handbook.

Bayreuth, December 1983 Otto Hutzinger

Contents

Aromatic Amines

L. Fishbein

Phosphate Esters

D.C.G. Muir

Phthalic Acid Esters

C.S. Giam, E. Atlas, M.A. Powers, Jr., and J.E. Leonard

Thallium

J. Schoer

List of Contributors

Dr. Elliot Atlas
Department of Chemistry
Texas A & M University
College Station,
TX 77843, USA

Dr. Lawrence Fishbein
Department of Health
and Human Services
Food and Drug Administration
National Center
for Toxicological Research
Jefferson, AR 72079, USA

Dr. C.S. Giam
Department of Chemistry
Texas A & M University
College Station,
TX 77843, USA

Dr. Jack E. Leonard
Department of Chemistry
Texas A & M University
College Station, TX 77843, USA

Dr. Derek C.G. Muir
Freshwater Institute
Department of Fisheries and Oceans
Winnipeg, Canada R3T 2N6

Dr. Mack A. Powers, Jr.
Department of Chemistry
Texas A & M University
College Station, TX 77843, USA

Jürgen Schoer
Institut für Sedimentforschung
Universität Heidelberg
D-6900 Heidelberg
Federal Republic of Germany

Volume 3, Part A: **Anthropogenic Compounds**

Volume 3, Part B: **Anthropogenic Compounds**

Aromatic Amines

L. Fishbein

Department of Health and Human Services
Food and Drug Administration
National Center for Toxicological Research
Jefferson, AR 72079, USA

Introduction

Aromatic amines represent a category of chemical agents of considerable importance as withnessed by their widespread use as intermediates in the manufacture of drugs, pesticides and plastics, as antioxidants in the preparation of rubber for the manufacture of tires and cables and as curing agents in the preparation of various plastics. In addition, they are widely used as intermediates in the preparation of dyes and pigments extensively employed to color textiles, leathers, rubber, printing inks, paints, lacquers, metal finishes, plastic and paper products, as well as in semi-permament coloring products [1–9].

Although aromatic amines can be acutely toxic, causing methemoglobinemia, acute hemolysis and dermatitis, the major health hazard to man is cancer of the urothelial tract, principally the urinary bladder, as a result of occupational exposure. Hence, particular concern is the growing realization that the epidemiology of aromatic amine carcinogenesis is essentially the epidemiology of human bladder cancer of industrial origins. The eight occupational categories in which increased bladder cancer incidence may be related to aromatic amine exposure are: chemical and dyestuff industries, textile dyeing and printing, pigment manufacture, rubber and cable making, gas workers, laboratory workers, rodent control, and patent fuel tar and pitch workers [8].

The most important aromatic amines from the standpoint of industrial epidemiology are benzidine and its congeners [e. g., 3,3′-dichlorobenzidine, 3,3′-dimethylbenzidine (o-tolidine), 3,3′-dimethoxybenzidine (o-dianisidine) and their derived dyes], 1- and 2-naphthylamines, 4-aminobiphenyl and diphenylamine [4–17]. Exposure to these agents, a number of which are carcinogenic and/or mutagenic [4–23] can occur via skin contact, inhalation or ingestion. The latent period between entry into work with an occupational carcinogen and development of bladder cancer is lengthy (18–22 years).

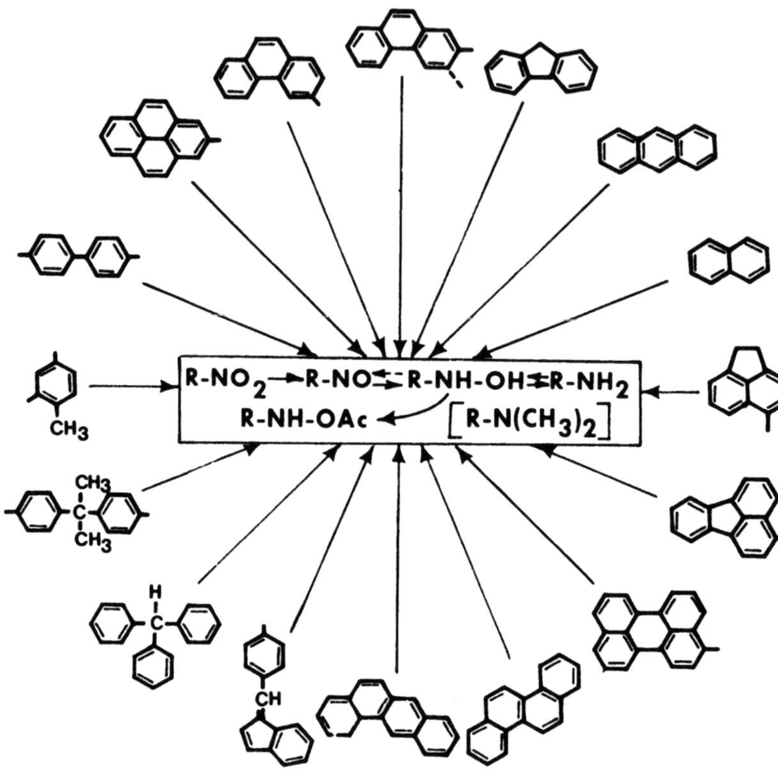

Fig. 1. Spectrum of typical hydrocarbons moieties present in carcinogenic aromatic amines

Figure 1 illustrates a spectrum of typical hydrocarbon moieties present in carcinogenic aromatic amines [23]. Metabolic activation of the carcinogenic aromatic amines proceeds by pathways involving either an initial hydroxylation of the amine group followed by further enzyme-mediated reactions [10, 11, 24–30], or ring-hydroxylation *ortho* to the amine [31–33]. A number of these activated aromatic amines bind covalently with nucleic acids and proteins [10, 24–30, 34–37]. It should also be noted that aromatic amines are metabolized to a wide variety of intermediate compounds via N- and ring oxidation as well as N-acetylation. The broad utility of the aromatic amines moreover, create the potential for their wide distribution in aquatic and terrestrial ecosystems. Additionally, not only are segments of the labor force directly exposed to these agents, but consumers, as well, may be in contact with derived dyes and unreacted aromatic amine precursors in clothing and through the use of packaged retail products, such as dyes for home use, art and craft supplies and hair dyes [38].

Other sources of aromatic amines are known to exist although their toxicological significance is, in many cases, as yet uncertain. For example, polycyclic aromatic amines are found in coal-derived petroleum substitutes (synfuels) [38–43], natural and synthetic crudes [44] and have been found to be the determinant mutagens in some coal derived materials [41, 45, 46]. These species include recognized or suspected carcinogens such as 1-aminonaphthalene, 2-aminonaphthalene, 4-aminobiphenyl, 1-aminoanthracene and 2-aminoanthracene [33, 47, 48].

Heterocyclic aromatic amines are produced by pyrolysis of proteins or amino acids [49, 50] or as a consequence of drying, boiling or broiling beef and fish [51–53]. Aromatic amines are also found in cigarette smoke [54, 55] to which large segments of the general public are exposed.

In contrast to classes of agents such as the metals and metalloids, aromatic hydrocarbons, chlorinated hydrocarbons etc., there is a paucity of information relating to the environmental fate of most members of the aromatic amines as a class. Accordingly, the principal objectives of this chapter are to focus primarily on a select number of the major aromatic amines: aniline, benzidine, 3,3′-dichlorobenzidine (and azo dyes and pigments derived from these amines), naphthylamines and 4,4′-methylenebis(2-chloroaniline) (MOCA) where there are more substantive data on their production, utility and environmental fate. Selected methods of analysis for some aromatic amines and azo dyes in the general and industrial environment have been reported [56–58] and aspects of environmental and metabolic transformations of primary aromatic amines and related compounds were recently reviewed [59].

Aniline

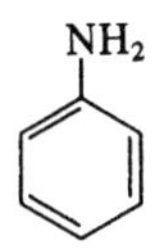

Aniline (and its hydrochloride and sulfate salts) occupy a position of singular importance among the aromatic amines because of its broad spectrum of utility *per se* as well as potential environmental significance resulting from these applications as well as due to its formation from N-phenyl carbamate and N-phenyl urea herbicides.

Production and Use

Aniline was produced for many years by the Bechamp reduction of nitrobenzene with iron and hydrochloric acid and until 1966, when amination of chlorbenzene with ammonia was used [1, 5]. Currently, aniline is produced in the United States, in several western European countries, and Japan by the catalytic hydrogenation of nitrobenzene carried out either in the vapor phase or in a solvent system [1, 5].

The total US production of aniline by eight companies in 1978 and 1979 amounted to 275 and 313,000 t respectively [5]. The annual capacity for aniline production in western Europe was approximately 500,000 t in 1979 with the U.K. and Federal Republic of Germany being the largest producing countries. Aniline is believed to be produced by three companies in Spain, two companies each in Belgium, Federal Republic of Germany and the U.K. and by one company each in France, Italy, Portugal and Switzerland [5]. The total production of aniline in Japan by four companies in 1979 was estimated to have been 79,100 t [5].

About 300 chemical products and intermediates are currently manufactured from aniline. While some are produced by a single process, many others, principally dyes and pharmaceuticals require several steps [1]. The major uses of aniline are in the polymer, rubber, agricultural and dye industries. The demand in the dye industry has decreased markedly in the US in the 1970's because of the increased use of synthetic fabrics and stricter controls by the Food and Drug Administration [1]. Although over 170 dyes can be made drom aniline and over 700 dyes prepared from aniline derivatives [5, 60], it should be noted that very few of these are now produced commercially in significant quantities [1]. Of 33 typical dyes derived from aniline [61], only 16 were reported to be produced commercially in the US in 1979. Those produced in the largest volumes were: Basic Orange 2 (275 t), Solvent Yellow 14 (150 t), Acid Red 1 (128 t), and Acid Black 1 (169 t) [5].

Aniline was used in the US in 1979 as follows: approximately 50% for production of 4,4'-methylenediphenyldiisocyanate (MDI) and polymethylene polyphenylisocyanate (so-called polymeric MDI), 27% for production of rubber chemicals, 5% for hydroquinone production, 3% for production of drugs and 9% for miscellaneous applications, including production of herbicides and fibers [5, 62].

Use of aniline in western Europe in 1979 was estimated to have been as follows: 60–65% for production of isocyanates (MDI and polymeric MDI), 20–22% for production of rubber chemicals, 9–10% for synthesis of dyes and intermediates and 4–10% for miscellaneous uses [5].

Approximately 50% of the aniline consumed in Japan in 1979 was for the production of isocyanates (principally MDI), 30% to produce rubber chemicals, 10% for dye manufacture, and 10% for miscellaneous applications including the production of pharmaceuticals [5].

Aniline condenses with formaldehyde to yield methylene dianilines and polymers which on phosgenation yield methylenebis(phenylisocyanates) and the polymeric isocyanates. The pure diisocyanate [MDI; 1,1'-methylenebis(4-isocyanatobenzene)] is used for the production of urethane elastomers [1]. The US production of polymeric MDI by three companies in 1979 was 221,000 t [5].

The rubber-processing agents produced from aniline include: antioxidants, antidegradants and vulcanization accelerators including aldehyde-aniline condensates (e.g., the n-butyraldehyde condensate), guanidines (e.g., 1,3-diphenylguanidine), thiazoles (e.g., 2,2'-dithiobisbenzothiazole and other derivatives of 2-mercaptobenzothiazole), and a variety of derivatives of diphenylamine [1, 5].

In the pharmaceutical industry, aniline is used in the manufacture of acetanilide (antifebrin) which is also used in the production of several sulfa drugs including sulfafurazole, sulfamethoxazole, and phenazopyridine [1, 5].

Metabolism and Toxicity

The mammalian metabolism of aniline results in a large number of products including: ortho-, para- and meta-aminophenols, acetanilide, ortho-hydroxyacetanilide, para-hydroxyacetanilide, aniline glucuronide, N-phenylhydroxylamine (N-hydroxyaniline), nitrosobenzene, in addition to glucuronic, sulfuric and mercapturic acid conjugates [5, 17, 63–70].

Metabolic conversion of aniline to urinary conjugates of para-aminophenol has been observed in man [64] and the urinary excretion of these metabolites has been found to reflect the absorption of aniline vapor through the skin and respiratory tract [71–74]. The methaemoglobinaemia produced in humans by aniline is believed to result from its N-hydroxylation [5].

Earlier studies indicated that the rat, mouse, rabbit, guinea pig, gerbil, hamster, cat and dog oxidize aniline to ortho- and para-aminophenol, which are excreted in the urine as various conjugates [63, 64]. The ratios of these isomers in the urine differed widely in the various species. Small amounts of free aniline, phenyl sulphamic acid and aniline-N-glucuronide were also found in the urine of some species after administration of aniline [63, 65], while meta-aminophenol was detected in trace amounts in the urine of dogs and rabbits [63]. Ortho-, and para-aminophenyl and para-acetylamino-phenylmercapturic acid are also excreted in rats and para-acetylaminophenol and para-acetylaminophenylmercapturic acid in rabbits [66].

Acetanilide has been detected in the urine of rabbits but not in the urine of dogs [64]. More recent studies of the metabolism of orally administered aniline (50 mg(kg) in rats, pigs, and sheep indicated that contrary to previous reports, N-acetylated derivatives were identified as the major 24 h urinary metabolites of aniline representing 76%, 85% and 82% of the urinary metabolites from the rat, pig and sheep respectively [70]. The double conjugate, N-acetyl-o-aminophenyl glucuronide was the major metabolite in the sheep and pig (60% and 66% respectively), whereas N-acetyl-p-aminophenyl sulfate was the major metabolite in the rat (56%). Minor urinary metabolites of aniline from these species included: O-conjugates of ortho- and para-aminophenol (ca 20%), acetanilide (ca 3%) and N-acetyl-p-aminophenol (ca 10%). N-Glucuronides and sulfates of aniline and free aniline were not detected as urinary metabolites.

While no phenyl hydroxylamine (N-hydroxyaniline) has been detected in the urine of animals treated with aniline, both phenylhydroxylamine and nitrosobenzene have been found in the blood of treated dogs and cats [5, 67]. The formation of phenylhydroxylamine from aniline appear to be the major cause of the methaemoglobinaemia that follows its administration [5, 67]. The N-hydroxylation of aniline by hepatic microsomal preparations from several species has been observed in vitro [67]. The proposed metabolic pathway of aniline in the rat is shown in Fig. 2 [17].

Single oral doses of 25–65 mg/person of aniline caused a dose-dependent increase in methaemoglobin formation while doses of 45–65 mg/person also produced a slight increase in serum bilirubin in several subjects [255]. There is limited evidence for the carcinogenicity of aniline hydrochloride in experimental animals [5]. Aniline hydrochloride tested by oral administration in rats in one experiment produced fibrosarcoma, sarcomas and haemangiosarcomas of the spleen or the peritoneal cavity. When tested in mice by dietary administration, aniline hydrochloride produced no carcinogenic effects. Aniline was inactive in bacterial and mammalian DNA repair assays, in tests for mitotic recombination with yeast and in cell transformation assays. It induced sister chromatid exchanges in cultured mammalian cells [5].

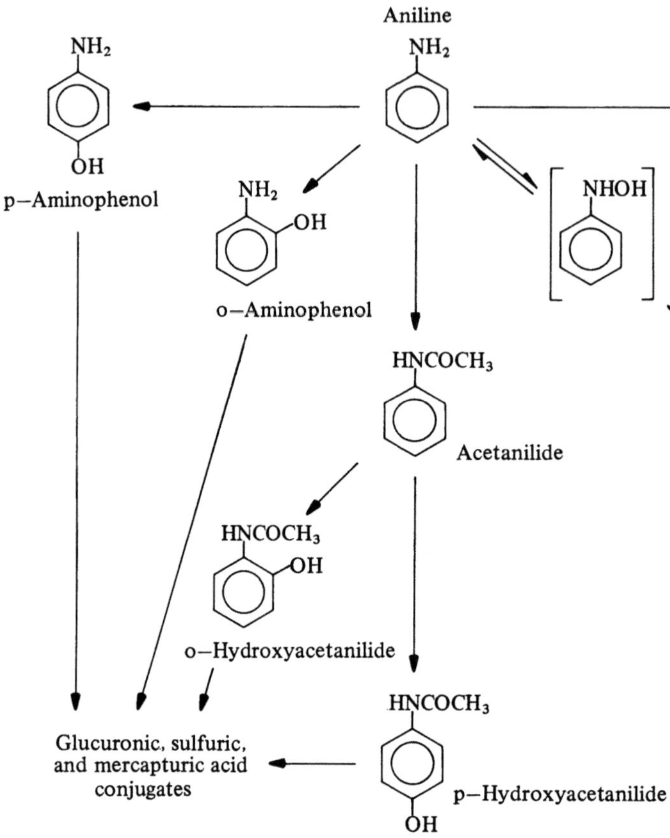

Fig. 2. Proposed metabolic pathway of aniline in the rat

Occurrence and Reactions

Aniline and a number of substituted anilines are introduced into the environment directly as industrial waste and indirectly by the combustion of plastics and urethane products or as reduced form of nitrobenzene products [75, 76]. The degradation of acyl anilide, N-phenylcarbamate and N-phenyl urea herbicides in soils [78–84], environmental water systems, and humans [81] is the most common source of aniline and its congeners [75]. Although aniline biodegradation products have no herbicidal properties [68], they are more toxic than the parent compounds and persist in the environment strongly bound to soil organic matter [80, 81].

(a) Air

Aniline undergoes partial air oxidation to such products as hydrazobenzene, azobenzene, azoxybenzene and phenazine [85]. The effect of aniline on photochemical reactions in atmospheric and synthetic air sample was reported [86]. Whereas aniline concentration of 0.81 ppm had negligible effect in a light traffic air sample containing negligible NO, 20 ppm in a heavy traffic sample caused a 70% decrease in the oxidant dosage, a 50% decrease in the NO_2 peak and a four-fold increase in the NO half-life. The mechanism of this inhibition is probably more than just hydrogen abstraction by hydroxyl radicals as indicated:

$$C_6H_5NH_2 + OH \rightarrow H_2O + C_6H_5NH \quad (1)$$

Further reactions which were proposed [86] to explain the above experimental results were:

$$C_6H_5NH_2 + O \rightarrow OH + C_6H_5NH \quad (2)$$

$$C_6H_5NH_2 + O \rightarrow C_6H_5NHOH \quad (3)$$

$$C_6H_5NH + NO_2 \rightarrow NH_2C_6H_4NO_2 \quad (4)$$

Hydrogen abstraction by reaction (2) is similar to that of reaction (2) and oxidation in reaction (3) is possible due to the unshared pair of electrons on the nitrogen atom of aniline. The hydroxyl radical produced by reaction (2) is scavenged by aniline in reaction (1). Scavenging of hydroxyl radicals and oxygen atoms by aniline would lead to retardation of hydrocarbon consumption since O, OH, and O_3 are the only species responsible for hydrocarbon oxidation, with O und OH being the most important one [86].

Aniline was earlier found to be the best inhibitor of those studied for the oxidation of NO [87, 88]. In synthetic automobile exhaust systems, aniline was found to inhibit ozone formation [89].

Aniline has been found in tobacco smoke at concentrations of 50.6 – 577 ng/cigarette [90].

(b) Water and Sediments

As noted earlier, since aniline is extensively used in the production of many products including dyestuffs, drugs, photographic chemicals, resins, and varnishes, there are many opportunities for aniline to enter into industrial effluents or wastewaters [91]. Once aniline is in those waters it can act as a pollutant in two ways. First, because it is biodegradable [92, 93], it exerts a carbonaceous oxygen demand that must be met; secondly, since it contains nitrogen in excess of the amount required in cell synthesis, ammonia nitrogen will be the end product of its biodegradation [91].

Aniline has been found in surface water samples taken from some rivers in the Federal Republic of Germany in concentrations of 0.5–3.7 μg/l [94] as well as in the rivers in the Netherlands, both in the Waal River [95] and in the Rhine delta [96]. In the River Rhine, aniline, 2-, 3-, and 4-monochloroaniline and 3,4-dichloroaniline were frequently detected and in 1979 their mean values were 3.0, 0.52, 0.22, and 0.40 μg/l respectively [97]. Microgram levels of aniline in river water and soil samples in the Federal Republic of Germany were attributed to industrial chemical waste [98].

In the US aniline has been identified at unspecified levels in surface water, finished drinking water and river water as well as in industrial effluents from oil shale recovery and oil refineries and from chemical and coal conversion plants [5, 99–101].

Aniline was found in composite samples of raw wastewater and final effluent (at levels of 36–480 ppb and 10–96 ppb respectively) from a dye manufacturing plant near the Back River in Massachusetts [101]. In addition a large variety of aniline derivatives were found in the above composite samples including: bromo, chloro-, dichloro-, dibromo-, tribromo-, nitro-, bromonitro-, and methylmethoxy anilines.

(c) Degradation and Biotransformation

Literature on the chemical degradation of organic chemicals in ground water and surface waters is extremely scant [102, 103]. In principle hydrolysis, oxydation and reduction processes can take place at the surfaces of e.g., clay materials which might exert a catalytic activity. Chemical degradation reactions may play a more pronounced role than biological degradation processes with regard to organic pollutants in surface waters as compared to ground waters [104, 105].

Aniline was found to have average concentrations after infiltration of Rhine water in the period of 1976–1978 of 2.3 µg/l (before infiltration) and 1.4 µg/l (1–12 months after bank filtration [103, 106]. The estimated half-life of aniline in river water was 2.3 days for each studied (Lobith/Gorinchem in the Netherlands) [103].

The reaction of aniline and several ring-substituted anilines with humate has been studied in aqueous solution [107]. The primary amines bind to humate in two phases. Initially a rapid, reversible equilibrium is established which may represent formation of imine linkages with the humate carbonyls. Subsequently, there is a slow reaction that is not readily reversed. The slow reaction is thought to represent 1,4-addition to quinone rings followed by tautomerization and oxidation to give an amino-substituted quinone. The slow reaction is considered not likely to be imine formation or other processes that require primary amines since N-methylaniline, a secondary amine also reacts with humate. These processes represent ways in which aromatic amines may be converted to latent forms with undetermined effects on the biosphere [107].

Aniline and its analogs yield colored complexes with basic Wyoming bentonite [108] indicating that charge-transfer complexes with clays may represent an additional reaction pathway, depending on soil conditions. However, lack of color development when aniline was absorbed on montmorillonite and hectorite in a nitrogen atmosphere suggests that molecular oxygen is required [109]. It is believed, however, that of greater importance, with respect to soil reactions, is probably the strong binding of aniline residues to soil organic matter [110]. Loss of CO_2 from these bound residues is very slow [111–113].

Aniline has been found to be partially degraded in sterile soil to azobenzene, azoxybenzene, phenazine, formanilide and acetanilide. Nitrobenzene, p-benzo-quinone and unidentified species were also possible products. Additionally, substantial bound residues may also have formed [110]. Freundlich adsorption constants showed: azobenzene > azoxybenzene > phenazine > aniline. Despite relatively low recoveries and limited aniline conversion in 1–3 days, soil catalyzed conversion of aniline or [d_5]aniline was evidenced by a 6 to 24 times greater product recovery in sterile soil than in sterile water alone, a process inhibited by $Na_2S_2O_4$. Hence, soil appears to catalyze aniline transformations by some unknown mechanism(s). If molecular oxygen were directly involved (considered plausible since aniline is easily oxidized), then addition of a reducing agent such as sodium dithionite might curtail aniline oxidation. This agent did in fact, considerably reduce the relative formation of azobenzene, azoxybenzene and phenazine [110].

Pillai et al. [110] in their studies of the possible pathways of soil catalyzed oxidation of aniline above noted the importance of those intermediates of aniline degradation *not* actually identified. For example, although hydroazobenzene (C_6H_5–NH–NHC_6H_5) was not detected, TLC and GLC analyses has demonstrated how readily it oxidizes to azobenzene [114].

Hydrazobenzene may form in soil via the coupling of two C_6H_5NH radicals. Thus the presence of azobenzene might indirectly furnish evidence of hydroazobenzene. Additionally, hydrazobenzene may undergo the acid-catalyzed benzidine rearrangement, viz., $C_6H_5NH–NH–C_6H_5 \rightarrow H_2N–C_6H_5–NH_2$.

Phenazine, as an aniline transformation product found in most of the sterile soil oxidation of aniline samples may be formed via the coupling of C_6H_5NH radicals. Thus, in this case C-N-C bonds would form rather than C–N–N–C bonds as with hydrazobenzene *viz.*,

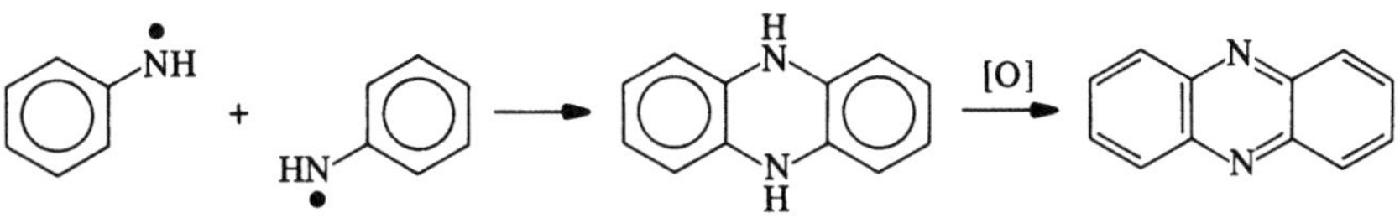

Phenazine is among known air oxidation products of aniline [85].

Although a variety of aniline transformation products were identified in the study of Pillai et al [110], it should be noted that most of the added aniline was unaccounted for. While inefficient extraction may account for some of this discrepancy, it was suggested that the isolated products indicate intermediate formation of reactive polar species that will largely form higher molecular weight polymers or react with soil organic matter leading to the formation of non-extractable soil bound residues [110]. Approximately 60% of[^{14}C]aniline became bound after 10 weeks in four non-sterile soils [112]. The results of Pillai et al. [110] suggest that chemical transformations may play a significant role in aniline bound residue production.

Biotransformation and toxicity of aniline and aniline derivatives in cyanobacteria was recently reported by Cerniglia et al [115]. Cyanobacteria (blue-green algae) and microalgae are ubiquitous and major contributors to the primary production of nutrients in marine habitats. *Oscillatoria* sp strain JCM and *Agmenellum quadruplicatum* strain PR-6 metabolized aniline to formanilide, acetanilide and p-aminophenol. The N-formyl and N-acetyl derivatives of aniline were the predominant metabolites. The formation of formanilide as a major metabolite suggests that cyanobacteria have the ability to N-formylate aromatic amines. This metabolic pathway has been previously reported for the microbial metabolism of aromatic amines [116].

Both cyanobacteria of the above strains formed small amounts of para-aminophenol from aniline. It was suggested that para-aminophenol could be formed via initial epoxidation followed by isomerization to the phenol [115]. These reactions had been previously described for the fungal [117, 118] and mammalian [119] metabolism of aniline. Alternatively, it is possible that p-aminophenol could be formed via the direct hydroxylation of aniline or by rearrangement of phenylhydroxylamine [120]. Cerniglia et al. [115] stressed that although their studies indicated that cyanobacteria metabolized aniline predominantly to metabolites that are less toxic, further studies are required to determine the metabolic significance of these reactions in cyanobacteria and their importance in aquatic ecosystems.

Although the biochemical transformation of anilines in soil [121–124], in the liver microsome system [125–127] and by pure cultures of microorganisms [128] have been studied, the pathway of aniline degradation by soil microorganisms is still largely unknown [129]. The conversion of aniline into pyrocatechol by a mutant strain of *Nocardia sp* bacteria was reported to require the presence of oxygen [129]. Accumulation studies in an $^{18}O_2$-atmosphere revealed that both hydroxyl groups of pyrocatechol were equally labeled, hence eliminating an epoxide intermediate formed by mono-oxygenation of aniline. It was suggested that a cyclic peroxide formed after dioxygenation is the more likely intermediate in aniline oxidation. This conclusion is in agreement with the observation that mono-oxygenation is the mechanism of biological oxidation of non-hydroxylated aromatic compounds in mammalian liver [130] whereas dioxygenation is often found in bacteria [131].

The effects of aniline after biodegradation in industrial waste waters was investigated with primary focus as to whether aniline degradation by activated sludge could be sufficiently complete to allow nitrification to proceed at a reasonable rate either in the same reactor or in a subsequent unit operation [91]. Under normal conditions autotrophic nitrifying bacteria grow so slowly that relatively long solids retention times (SRT's) are needed for efficient nitrification. The study of Joel and Grady [91] suggests that completely mixed activated sludge systems should have no difficulty in degrading aniline-containing waste waters sufficiently to allow nitrification of the effluent.

Biodegradation studies of seven aliphatic amines, cyclohexylamine, morpholine and aniline with river mud bacteria, treatment plant sludge bacteria and adapted treatment plant sludge bacteria showed that only aniline underwent a primary biodegradation with all the different inocula and concentrations [132].

Chlorination is the most common disinfection process in water treatment and compounds resulting from the chlorination of natural and process waters have been characterized [133]. The fate of aniline, N,N-dimethylaniline and benzidine in chlorinated waters was investigated under conditions which were controlled to approximate the process chlorination of raw water supplies and wastewater secondary effluents [134].

For the monophenylamines, ring chlorination was a significant depletion pathway. Aniline yielded principally ortho-chloroaniline. At a 1:1 (Cl_2)/(aniline) ratio,

19.13 mg aniline were depleted and 3.83 mg of ortho-chloroaniline (20.0%) were produced. In general, as $(Cl_2)/(\text{aniline})$ ratios decreased, the percent yields of the chlorinated products decreased suggesting that ring chlorination of aniline was a less favored reaction pathway at higher relative aniline concentrations. If aniline/chlorine mixtures were allowed to stand overnight, a reddish-brown precipitate formed which was not structurally identified [134].

Chlorination of aniline in distilled water or secondary effluent produced a faint violet color, presumably due to the formation of indoaniline, a colored end product which had been previously identified from the reaction of aniline with concentrated NaOCl solutions, equations 1 and [135].

$$C_6H_5NH_2 + NaOCl \rightarrow p\text{-}OH\text{-}C_6H_4NH_2 + NaCl \quad (1)$$

$$\begin{aligned} &C_6H_5NH_2 + p\text{-}OH\text{-}C_6H_4\text{-}NH_2 + 2NaOCl \rightarrow \\ &\quad HO\text{-}C_6H_4\text{-}N{=}C_6H_4{=}NH + 2NaCl + 2H_2O \end{aligned} \quad (2)$$

Indoaniline.

By analogy to the formation of chloramines from the reaction of HOCl and NH_3, organic amines might be expected to form organic chloramines, *viz.*,

$$R\text{-}NH_2 + Cl_2 \rightarrow R\text{-}NHCl + HCl.$$

Sodium sulfite is known to destroy chloramines. However, when aniline-chlorination mixtures were treated with Na_2SO_3, no change in chromatographic analysis or in solution color was observed, suggesting the absence of organic chloramines.

(d) Aniline Degradation Products of Pesticides

The degradation of acylanilide, N-phenyl carbamate and N-phenyl urea herbicides in soil was noted earlier in the chapter as environmental sources of aniline and its congeners, primarily halogenated anilines [78–84, 110, 121, 136–142]. The aniline degradation products of a number of N-phenyl carbamate and N-phenylurea herbicides are shown in Table 1 [76].

Mechanisms of substituted aniline transformation in soil, by chemical systems and by isolated soil microorganisms have been widely studied. Several different transformation products have been found, but complete mineralization has only been demonstrated for unsubstituted aniline [129, 143] and 3-chloroaniline [79]. It should be noted, however, that 4-chloroaniline and 3,4-dichloroaniline represent the major portion of metabolites of aniline-based pesticides in the environment as was shown for the Federal Republic of Germany [144, 145].

Four major modes of transformation of halogenated anilines in soil and by microorganisms isolated from soil have been described. These include:

(a) condensation of two molecules which leads to the formation of both symmetrical and asymmetrical azobenzenes [140] and the condensation of three molecules with the formation of halogen-substituted aniline azobenzene [146];

Table 1. Decomposition of certain herbicides to aniline degradation products

I

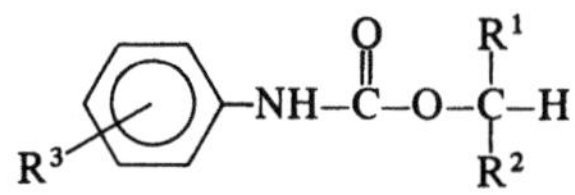

N–Phenylcarbamate

II

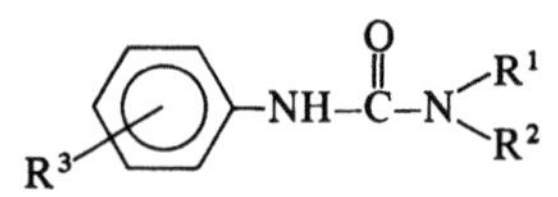

N–Phenylurea

Class	Common name	R^1	R^2	R^3	Degradation product
I	Propham (IPC)	$-CH_3$	CH_3	$-H$	
I	Carbetamide	$-CH_3$	$-\overset{\overset{O}{\|}}{C}NHCH_5CH_3$	$-H$	aniline
II	Fenuron	$-CH_3$	$-CH_3$	$-H$	
II	Siduron	$-H$	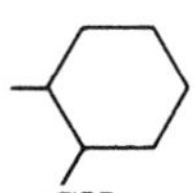 CH_3	$-H$	
I	Chloropropham	$-CH_3$	$-CH_3$	3-Cl	3-chloroaniline
I	Barban	$-C{\equiv}CCH_2Cl$	$-H$	3-Cl	
II	Monolinuron	$-CH_3$	$-OCH_3$	4-Cl	
II	Buturon	$-CH_3$	$-CH(CH_3)C{\equiv}CH$	4-Cl	4-chloroaniline
II	Monuron	$-CH_3$	$-CH_3$	4-Cl	
II	Metabromuron	$-CH_3$	$-OCH_3$	4-Br	4-bromoaniline
II	Chlorotoluron	$-CH_3$	$-CH_3$	3-Cl, $4-CH_3$	3-chloro-4-methylaniline

(b) condensation of two molecules of a primarily hydroxylated 4-chloroaniline with the formation of 7-chloro-2-amino-3H-phenoazin-3-one [147];
(c) acylation resulting in a formylated [123], acetylated [122, 148–150] or malonic acid conjugated [151] derivative; and
(d) oxidation of the amino to a nitro group [152].

Bordeleau et al. [114] using an *in vitro* peroxidase system, suggested that chlorohydroxylamines and chlorohydrazobenzenes were labile intermediates in the oxidation of chloraniline to chloroazobenzenes. Figures 3–5 illustrate the transformation of mono-, di-, and tri-substituted anilines respectively by *G. candidum* L-3 peroxidase and aniline oxidase. However, the role of soil peroxidase in such reactions has been questioned by Burge [153]. Bacteria containing a nitrate reduc-

Fig. 3. Transformations of monosubstituted anilines by G. *candidum* L-3 peroxidase and aniline oxidase. Numbers in parentheses refer to chemical names: (1) aniline, (2) 2-nitroaniline, (3) 3-nitroaniline, (4) 4-nitroaniline, (5) 2-fluoroaniline, (6) 3-fluoroaniline, (7) 4-fluoroaniline, (8) 2-chloroaniline, (9) 3-chloroaniline, (10) 4-chloroaniline, (11) 2-bromoaniline, (12) 3-bromoaniline, (13) 4-bromoaniline, (14) 2-iodoaniline, (15) 3-iodoaniline, (16) 4-iodoaniline, (17) 2-methylaniline, (18) 3-methylaniline, (19) 4-methylaniline, (20) 2-methoxyaniline, (21) 3-methoxyaniline, (22) 4-methoxyaniline

Transformation by peroxidase ← → *Transformation by aniline oxidase*

Azobenzene	Activity	ANILINE	Activity	Azobenzene
C_6H_5-N=N-C_6H_5	960	C_6H_5-NH_2 (1)	1140	C_6H_5-N=N-C_6H_5
	0	2-NO_2-C_6H_4-NH_2 (2)	0	
	0	3-NO_2-C_6H_4-NH_2 (3)	0	
	0	O_2N-C_6H_4-NH_2 (4)	0	
2-F-C_6H_4-N=N-C_6H_4-2-F	157	2-F-C_6H_4-NH_2 (5)	0	
3-F-C_6H_4-N=N-C_6H_4-3-F	104	3-F-C_6H_4-NH_2 (6)	0	
F-C_6H_4-N=N-C_6H_4-F	190	F-C_6H_4-NH_2 (7)	92	F-C_6H_4-N=N-C_6H_4-F
2-Cl-C_6H_4-N=N-C_6H_4-2-Cl	230	2-Cl-C_6H_4-NH_2 (8)	0	
3-Cl-C_6H_4-N=N-C_6H_4-3-Cl	205	3-Cl-C_6H_4-NH_2 (9)	0	
Cl-C_6H_4-N=N-C_6H_4-Cl	540	Cl-C_6H_4-NH_2 (10)	534	Cl-C_6H_4-N=N-C_6H_4-Cl
2-Br-C_6H_4-N=N-C_6H_4-2-Br	240	2-Br-C_6H_4-NH_2 (11)	0	
3-Br-C_6H_4-N=N-C_6H_4-3-Br	218	3-Br-C_6H_4-NH_2 (12)	0	
Br-C_6H_4-N=N-C_6H_4-Br	628	Br-C_6H_4-NH_2 (13)	785	Br-C_6H_4-N=N-C_6H_4-Br
2-I-C_6H_4-N=N-C_6H_4-2-I	270	2-I-C_6H_4-NH_2 (14)	0	
3-I-C_6H_4-N=N-C_6H_4-3-I	235	3-I-C_6H_4-NH_2 (15)	0	
I-C_6H_4-N=N-C_6H_4-I	760	I-C_6H_4-NH_2 (16)	1080	I-C_6H_4-N=N-C_6H_4-I
2-CH_3-C_6H_4-N=N-C_6H_4-2-CH_3	490	2-CH_3-C_6H_4-NH_2 (17)	940	2-CH_3-C_6H_4-N=N-C_6H_4-2-CH_3
3-CH_3-C_6H_4-N=N-C_6H_4-3-CH_3	320	3-CH_3-C_6H_4-NH_2 (18)	338	3-CH_3-C_6H_4-N=N-C_6H_4-3-CH_3
H_3C-C_6H_4-N=N-C_6H_4-CH_3	1530	H_3C-C_6H_4-NH_2 (19)	2080	H_3C-C_6H_4-N=N-C_6H_4-CH_3
None	1290	2-OCH_3-C_6H_4-NH_2 (20)	1985	None
None	392	3-OCH_3-C_6H_4-NH_2 (21)	614	None
None	2120	H_3CO-C_6H_4-NH_2 (22)	9150	None

Transformation by peroxidase ← → *Transformation by aniline oxidase*

Azobenzene	Activity	ANILINE	Activity	Azobenzene
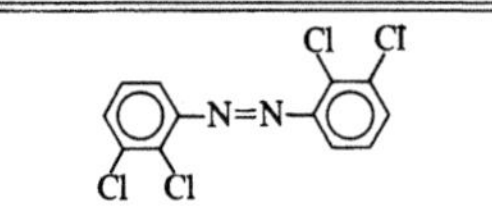	118	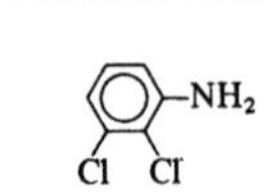 (1)	0	
	0	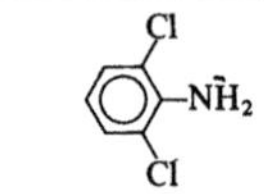 (2)	0	
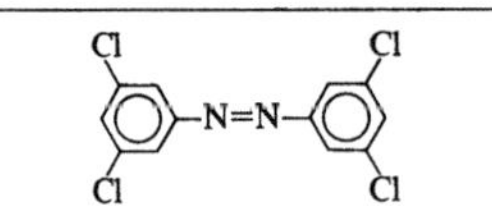	160	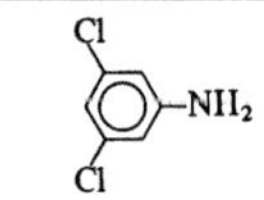 (3)	0	
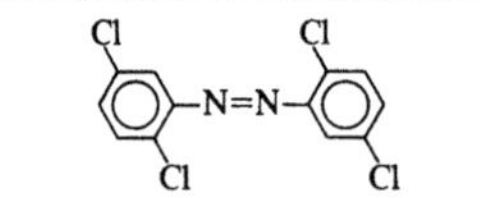	120	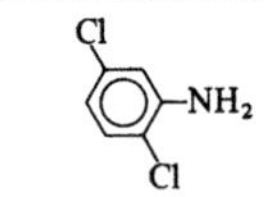 (4)	0	
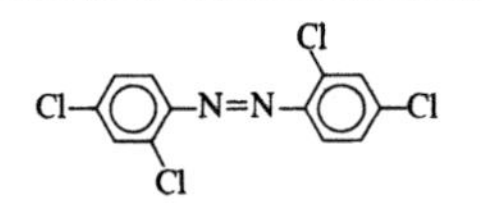	528	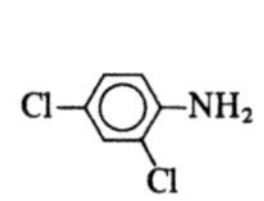 (5)	92	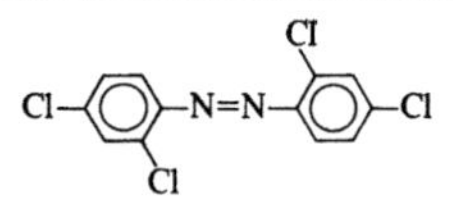
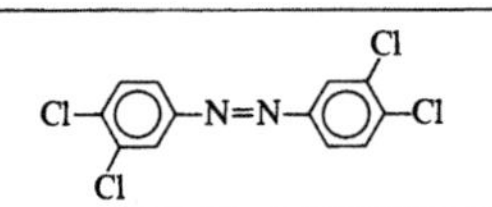	510	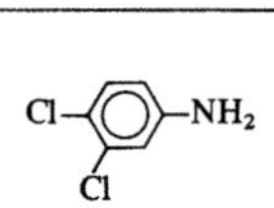 (6)	118	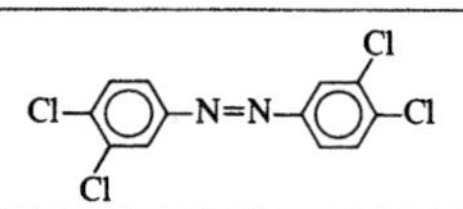
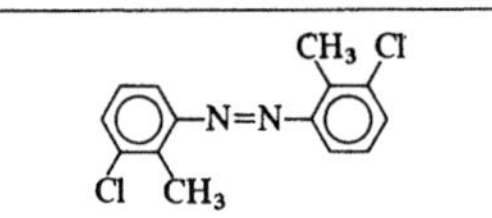	490	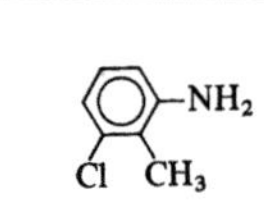 (7)	182	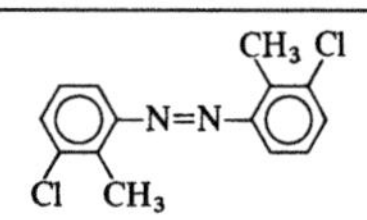
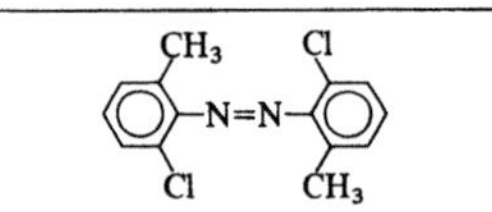	92	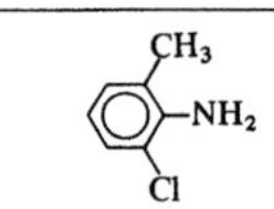 (8)	0	
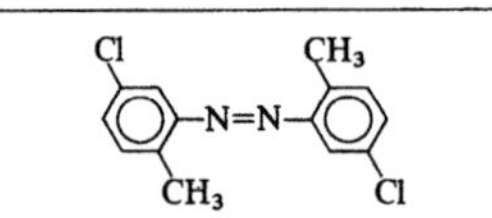	512	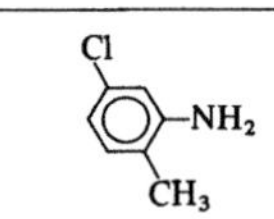 (9)	240	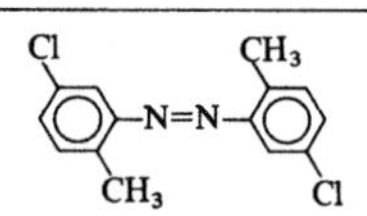
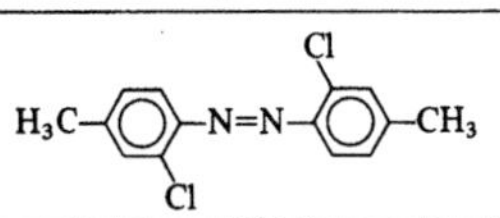	812	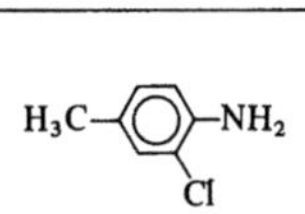 (10)	195	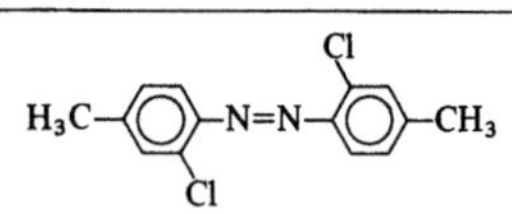
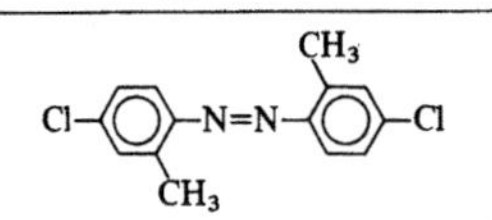	1205	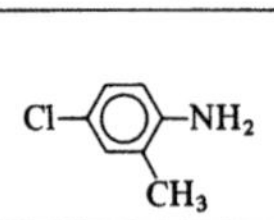 (11)	1420	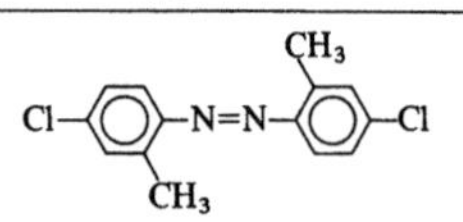
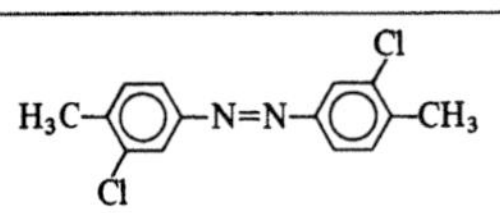	928	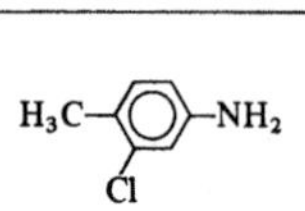 (12)	1270	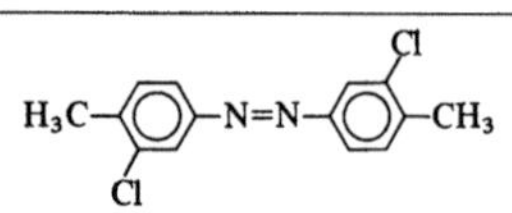

Fig. 4. Transformations of disubstituted anilines by G. *candidum* L-3 peroxidase and aniline oxidase. Numbers in parentheses refer to chemical names: (1) 2,3-dichloroaniline, (2) 2,6-dichloroaniline, (3) 3,5-dichloroaniline, (4) 2,5-dichloroaniline, (5) 2,4-dichloroaniline, (6) 3,4-dichloroaniline, (7) 2-methyl,3-chloroaniline, (8) 2-chloro,6-methylaniline, (9) 2-methyl,5-chloroaniline, (10) 2-chloro,4-methylaniline, (11) 2-methyl,4-chloroaniline, (12) 3-chloro,4-methylaniline

Transformation by peroxidase Azobenzene	Activity	ANILINE	Activity	*Transformation by aniline oxidase* Azobenzene
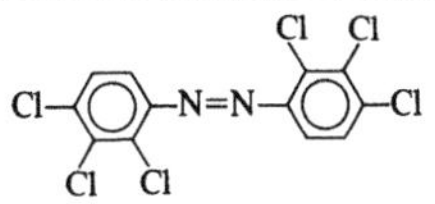	470	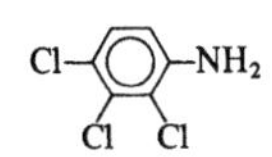 (1)	0	
	0	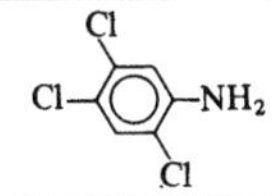 (2)	0	
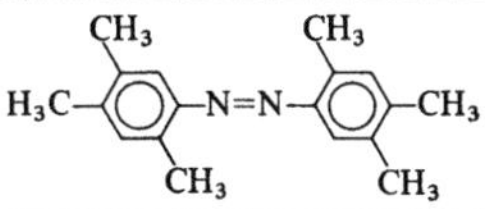	4750	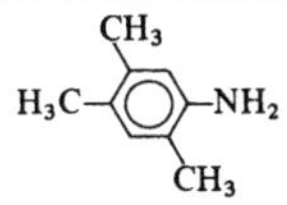 (3)	9200	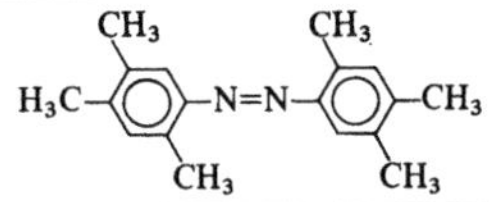
	0	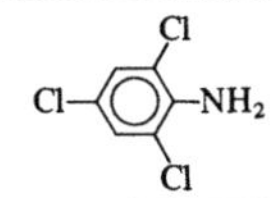 (4)	0	
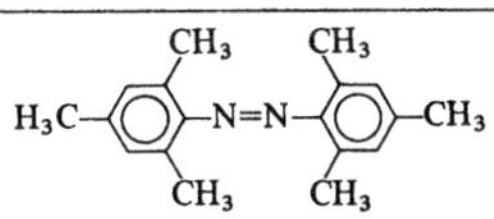	4700	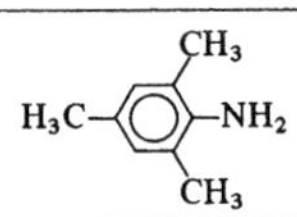 (5)	8750	None
None	12700	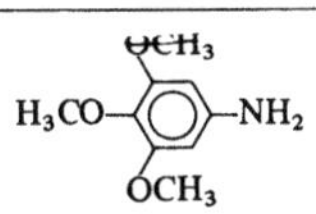 (6)	22300	None
	0	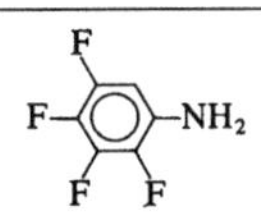 (7)	0	
	0	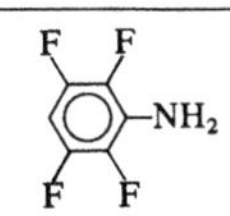 (8)	0	
	0	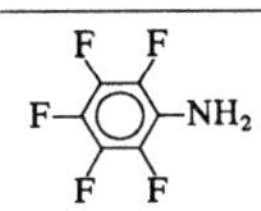 (9)	0	

Fig. 5. Transformations of tri-, tetra-, and penta-substituted anilines by G. *candidum* L-3 peroxidase and aniline oxidase. Numbers in parentheses refer to chemical names: (1) 2,3,4-trichloroaniline, (2) 2,4,5-trichloroaniline, (3) 2,4,5-trimethylaniline, (4) 2,4,6-trichloroaniline, (5) 2,4,6-trimethylaniline, (6) 3,4,5-trimethoxyaniline, (7) 2,3,4,5-tetrafluoroaniline, (8) 2,3,5,6-tetrafluoroaniline, (9) 2,3,4,5,6-pentafluoroaniline

tase enzyme system may, in a somewhat acidic reducing environment, produce nitrite ions that then react with anilines to yield diazonium cations. These labile intermediates were linked to formation of triazenes, azobenzenes and biphenyls from aniline precursors [154]. The hydroxylamine, nitroso, nitro [122], and *ortho*-hydroxy [155] derivatives are among the non-conjugated chloroaniline metabolites isolated thus far [110]. The ultimate degradation products of aniline are CO_2 and NH_3.

Benzidine

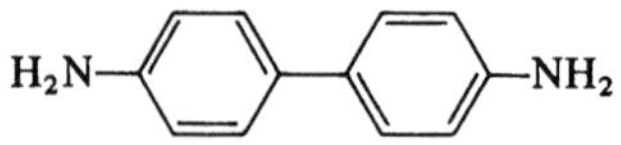

Benzidine and a number of its salts (principally the hydrochloride, dihydrochlorides und sulfates), derivatives and analogs have occupied a paramount position among the aromatic amines with regard to their applications and their potential risk to health.

Production and Use

Benzidine was initially prepared in 1845 via the reduction of azobenzene with ammonium sulphide followed by treatment of the hydrazobenzene with sulphuric acid treatment with a base to release the free benzidine [6]. Commercial manufacture involves the alkaline reduction of nitrobenzene in either one stage to hydrazobenzene, or stepwise, changing conditions at the azoxybenzene or azobenzene stages. Although several methods for the reduction of nitrobenzene can be used (e. g., zinc or sodium amalgams, iron, the electrolytic cathode, zinc metal or methanol), the key reaction in all routes of benzidine synthesis is the rearrangement of the intermediate hydrazobenzene to benzidine with hydrochloric acid. The resulting benzidine hydrochloride is generally used for subsequent reaction rather than converted to the free base [2, 6].

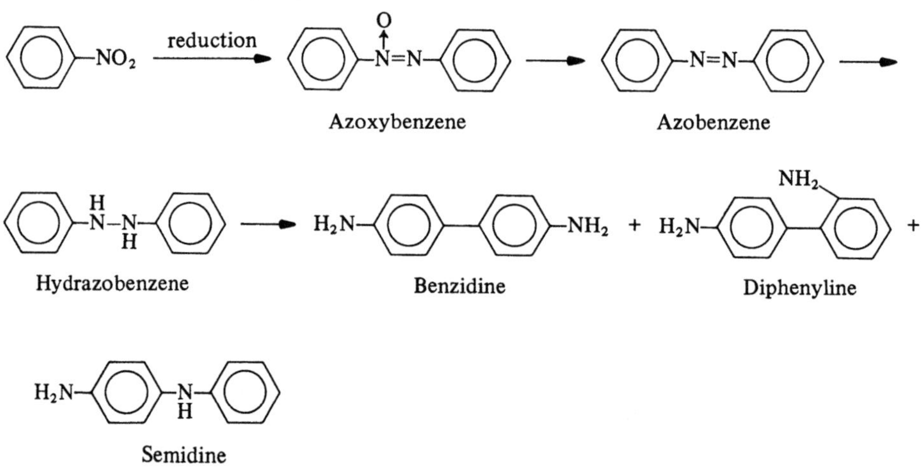

Benzidine and its sulfate were produced commercially in the US for at least 60 years. Currently, benzidine dihydrochloride is the only form produced commercially in the US (using hydrazobenzene as the starting material) and this only as an *unisolated* intermediate which is further converted by the single manufacturer to a variety of dyes [6, 156, 157]. One US company in 1977 reported its production of benzidine dihydrochloride for captive production was in the range of 45.4–454 t [6, 157]. Commercial production of benzidine in the U. S. was last reported (by a single company) in 1976 and that would appear to be the last year of large-scale production of benzidine in the U. S. [6]. An estimated 500 t of benzidine were produced in 1974 and 700 t in 1972 [2, 6]. This is in sharp contrast to the 1,800 t of benzidine produced in the US in 1948 [6, 158]. A total of 4.1 t of benzidine were imported into the US in 1980 [6].

In western Europe, one company in France is believed to produce benzidine dihydrochloride as an unisolated intermediate for dyes [6]. The following countries which are believed to manufacture benzidine-based dyes may have plants in which benzidine and/or its salts are produced as unisolated dye intermediates: Argentina, Brazil, India, Mexico, the People's Republic of China, Poland, Romania, South Korea, and the USSR [6]. Commercial production of benzidine in Japan was believed to have been terminated in about 1966 [6].

The most important and by far the largest use of benzidine is in the production of dyes and pigments (principally azo dyes, sulfur dyes, fast colour salts, naphthols and other dyes and dyeing compounds) [2, 6–8, 156–157]. Over 250 dyes and pigments can be prepared from benzidine based primarily via the coupling of tetrazotized benzidines with phenols and aromatic amines [6, 156, 157, 159, 160]. (The distinction between the term "dye" and "pigment" as stated by the dye industry is that dyes are soluble and pigments are insoluble in the medium in which they are used. However, where solubility is not an important consideration, the word "dye" is sometimes used to refer to dyes and pigments collectively.) [156].

Only 16 (including the major volume dyes Direct Black 38, Direct Blue 6 and Direct Brown 95) out of a list of 232 benzidine dyes are being produced in the US or being imported and only one benzidine-based pigment, Pigment Red 39 which had been produced is no longer produced commercially in the US [6, 161]. Data on US production (or sales), imports, and principal uses of 13 dyes during the period 1971–1979 have been reported [6].

The general use of benzidine and benzidine-based dyes have decreased dramatically because of the recognition of benzidine as a human carcinogen [6, 7, 10, 11]. For example, in 1948, 14 million kg of benzidine-based dyes were produced in the US representing 21% of all dyes manufactured and almost all of the direct dyes on the market that year [6, 158]. In 1977, 600 t of benzidine-based dyes were reported produced in the US and approximately 35 t imported and that benzidine-based dyes are now produced annually in the US at about 230 kg [162]. No insoluble benzidine-based pigments have been produced in, or imported into the US since 1974 [162].

Benzidine or its salts have also been widely used in the following minor applications: as a hardener for polyurethane in the rubber, adhesives and plastics industries; for the detection of hydrogen peroxide in milk and in the past was extensively used (as well as o-toluidine and o-dianisidine) in analytical and clinical chemistry for the detection of chlorinated organic pesticides using TLC, nicotine, sugars, occult blood, chlorine, or pyridine in water, bacterial cytochromes, naphthalene sulphonic acid in detergents and meta- and para-cresols [2, 6, 156–158, 163-164].

Human exposure of benzidine or its dyes can occur in three ways:

(1) workers can be exposed during the production of benzidine or benzidine based dyes in manufacturing plants or during the dyeing of textiles, paper, leather, plastics, etc. in dye applicator plants;

(2) consumers, principally those in arts and crafts can be exposed via inhalation from home kits and by absorption from use of textiles and coloring instruments colored with the dyes and

(3) segments of the general public who are potentially exposed to benzidine contaminants in the water supply.

Metabolism and Toxicity

Benzidine may enter the body by percutaneous absorption, ingestion or inhalation with percutaneous absorption being the primary route following occupational exposure [6]. Occupational exposure to benzidine has been strongly associated with bladder cancer in numerous case reports from many countries and this association has been recognized for about 60 years [6, 165–167].

Benzidine and its salts have been recognized as carcinogenic by the following 13 countries by regulation or guidelines: Australia, Belgium, Finland, the Federal Republic of Germany, Italy, The Netherlands, Poland, Romania, Sweden, Switzerland, the United Kingdom, the USA and Yugoslavia, Those countries which also designate it as a skin irritant are Australia, Belgium, Italy, The Netherlands, Poland, Romania, Switzerland, and the USA [6]. IARC [6] has concluded that there is sufficient evidence that benzidine is carcinogenic to man as well as to mice, rats, hamsters and dogs.

Benzidine and urine from rats fed benzidine are mutagenic to *Salmonella typhimurium* TA 1538, TA 98 and TA 100 with metabolic activation [6, 18, 168–169]. Benzidine is mutagenic to *Drosophila melanogaster* [6, 170], inhibits DNA synthesis in HeLa cells and in renal and hepatic cells in mice *in vivo* [6, 171, 172], induces unscheduled DNA synthesis in HeLa cells and in rat hepatocytes [6, 173, 174] transformed Syrian hamster emryo cells [6, 175] and was positive in the BHK21 clone-13 cell systems [6, 19].

An increase in the number of sister chromatid exchanges was reported in peripheral blood lymphocytes of 15 subjects occupationally exposed to benzidine [176]. However, the significance of this effect was questioned by IARC [6].

Following application of 100 mg benzidine to the skin of human volunteers, <0.02 mg benzidine and metabolites were detected in the urine. Concentrations of 0.27 to 1.60 mg/l of benzidine were measured in the urine of workers in a chemical plant manufacturing benzidine and substituted benzidines [6, 177]. In benzidine exposed workers, 3 hydroxybenzidine conjugates were found to constitute 80–90% of the urinary constituents while 5–10% was excreted as diacetylbenzidine, 1–5% monoacetyl benzidine and 4–6% as the unmetabolized parent compound [6, 178].

The major routes of excretion of intraperitoneally or intravenously administered benzidine in experimental animals are urine and feces [6, 179]. The materials excreted in the urine are predominantly metabolites of benzidine involving N-acetylation, 3-hydroxylation and sulfate and glucuronide conjugations [6, 180–183].

Occurrence and Reactions

(a) Occupational Exposure

The National Occupational Hazard Survey estimated that about 700 people were occupationally exposed to benzidine [6, 158]. However, it should be noted that many more individuals can be potentially exposed as a result of manufacturing or of using benzidine-based dyes. Twenty-six US produced benzidine based dyes were found to contain < 1–20 mg/kg benzidine and one dye contained 270 mg/kg. Eight of 33 benzidine-based dye samples from Belgium, Egypt, India, The Netherlands, Poland, Romania, and South Korea were found to contain 38–1,254 mg/kg of benzidine and the others had 24 mg/kg or less [6, 158].

Information on occupational exposure levels in benzidine plants is extremely limited. One plant had airborne concentrations of $<0.007\ mg/m^3$ at four locations, $0.152\ mg/m^3$ at a salting-out tub, $0.072–0.415\ mg/m^3$ at a filter press and $17.60\ mg/m^3$ where benzidine was shovelled into drums [6, 184].

(b) Air

Although no data have been reported on the distribution of benzidine in air, Radding et al. [185] considered the probable principal chemical reactions of benzidine in air to be photolysis and oxidation by ozone. Lacking any reaction data, it was estimated crudely that the half-life of benzidine in air is approximately one day.

(c) Water and Sediments

Benzidine may enter the environment from benzidine production facilities from downstream chemical processing (primarily dye manufacture and application) and from the use of products containing benzidine or benzidine derivatives. While industrial measurements in the past indicated that discharges of benzidine in wastewater from production facilities usually did not exceed one pound per day, there have been significant episodes of accidental releases [7, 186]. Benzidine is also believed to have been present in sludge from industrial waste pretreatment plants with land disposal being the final method of disposal of this sludge in the past [7, 186].

While the levels of benzidine emissions from downstream processes are not well defined it is stressed that free benzidine is present in the benzidine-derived azo dyes. Quality control specification in the past have required that the level not exceed 20 ppm and in practice the level of benzidine in azo dyes is usually below 10 ppm [186].

It has been estimated that at the 300 users facilities a total of 450 pounds/year or about 1.5 pounds/year per facility of benzidine is discharged, assuming all of the free benzidine was discharged into the liquid effluent. It has also been estimated that based on a 1972 production figure of 4,700 t, a release rate of 9000 kg of benzidine/year could be calculated. This figure probably includes atmospheric processes and fugitive emissions from vapor pressure or other physical properties [187].

Benzidine has been detected in oil refinery, municipal and industrial effluents and in the surface water [99], as well as in river water and in raw sewage effluents [100] in the US.

Benzidine-based or congener-based dyes and pigments discharged into waters from manufacturing facilities are believed to be chemically reduced to the parent compound if either H_2S or SO_2 are present in the receiving waters. This was believed to account for benzidine concentrations of up to 233 μg/l which were found downstream from a dye plant on the Sumida River in Japan [6, 156, 188]. Effluents from factories where textiles are dyed with benzidine-based dyes were found to contain an average concentration of 3.5 μg/l benzidine; those from a leather factory and another manufacturing plant with heavy benzidine-dye use contained 0.25 and 3.5 μg/l benzidine respectively [6, 156].

(d) Degradation and Biotransformation

In soil, benzidine is postulated to be immobilized rapidly by adsorption to humic material [185] and clays [189] and the Fe_3, Al_3 and Cu_2 ions which are readily available in some clay soils are believed to oxidize benzidine very rapidly [187]. The chemical oxidation in soils can occur via the formation of a free radical species of benzidine that is stabilized against further oxidation by adsorption to clay. Electron transfer is promoted by Fe^{3+}, a faster reaction than when O_2 alone is the oxidizing agent [190].

Batch equilibrium sorption isotherms were determined for benzidine in 14 soil and sediment samples [191]. The sorption isotherms were nonlinear and were well represented by the Freundlich equation with 1/n values of approximately 0.5. Sorption was largely controlled by the pH of the aqueous phase. Benzidine sorption was corrected for sorption of the neutral species by organic matter, and sorption of the ionized species was highly correlated with surface area of the soil or sediment [191].

Activated sludge has been shown to deplete biphenyl amines and monophenylamines (including benzidine, aniline, and N,N-dimethylaniline) [134, 192]. The aromatic amines were found to be quite toxic to activated sludges, thus requiring the use of acclimated sludges. Although depletion of the aromatic amines was demonstrated, the oxidation products were not characterized hence the effective removal of toxic or carcinogenic substances cannot be stated [134, 192].

Acclimated extended aeration sludges were found to completely oxidize continuous doses of 1 mg/l benzidine. Less complete oxidation occurred at higher dose levels and intermediate oxidation products began to accumulate in the system. Even at the highest level of benzidine (30 mg/l) no marked interference with carbonaceous removal efficiency was noted, since the increased COD of this effluent could be related to the presence of undegraded benzidine [193].

In contrast to the monophenylamines (e.g., aniline) when benzidine was exposed to chlorinated waters, the primary depletion route was observed to produce an immediate polymeric precipitate with no ring chlorination. The depletion of benzidine in contact with chlorine was observed to be greater at low $(Cl_2)/(amine)$ ratios than for monophenyl anilines. From the above results it appeared that 80% of the benzidine present in waste streams could be removed by chlorination at 1:1 molar ratio. Although the toxicity and carcinogenicity of the polymeric product(s) is unknown, it appears unlikely that this product would retain the carcinogenicity of benzidine [134].

The effects of ozonation on the mutagenicity of benzidine has been investigated by Caulfield et al [194]. When benzidine was dissolved in water at a concentration of 100 μg/ml and ozonated, there was a transient increase in the mutagenic index due to the formation of an intermediate which no longer required activation by an S-9 microsomal fraction to be mutagenic for *S. typhimurium* TA 98 tester strain. Upon further ozonation, the mutagenicity of benzidine and its ozonation products were eliminated. In contrast, oxygenation had no effect on the mutagenicity of benzidine. Earlier studies have shown benzidine to be activated by ozonation to an intermediate not requiring homogenate activation in the *Salmonella*-microsome assay system [195, 196].

The accumulation and bioconcentration of benzidine as studied in a laboratory model ecosystem was closely correlated with its octanol/water partition coefficient (23) and water solubility (500 ppm at 25 °C). Benzidine was much more rapidly metabolized than benzo-(alpha)-pyrene in *Oedogonium*, Daphnia, Physa, Corlex, and Gambusia and the parent compound was stored in amounts ranging from 8% in alga and daphnia to 12% in fish and 13% in snail. Biomagnification was much lower than with benzo-(α)-pyrene (fish, 55; daphnia, 293; mosquito, 456; snail, 645; and alga 2617). The degradative pathways of benzidine were through N-methylation and acetylation of the amino groups. Relatively large amounts of polar radioactivity were formed by the various organisms [197].

Benzidine-Based Dyes

Benzidine-based dyes are defined as dyes that contain benzidine attached to other substituents by diazo linkages. Benzidine-based dyes may contain residual amounts of benzidine as well as other substances such as 4-aminobiphenyl, an OSHA regulated carcinogen [4, 8, 156–158, 198–202]. Thousands of workers involved in the production of dyes, textile, paper and leather goods potentially are exposed to dyes based on benzidine. The National Institute of Occupational Safety and Health (NIOSH) and the Occupational Safety and Health Administration (OSHA) recently reported that three major benzidine-based azo dyes (Direct Black 38, Direct Blue 6 and Direct Brown 95) (Fig. 6) were carcinogenic in animals, in which they were converted to benzidine [202].

NIOSH estimated that about 79,000 workers in 63 occupations were potentially exposed to benzidine-based dyes in the period 1972–1974 [8, 200–203]. Workers are occupationally to Direct Black 38, Direct Blue 6 and Direct Brown 95 in a variety of industries including: paper and allied products, petroleum and related industries, rubber and plastic products, leather and leather products. In addition the textile industry accounts for a substantial occupational exposure. It is estimated that 25% of the benzidine derived azo dyes are applied to textiles, 40% to paper, 15% to leather and the remainder to diverse applications [204].

Production and Uses of Direct Black 38, Direct Blue 6 and Direct Brown 95

Direct Black 38 was first synthesized in 1898 and is produced commercially by:
(1) coupling diazotized benzidine with 1 mole H acid (8-amino-1-naphthol-3,6-disulphonic acid) under acid conditions,
(2) reacting the resulting product with 1 mole diazotized aniline under alkaline conditions,

DIRECT BLUE 6

DIRECT BLACK 38

DIRECT BROWN 95

COPPER COMPLEX DERIVED FROM

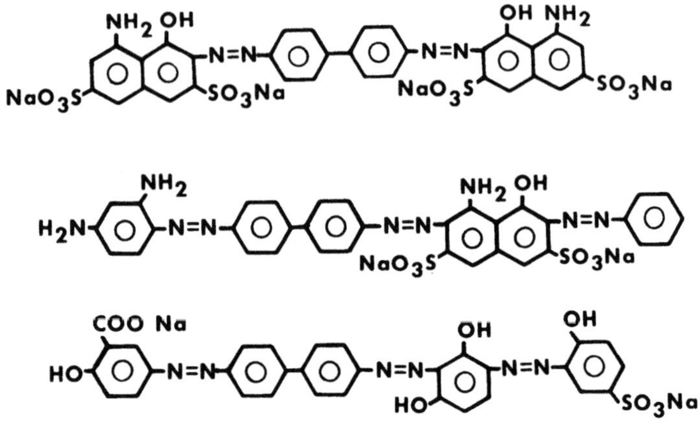

Fig. 6. Benzidine derived dyes

(3) coupling this intermediate with *meta*-phenylenediamine, and
(4) neutralization with sodium hydroxide [6].
In 1976, US production of Direct Black 38 by 5 companies amounted to 1.71 million kg down sharply from the 3.1 million kg produced in 1973. It is presently produced commercially by only one company in the US, whose production in 1978 amounted to 374 thousand kg [199] which was the largest quality of a benzidine-based dye produced in the US in that year [6]. US imports of Direct Black 38 in 1980 were nearly 95 thousand kg down from the 149 thousand kg imported in 1979 [6]. Direct Black 38 (as well as Direct Blue 6 and Direct Brown 95) are believed to be produced commercially by one company in France although it may be imported into other Western European countries from India or from eastern Europe [6]. Although benzidine-based dyes are also believed to be produced commercially in Argentina, Brazil, India, Mexico, the People's Republic of China, Poland, Romania and the USSR, it is not known whether Direct Black 38 (or Direct Blue 6 and Direct Brown 95) are among the dyes produced.

Direct Black 38 can be used to: (1) dye cellulose, wool, silk, bast and hog's hair; (2) print cellulose, wool and silk; (3) dye leather, plastics, and wood flour used as a resin filler; (4) stain wool, silk, acetate, nylon, wood and biological materials and (5) produce aqueous inks [6, 205]. It has reportedly been used in hair dyes [6, 206].

Direct Black 38 was being used commercially in the US in 1980 to dye textiles, leather and paper, while in Japan, 60% of the Direct Black 38 is used for dyeing fibers, 20% for dyeing paper and 20% for dyeing leather [6].

NIOSH estimated that approximately 13,000 US workers are exposed to Direct Black 38 [6, 200].

Direct Blue 6 was first synthesized in 1890 and is prepared commercially by coupling diazotized benzidine with 2 moles H-acid (gamma-amino-1-naphthol-3,6-disulfonic acid) under alkaline condition [6]. In both 1971 and 1976 US production of Direct Blue 6 amounted to 148 thousand kg [6, 200]. It is presently produced commercially in the US by one company whose production in 1978 totalled 28 thousand kg making it the seventh largest volume benzidine-based dye produced in that year [6, 200]. Two thousand kg of Direct Blue 6 were imported into the US in 1978 [6].

Direct Blue 6 can be used to: (1) dye cellulose and silk; (2) stain silk, wool and nylon fibers; (3) print cellulose fabrics; (4) dye leather and paper; (5) stain biological materials; and (6) produce aqueous writing inks [6, 210]. Direct Blue 6 has reportedly been used in hair dyes [6, 206].

NIOSH estimated in 1980 that approximately 800 US workers were exposed to Direct Blue 6 [200].

Direct Brown 95 was first synthesized in 1931 and is prepared commercially by: (1) coupling diazotized 2-amino-1-phenol-4-sulphonic acid with resorcinol; (2) coupling 1 mole of the resulting intermediate with 1 mole diazotized benzidine; (3) coupling the resultant intermediate with salicylic acid; and (4) forming a copper complex with a copper salt.

In 1976 US production of Direct Brown 95 by four companies totalled 270 thousand kg, a slight increase over the 257 t produced in 1973. It is presently produced commercially by one company in the US whose production in 1978 totalled 34.5 t making it the fifth largest volume benzidine-based dye produced in the US in that year [6, 200]. US imported 10.9 t of Direct Brown 95 in 1980.

Direct Brown 95 can be used to: (1) dye cellulose and silk fibers; (2) stain wool, acetate and nylon fibers; (3) print cellulose and silk fabrics; (4) dye leather, paper and casein-formaldehyde plastics; and (5) produce its heavy metal salts which can be used as pigments [6, 158, 205]. In Japan, 70% of Direct Brown 95 is used for dyeing leather and 30% for dyeing paper [6].

NIOSH estimated that approximately 700 US workers are exposed to Direct Brown 95. The potential for exposure of the above benzidine-based dyes is greatest in two areas of operation: initial process reactions and filter press and blender units. The greatest potential for exposure (either dermal or via inhalation) is in the latter operations primarily because of the physical nature of the dye in the form of press cake or powder [200].

Metabolism and Toxicity

The benzidine-based dyes may enter the body by percutaneous absorption, ingestion or inhalation with percutaneous absorption being the primary route following occupational exposure [6]. Benzidine-derived azo dyes may be degraded metabolically in the gut or liver in man to free benzidine or monoacetylbenzidine [6, 207]. Environmental and urine samples were collected at six factories where workers were potentially exposed to benzidine-based dyes (two benzidine-based dye manufacturers, two textile-dyeing plants, a leather dyeing plant and a mill where paper was dyed). Monoacetylbenzidine was detected in the urine of 2/8 workers at one of the dye-manufacturing plants at levels of 3 and 7 ppb. At the second factory, 4 workers exposed to average levels of 7.9, 5.2, 11.7 and 17.4 mg total particulate/m^3 had corresponding urinary concentrations of 52, 11, 10 and 112 ppb benzidine; 590, 248 and 22 ppb monoacetylbenzidine were detected in urine samples containing 112, 52 and 11 ppb benzidine. Traces of diacetylbenzidine, *ortho*-tolidine and *ortho*-dianisidine were also detected [6].

Although the epidemiological data were inadequate to evaluate the carcinogenicity to man of Direct Black 38, Direct Blue 6 or Direct Brown 98, they, together with the presence of benzidine in the urine of exposed workers indicated that those exposures were strongly associated with the occurrence of bladder cancer. Base reports and other epidemiological studies, according to IARC, support the existence of such a realationship [6]. Additionally, IARC has deemed that there is sufficient evidence that commercial Direct Black 38, Direct Blue 6 and Direct Brown 95 are carcinogenic to rats [6].

Direct Black 38 was mutagenic in *S. typhimurium* strains TA 98 und TA 100 only when tested in the presence of a mouse liver metabolic activation system [208]. Monoacetylbenzidine, a major metabolite of benzidine and urine from hamsters given Direct Black 38 (100 mg/kg b.w.) was mutagenic for *S. typhimurium* strain TA 1538 only when tested in the presence of metabolic activation [209, 210]. Urine from rats given 500 mg/kg b.w. Direct Black 38 was also mutagenic for *S. typhimurium* strains TA 98 und TA 100 in the presence of metabolic activation [211].

Occurrence and Reactions

Direct Black 38 has been detected in the workplace air of a paper-dyeing facility, at total airborne particulate levels of 1.6–5.1 mg/m^3 (0.05–0.16 ppm), and a textile-dyeing operation at unspecified levels [6, 158]. Direct Blue 6 has been detected in the workplace air of a textile-dyeing operation at total airborne particulate levels of 1.20–3.94 mg/m^3 [6, 158]. Total airborne particulate levels of Direct Brown 95 detected in the workplace air of a textile-dyeing operation were 1.3–1.54 mg/m^3; those in a leather-dyeing operation, 1.12–14.72 mg/m^3; and those in a paper-dyeing facility, 0.17–3.30 mg/m^3 [6, 158].

There is a paucity of data concerning the environmental fate of the benzidine-based dyes. Benzidine is an environmentally significant degradation product of benzidine-based dyes as supported by the finding of levels ranging from 0.082 to 0.233 ppm in the Sumida River, Japan which at the time was receiving large quantities of wastewaters from dye and pigment plants. Levels of total aromatic amines in the river were reported as 0.205 to 0.565 ppm [188].

It is generally conceded that calculation of the environmental release of dyes is extremely difficult. Consideration of annual production and import data on dyestuffs fails to account for the dyes which are part of dyed materials and articles. Additionally, wastewaters and sludges containing some proportion of annual production and imports, the disposal of all dyed materials (many of which were produced several to many years ago) must be accounted for [204].

A study of Yoshida et al. [211] disclosed that a benzidine-based dye on cotton cloth lost color in 72 hours when incubated with river water, ostensibly releasing free benzidine which resisted further bacterial degradation.

The study of Hitz et al. [212] showed that these benzidine-based dyes with their high affinity for cellulosic materials are adsorbed to sludges in biological treatment. This could potentially lead to additional environmental release when the sludges are disposed of, used as a soil supplement or incinerated. Anerobic digestion of the dye-containing sludge might also release aromatic amines and other degradation products [204]. In this regard, it should be noted that Cerniglia et al.

[213] recently reported that anaerobic bacterial suspensions isolated from human, monkey and rat feces and intestinal contents metabolized Direct Black 38 as well as 3,3′-dimethylbenzidine-based dye, Direct Red 2 and a 3,3′-dimethoxybenzidine-based dye, Direct Blue 15 to their respective carcinogenic free amines.

The US Environmental Protection Agency further cautioned [204] that the classical emphasis on decolorization of dye wastes without attention to the degradation products possibly formed is also a serious potential concern, particularly if these wastes are chlorinated [214].

3,3′-Dichlorobenzidine (DCB)

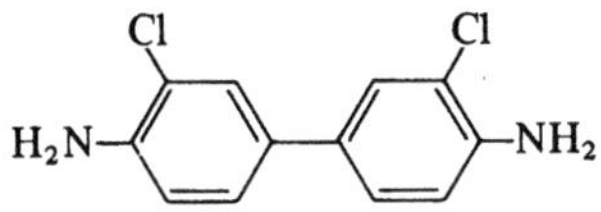

3,3′-dichlorobenzidine has been used in the color industry in England, Germany, Switzerland and in the United States for at least 60 years. It is an important starting material for the synthesis of diarylide yellow to red azo pigments that are used in printing inks, rubber, plastics, crayons, paints and textiles [6, 215].

Production and Use

3,3′-Dichlorobenzidine was first synthesized from *ortho*-chloronitrobenzene in 1900, and is produced commercially by alkaline reduction of *ortho*-chloronitrobenzene and rearrangement of the resulting 2,2′-dichlorohydrazobenzene with hydrochloric acid to produce the DCB dihydrochloride salt (the commercially available form) [2, 6]. An additional procedure is via the chlorination of diacetylbenzidine with hypochlorite followed by hydrolysis of the diamide to the diamine using HCl [216].

3,3′-Dichlorobenzidine has been produced commercially in the US since 1937, in western Europe since 1932 and in Japan since about 1955. At the present time two US companies produce commercial quantities of 3,3′-dichlorobenzidine dihydrochloride and the production of one of these companies in 1977 was 0.454–4.540 t [157]. Combined US production of 3,3′-dichlorobenzidine and its salts by three US companies in 1971 totalled 1.600 t US imports of 3,3′-dichlorobenzidine and its salts totalled 147 t in 1980 sharply reduced from the 658 t imported in 1971 [6].

3,3′-Dichlorobenzidine and its salts are believed to be produced by one company each in France, the Federal Republic of Germany, Italy and the UK with the total production estimated in 1979 to have been 100 t to 500 t [6]. Commercial production of 3,3′-dichlorobenzidine and its salts by the two producing companies in Japan was about 1,800 t in 1980 while in 1975 production was only about 870 t and about 100 t were exported [6].

The major use of 3,3′-dichlorobenzidine (in the form of its dihydrochloride) is as an intermediate for the production of organic pigments. While 15 organic dyes and pigments can be prepared from this amine, eight 3,3′-dichlorobenzidine-based pigments were produced in the US. In 1975, US production quantities of these 4 pigments were: C.I. Pigment Yellow 12, 2,7 million kg; Pigment Yellow 14, 0.87, kg; Pigment Yellow 11 million 187 thousand kg and Pigment Yellow 13, 108 thousand pounds. These and Pigment Yellow 55 are widely used in printing inks, plastics, elastomers and in textile printing [7, 217].

Of the remaining three, 3,3′-dichlorobenzidine based pigments produced in the US in 1975, C.I. Pigment Orange, 94 thousand kg are used mainly to color rubber products, printing inks, paper and all types of plastic, while C.I. Pigment Orânge 34 (446 thousand kg) are used to decorate finishes, transparent metal coatings, tin printing, foil lacquers, vinyl, polyethylene, and polypropylene plastics and textile printing inks. C.I. Pigment Red 38 are used primarily in rubber, plastics, inks and paints [217]. In general, the 3,3′-dichlorobenzidine-based pigments have maintained a fairly steady production/sales level in the US during the period 1971–1979, and in some cases, e.g., Pigment Yellow 12 and Pigment Yellow 13 have shown considerable growth [6]. In Japan in 1980 essentially all of the estimated 1,500 t of 3,3′-dichlorobenzidine dihydrochloride used was for the production of pigments. 3,3′-Dichlorobenzidine has been used alone or in blends with 4,4′-methylenebis(2-chloroaniline) as a curing agent for liquid-castable polyurethane elastomers [6, 7].

In 1973, 18 US companies were using 3,3′-dichlorobenzidine and 166 workers were potentially exposed [6, 218]. In 1974, NIOSH estimated that approximately 1,000 people were exposed to DCB. However, no data were available on the number of workers exposed during the actual manufacturing of the compound as well as those involved in processing, storage and handling and repackaging operations.

Metabolism and Toxicity

Although occupational exposure to 3,3′-dichlorobenzidine has and probably still does occur during its conversion to derived pigments and rubber workers were formerly and may still be exposed to 3,3′-dichlorobenzidine used for curing polyurethane elastomers, the epidemiological data are thus far deemed inadequate to evaluate the carcinogenicity of this amine in man [6].

DCB following its oral administration produced liver-cell tumors in mice, hepatocellular carcinomas in dogs, mammary and Zymbal-gland tumors in rats and carcinomas of the urinary bladder in hamsters and dogs [6].

DCB is mutagenic to *S. typhimurium* TA 1538, TA 98 and TA 100 in the presence and (for some strains) absence of a liver activation system [6, 18, 20, 219]. 3,3′-Dichlorobenzidine is believed to be rapidly degraded *in vivo* [6, 178, 179].

DCB is readily absorbed following oral administration in the rat and then metabolized and excreted mainly via the feces. DCB metabolites recovered in the urine were mostly unconjugated compounds, whereas most of the metabolites in the bile were conjugates. It was suggested that DCB was metabolized to reactive metabolites which bind covalently to cellular macromolecules. The metabolic transformation of DCB may involve N-acetylation, N-hydroxylation, and ring-hydroxylation followed by O-esterification or O-glucuronidation [220].

Occurrence and Reactions

(a) Occupational

A study in 1970 of worker exposure to 3,3′-dichlorobenzidine in a pigment manufacturing plant in Japan showed that concentrations of 3,3′-dichlorobenzidine in the air reached a level of 25 μg/m³ (2 ppb) within 10 min of charging reaction vessels and dropped to 2 μg/m³ (0.2 ppb) within 20 min [6, 221].

(b) Water and Sediments

With current work practices, effluents containing 3,3′-dichlorobenzidine may be discharged directly into receiving waters. Moreover, the discharge of DCB-pigment wastes into receiving waters constitutes an additional source of DCB into the environment, since free unreacted DCB is reported to be present in these pigments [222].

DCB absorbs extensively to a variety of aquatic sediments and becomes more tightly bound with time [223]. Purge wells and seepage water near a waste disposal lagoon receiving wastes from the manufacture of DCB were reported to contain 0.13–0.27 mg/l DCB. In Japan, the Sumida River which receives waste effluents from several dye and pigment factories was also found to contain DCB [6, 221]. DCB has also been detected in unspecified amounts in oil refinery effluents, industrial effluents and surface water [99].

Adsorption of DCB to natural aquatic sediments was inhibited under alkaline (pH 9) conditions by 30–50%. Desorption of DCB from sediments was very low confirming the affinity for aquatic sediments [224].

(c) Degradation and Biotransformation

3,3′-Dichlorobenzidine is very rapidly photodegraded in aqueous solution to yield monochlorobenzidine (MCB), benzidine, and a number of brightly colored water insoluble materials. Dissappearance quantum yields for DCB and MCB are high, whereas that for benzidine is much lower. Transient products are generated upon photolysis of DCB in acidic solutions or on treatment of DCB with chlorine water. The photolability of DCB is much lower in organic solvents and the mechanism of dechlorination does not involve simple carbon-chlorine bond homolysis [225]. From environmental considerations, the action of sunlight on DCB will lead to its degradation but not necessariliy to its detoxification since benzidine, a relatively photostable carcinogen, is one of the products. Additionally, the inertness of DCB in hydrocarbon solvents coupled with its high octanol/water coefficient might lead to enhanced stability in water contaminated with hydrocarbons [225].

Bioconcentration, elimination and metabolism studies in bluegill sunfish revealed that [^{14}C]-DCB was rapidly accumulated by the fish from water containing 5 ppb or 0.1 ppm of the chemical. Based on total ^{14}C residues, bioconcentration factors of 495 to 507 were observed in the whole fish with equilibria achieved in 96 to 168 hours. The ^{14}C residues were distributed in both the edible and nonedible portions. Bioconcentration was 12- to 43-fold higher in the nonedible portions. No evidence of toxicity was found. Elimination of the residues was progressive in the depuration phase of the studies, with 60–73% of the residues in the edible portion and 96% of those in the nonedible portions eliminated after 14 days exposure to fresh water. The only metabolite detected in the fish was an acid-labile conjugate of DCB which appears to be an N-glucuronide. It was cautioned that the ability of DCB to concentrate in aquatic organisms may present a direct hazard to human health through the consumption of contaminated fish [222].

1- and 2-Naphthylamines

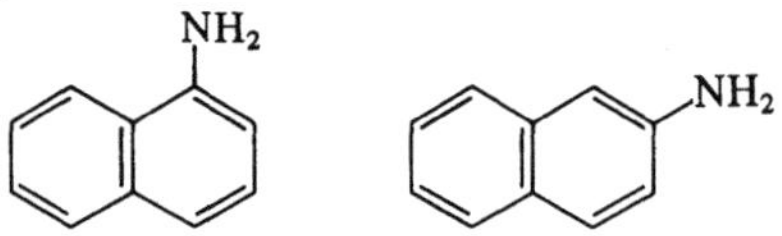

1- and 2-Naphthylamines have been produced commercially for at least 50 years until US production of the latter was stopped in early 1972. 2-Naphthylamine is recognized as a human carcinogen producing bladder cancer on prolonged exposure. Both 1- and 2-naphthylamines had been widely used in the production of dyes, chemical intermediates, antioxidants and pesticides [7, 8, 10, 11, 216, 226].

Production and Use

Although 1-naphthylamine was produced from 1-nitronaphthalene by reduction with iron and hydrochloric acid in the past, in recent years it has been made exclusively by catalytic hydrogenation of 1-nitronaphthalene using a nickel catalyst [3, 216].

Production data on 1-naphthylamine is extremely scant. Although in 1963 US production of 1-naphthylamine was estimated at more than 500 t, by 1972 only one US company was still producing the chemical, and no other data are available on its production in any other countries [216]. US imports of 1-naphthylamine have grown from 2 t in 1967 to 27 t in 1971. Although 1-naphthylamine can be used as an intermediate in the production of 150 dyes, it has actually been used in the preparation of about 50 dyes [216]. It is also used as an intermediate in the production of 1-naphthylamine-4-sulfonic acid (naphthionic acid) which is used to make 1-naphthol-4-sulfonic acid, a coupling agent for azo dyes. 1-Naphthylamine is used to produce the herbicide N-1-naphthylphthalamic acid as well as the rodenticide ANTU (1-naphthylthiourea), and the rubber antioxidants N-phenyl-1-naphthylamine and the aldol-1-naphthylamine condensate [3, 216]. Although no production data are available for the herbicide N-1-naphthylphthalamic acid, one survey estimated the total US farm usage in 1966 to be 450 t [227]. As of August, 1968, the acid and its sodium salt were approved in the US for use on 5 fruits, 8 vegetables and on cotton and peanuts.

2-Naphthylamine was generally prepared via the reaction of 2-naphthol with ammonia and ammonium sulfite. It is no longer produced in the US and in 1967 US imports of the amine amounted to 17.4 thousand kg [216]. Data on the production of 2-naphthylamine in countries other than the US are not available.

2-Naphthylamine has been used principally as an intermediate in the production of twenty-one dyes and antioxidants. Although the antioxidant N-phenyl-2-naphthylamine and another antioxidant derived from it, N-phenyl-2-naphthylamine-acetone condensate, may have been made from 2-naphthylamine in the US in the past, it is probable that in recent years these products have been manufactured by the condensation of aniline and 2-naphthol [216].

Human exposure to 1- and 2-naphthylamines in the past was primarily via inhalation and dermal routes. NIOSH estimated that approximately 35,000 and 500 workers were exposed to 1-naphthylamine and 2-naphthylamine respectively, in the period 1972–1974 [162].

Metabolism and Toxicity

An association between the manufacture or handling of 2-naphthylamine and the induction of bladder cancer has long been recognized [256–258]. Bladder cancer has been observed in dogs (with commercial 2-naphthylamine [259–260] and purified 2-naphthylamine [261–263], hamster [264, 265], rhesus monkey [266], but not the mouse [267]. There is conflicting evidence on the suspectibility of the rat [267–269].

The data on the carcinogenicity of 1-naphthylamine in humans [256] and experimental animals [260] are not clear primarily due to the fact that commercial 1-naphthylamine contained 3–6% of 2-naphthylamine until the last decade when new production methods dramatically reduced the levels of the carcinogenic 2-naphthylamine [263]. Other studies in dogs [270, 271] and in hamsters [264] failed to demonstrate the induction of bladder cancer by 1-naphthylamine. In recent lifetime carcinogenicity study of 1- and 2-naphthylamine in dogs [263] all dogs which received pure 2-naphthylamine developed transitional cell carcinomas of the bladder with 34 months. No carcinogenic effect of 1-naphthylamine was observed when it was administered alone to dogs.

Carcinogenicity of 2-naphthylamine is attributed to its metabolism in the liver to the N-hydroxy and nitroso-derivatives which are excreted in the urine either as the conjugates or as the metabolite itself. The conjugates are unstable in the bladder and hydrolyze yielding the original oxidation products [272]. The N-hydroxy derivatives of both 1- and 2-naphthylamine are carcinogenic. The N-hydroxy derivative of 1-naphthylamine is excreted in the urine [273] is considered stable and unreactive with low molecular weight urinary nucleophiles [274] and is more potent at sites of direct application than the 2-hydroxy derivative [273, 274]. Both 1- and 2-naphthylamine are positive in *in vitro* mutagenicity assays which incorporate liver microsomal preparations for metabolism [275].

Occurrence and Reactions

1- and 2-Naphthylamines have been found in coal-tar and in the pyrolysis of nitrogen-containing matter (e.g., pyrolysis of 2-glutamic acid and L-leucine at 700 °C) [216]. Gas retort houses have been shown to have atmospheric levels of 2-naphthylamine sufficient to give a human exposure of 0.2 μg per day [216]. Cigarette smoke has been reported to contain 0.03 μg and 0.02 μg/per cigarette respectively of 1- and 2-naphthylamine which is equivalent to 1 μg/50 cigarettes, the daily ratio of a heavy smoker. A man smoking 30 cigarettes per day could absorb 200 μg of 2-naphthylamine per year [216, 228, 229].

The mainstream smoke of one US 85 mm cigarette, without filter tip, was found to contain 4.3 ng of 1-naphthylamine, 102 ng of aniline, 61 ng of 2-, 3- and 4-tolidine, 55.8 ng of ethyl and dimethylaniline, 6.9 ng of 2-, 3-, and 4-aminobiphenyl and 5.8 ng of 2-methyl-1-naphthylamine. Side stream smoke contained levels of aromatic amines 20 to 68 times higher than those in the mainstream smoke [230].

1- and 2-Naphthylamines were found in the waste streams in the Sumida River, Japan, from plants where they were produced and used and have been reported to be present in the effluent from certain dyestuffs factories [188].

Phenyl-2-Naphthylamine (PBNA)

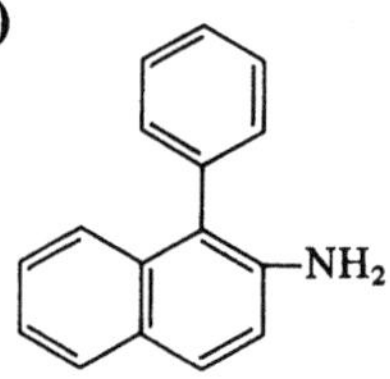

Production and Use

Since the 1950's, 2-naphthylamine has been replaced as an antioxidant in rubber by phenyl-2-naphthylamine and phenyl-1-naphthylamine, which until recently was regarded as safe substitutes [158, 231]. It is believed that PBNA was made by at least one manufacturer in the US by the condensation of 2-naphthylamine and aniline. Three US companies reported the production of 2,240 t of PBNA in 1973, 1,370 t in 1974, and 708.7 t in 1975. It is believed that no US companies are currently producing PBNA. In Japan, the annual production of PBNA declined from 5,060 t in 1971 to 2,250 t in 1974, and is currently being produced by four Japanese companies [8, 231].

PBNA at levels of 1–2% is used primarily as an antioxidant in rubber processing [238] and in polymers, greases, lubricating and transformer oils [231]. It is also employed as a stabilizer in electrical-insulating silicone enamels [232] and as a chemical intermediate in the production of the rubber antioxidant, N-phenyl-2-naphthylamine acetone condensate, and in the production of seven dyes [233] of which only one, C. I. Acid Blue 98, appears to be of commercial importance worldwide [234]. PBNA has been described as an effective heat and light stabilizer in a number of applications as well as a vulcanization accelerator and polymerization inhibitor. It has been reported that commercial samples of PBNA in the US were contaminated with 20–30 mg/kg of the carcinogen 2-naphthylamine [235]. While PBNA samples in the UK in the past contained 15–50 mg/kg 2-naphthylamine, levels of less than 1 mg/kg have been reported in recent years in one commercial product. Commercial samples of PBNA in Japan have been found to contain aniline, 2-naphthol, and 2-naphthylamine as impurities [231].

NIOSH has estimated that 15,000 workers in the US are potentially exposed to PBNA during its manufacture and use [236].

Metabolism and Toxicity

PBNA is a suspect carcinogen because of its possible breakdown to 2-naphthylamine, a potent human bladder carcinogen [231]. Earlier reports suggested that PBNA did not cause cancer in exposed workers. These reports were based on epidemiological evidence that there was no increase in cancer incidence in workers engaged in the large-scale production of PBNA in a number of countries for many years [231, 236].

A statistically significant increase in the incidence of all tumors, particularly hepatomas in male mice, has been found in animals administered PBNA orally or by single subcutaneous injection [238]. 2-Naphthylamine has been found in the urine of workers who had been exposed to PBNA, and in the urine of healthy human volunteers to whom PBNA was administered [239], as well as in urine of dogs administered PBNA orally [240].

Occurrence and Reactions

There is a paucity of data relating to the occurrence of PBNA in the environment. PBNA was detected in the air in several synthetic rubber plants in the USSR [231]. When samples of 1,3-butadiene rubber were heated to 220 °C, the evolved fumes contained 0.012 mg/g (22 mg/kg) of PBNA [240, 241].

(a) Biotransformation and Degradation

PBNA was biotransformed into 2-naphthylamine in a laboratory sewage sludge system. 2-Naphthylamine appeared to accumulate in the aqueous phase while the undergraded PBNA appeared to be associated with the sludge solids [242]. Substantial transformation of PBNA to BNA occurred especially at the 2 ppm level where approximately 36% of the PBNA was converted to BNA. This conversion appeared to be dependent only on the microbial activity of the sludge since 2-naphthylamine was not detected in the sterilized sludge with 2 ppm added PBNA. At the 5 and 8 ppm levels, the conversions of PBNA to 2-naphthylamine were 25 and 16% respectively. Unlike PBNA, the majority of 2-naphthylamine was found in the aqueous phase and very little was associated with the solids. The large amounts of 2-naphthylamine found in the aqueous phase may suggest that further biodegradation of 2-naphthylamine is limited [242]. It should also be noted that 1-naphthylamine was not biodegraded in an activated sludge system that was acclimated with PBNA [243]. 2-Naphthylamine has also been shown to inhibit the oxygen uptake of sludge obtained from a water treatment plant [244].

4,4-Methylenebis(2-Chloroaniline)(MOCA)

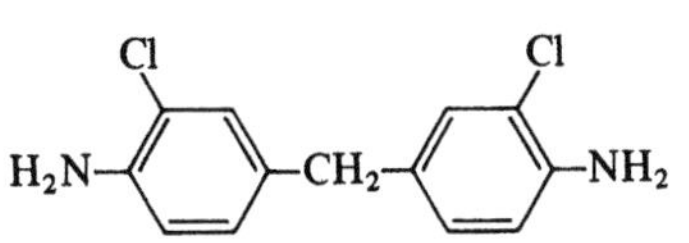

Production and Use

MOCA is prepared commercially by the condensation of formaldehyde with 2 moles of *ortho*-chloroaniline, traces of which may remain in the commercial product at levels of approximately 0.9% [245]. MOCA was produced in quantity until 1973, when OSHA promulgated an emergency temporary standard for MOCA and 13 other agents. US production figures for 1970 and 1972 are estimated to have been approximately 2,000 and 3,500 t respectively [216]. Although no production data are available for MOCA produced in countries other than the US, the U.K. was reported to have one producer in 1970 and Japan three producers in 1972 [216]. MOCA has been used principally as a curing agent for isocyanate-containing polymers. It has also been employed in conjunction with various other aromatic amines, such as 3,3′-dichlorobenzidine or 4,4′-methylenedianilide, to prepare curing agents. MOCA was perhaps the most widely used agent for curing liquid-castable polyurethane elastomers, suitable for molded mechanical articles, and for potting and encapsulating purposes [246]. Perhaps less than 1% of the total MOCA consumed is believed to have been used as a curing agent for epoxy and epoxy-urethane resin blends [216].

It has been estimated that approximately 18,000 workers have been potentially exposed to MOCA in the United States [247].

Metabolism and Toxicity

IARC [216] stated in 1974 that there are no conclusive epidemiological studies on which an evaluation of the carcinogenic risk of MOCA can be based. MOCA has induced lung and liver tumors in rats and mice [248, 249] and urinary bladder tumors in dogs [246].

MOCA is mutagenic in *S. typhimurium* [250], and was demonstrated to be genotoxic when tested in hepatocytes of rat, mouse and hamster. DNA repair was elicited in rat and hamster hepatocytes in primary culture exposed to 10^{-5} M MOCA (aniline was negative) [251].

Occurrence and Reactions

As in the case for all aromatic amines discussed in this chapter, MOCA can be present in air as an aerosol, a vapor or a combination of the two. In operations where it is packaged or transferred in bulk, it is most likely to be generated primarily as an aerosol. During compounding, MCOA is melted at approximately 121 °C, to be mixed with the formulation as a curing agent for polymer and epoxy resin systems containing isocyanates [252].

A recent study demonstrated that MOCA was not limited to the workers who handle the compound but MOCA contamination of soil and homes was shown in the area surrounding the manufacturing plant at Adrian, Michigan [253]. In 1979, the State of Michigan, found MOCA levels ranging from 1.6 up to 2,000 ppm as monitored from a variety of different samples at and around a company site in Adrian. The minimum concentrations of MOCA in the industrial lagoon sediments was 1,600 ppm, soil sampling taken near two monitoring wells showed concentrations of 1.6 and 6.3 ppm while drain sludge showed extremely high concentrations of 2,000 ppm. Ambient air levels of MOCA were reported to be <0.1–22 $\mu g/m^2$ [254].

References

1. Northcott, J.: Aniline and its derivatives, in: Encyclopedia of Chemical Technology (R.E. Kirk and D.F. Othmer (eds)), 3rd ed., Vol. 2, John Wiley & Sons, New York (1978), pp. 309–321
2. Ferber, K.H.J.: Benzidine and related biphenyldiamines, in: Encyclopedia of Chemical Technology (R.E. Kirk and D.F. Othmer (eds)), 3rd ed., Vol. 3, John Wiley & Sons, New York (1978), pp. 772–778
3. Dressler, H.: Naphthalene derivatives, in: Encyclopedia of Chemical Technology (R.E. Kirk and D.F. Othmer (eds)), 3rd ed., Vol. 2, John Wiley & Sons, New York (1978), pp. 309–321
4. IARC: Monographs on the Evaluation of Carcinogenic Risk of Chemicals to Man, Vol. 16, Some Aromatic Amines and Related Nitro Compounds – Hair Dyes, Colouring Agents and Miscellaneous Industrial Chemicals, International Agency for Research on Cancer, Lyon (1978), pp. 51–142
5. IARC: Monographs on the Evaluation of Carcinogenic Risk of Chemicals to Man, Vol. 27, Some Aromatic Amines, Anthraquinones and Nitroso Compounds and Inorganic Fluorides Used in Drinking Water and Dental Preparations, International Agency for Research on Cancer, Lyon (1982), pp. 39–188
6. IARC: Monographs on the Evaluation of Carcinogenic Risk of Chemicals to Man, Vol. 29, Some Industrial Chemicals and Dyestuffs. International Agency for Research on Cancer, Lyon (1982), pp. 149–183, 239–256, 295–300, 296–310, 311–320, 321–330
7. Fishbein, L.: Potential Industrial Carcinogens and Mutagens, Elsevier, Amsterdam (1979), pp. 256–416
8. Fishbein, L.: Aromatic amines of major industrial importance, use and occurrence, in: Environmental Carcinogens Selected Methods of Analysis, Vol. 4. Some Aromatic Amines and Azo Dyes in the General and Industrial Environment (H. Egan, L. Fishbein, M. Castegnaro, I.K. O'Neill, H. Bartsch (eds)), International Agency for Research on Cancer, Lyon (1981), pp. 51–74
9. Veys, C.A.: Ann. Occup. Hyg. *15*, 11–15 (1972)
10. Kriek, E.: Aromatic amines and related compounds as carcinogenic hazards to man, in: Environmental Carcinogenesis (P. Emmelot and E. Kriek (eds.)) Elsevier/North Holland Biomedical Press, Amsterdam (1979), pp. 143–164
11. Parkes, H.G.: Aromatic amines, in: Chemical Carcinogens (C.E. Searle (ed)) American Chemical Society Monograph No. 173, Washington, DC (1976), pp. 462–480
12. Lower, G.M., et al.: Environ. Health Persp. *29*, 71 (1979)
13. Yashida, O., Miyakoga, M.: Etiology of bladder cancer, in: Metabolic Aspects in Analytical and Experimental Epidemiology of Cancer (W. Nakahara, T. Hirayana, K. Wishroka, and A. Sugans (eds.)), University Park Press, Baltimore, MD (1973), pp. 31–39
14. Case, R.A.M., Hosker, M.E.: Brit. J. Ind. Med *11*, 75 (1954)
15. Clayson, D.G.: Brit. Med. Bull. *20*, 115 (1964)
16. Boyland, E.: Brit. Med. Bull. *14*, 153 (1958)
17. Wilmer, J.L., et al.: Environ. Mutagenesis *3*, 627 (1981)
18. Anderson, D., Styles, J.A.: Brit. J. Cancer *37*, 924 (1978)
19. Ashby, J., et al.: Brit. J. Cancer *38*, 34 (1978)
20. Lazear, E.J., Louie, S.C.: Cancer Lett. *4*, 21 (1977)
21. Garner, R.G., et al.: Cancer Lett. *1*, 39 (1975)
22. Rao, T.K. et al.: Arch. Toxicol. *49*, 185 (1982)
23. Miller, E.C., Miller, J.A.: Pharmacol. Res. *18*, 805 (1966)
24. Miller, J.A.: Cancer Res. *30*, 559 (1970)
25. Miller, E.C., Miller, J.A.: Biochemical mechanisms of chemical carcinogenesis, in: The Molecular Biology of Cancer (H. Bush (ed.)) Academic Press, New York (1974), pp. 376–402
26. Miller, E.C., Miller, J.A.: The metabolism of chemical carcinogens to reactive electrophiles and their possible mechanisms of action in carcinogenesis, in: Chemical Carcinogens (C. E. Searle (ed.)) ACS Monograph 173, Washington, DC (1976), pp. 736–762

27. Irving, E.C.: Species and tissue variations in the metabolic activation of aromatic amines, in: Carcinogens: Identification and Mechanisms of Action (A. C. Grffin and C.R. Shaw (eds.)) Raven Press, New York (1979) 211–227
28. Irving, C.C.: Metabolic activation of N-hydroxy compounds by conjugation, in: Biological Oxidation of Nitrogen in Organic Molecules (J.W. Bridges, J.W. Gorrod, and D. V. Parke (eds.)) Halsted Press, New York (1971), pp. 75–86
29. Uehleke, H.: N-Hydroxylation, in: Biological Oxidation of Nitrogen in Organic Molecules (J.W. Bridges, J.W. Gorrod and D.V. Parke (eds.)) Halsted Press, New York (1971), pp. 15–26
30. Kadlubar, F.F., et al.: Cancer Res. *37*, 805 (1977)
31. Clayson, D.B.: Brit. J. Cancer *7*, 460 (1953)
32. Boyland E.: The biochemistry of aromatic hydrocarbons, amines and urethane, in: Physico-chemical Mechanisms of Carcinogenesis (E.D. Bergman and B. Pullman (eds.)) Vol. 1, Israel Acad. Sci. and Humanities, Jerusalem (1969), pp. 25–44
33. Clayson, D.B., Garner, R.C.: Carcinogenic aromatic amines and related compounds, in: Chemical Carcinogens (C.E. Searle (ed.)) American Chemical Society, Washington, DC (1976), pp. 366–461
34. Roberts, J.J., Warwick, G.P.: Int. J. Cancer *1*, 179–196 (1966)
35. Miller, J.A., Miller, E.C.: Metabolic activation of carcinogenic aromatic amines and amides via N-hydroxylation and N-hydroxyesterification and its relationship to ultimate carcinogens as electrophilic reactants, in: Physico-Chemical Mechanisms of Carcinogenesis (E.D. Bergman and B. Pullman (eds.)) Israel Acad. of Sci. and Humanities (1969), pp. 237–261
36. Kriek, E., Westra, J.G.: Metabolic activation of aromatic amines and amides and interactions with nucleic acids, in: Chemical Carcinogens and DNA (P.L. Crover (ed.)) CRC Press, Boca Raton, FL (1979), pp. 1–28
37. Martin, C.N, et al.: Cancer Res. *42*, 2678–2686 (1982)
38. Jones, T.C.: Preliminary risk assessment. Phase I. Benzidine, its congeners and their derivative dyes and pigments. EPA Publ. No. 560/11-80-019, US Environmental Protection Agency, Washington, DC (1980)
39. Guerin, M.R., et al.: Environ. Res. *23*, 42 (1980)
40. Wilson, B.W., et al.: Mutat. Res. *79*, 193 (1980)
41. Guerin, M.R., et al.: Fuel. *60*, 282 (1982)
42. Buchanan, M.V., et al.: In: Chemical Analysis and Biological Fate: Polynuclear Aromatic Hydrocarbons (M. Cooke and A.J. Dennis (eds.)), Battelle Press, Columbus, OH (1981), pp. 133–144
43. Haugen, D.A., et al.: Anal. Chem. *54*, 32 (1982)
44. Tomkins, B.A., Ho, C.H.: Anal. Chem. *54*, 91 (1982)
45. Pelroy, R. A., Gandolfi, A.: Mutat. Res. *72*, 329 (1980)
46. Wilson, B.W., et al.: Proc. 29th Conf. Mass Spectrometry and Allied Topics. Abst. No. NPMO85 (1981)
47. Wilson, B.W., et al.: Proc., 2nd USDOE Environm. Control Symp. *2*, 415 (1980)
48. Hoffmann, D., Wynder, E.L.: in: Chemical Carcinogenesis (C.E. Searle (ed.)) Amer. Chem. Soc., Washington, DC (1976) pp. 366–462
49. Sugimura, T., et al.: Mutagen-Carcinogens in food with special reference to highly mutagenic pyrolytic production in broiled food, in: Origins of Human Cancer (H.H. Hiatt, J.D. Watson and J.A. Winsten (eds.)) Cold Spring Harbor Laboratory, Cold Spring Harbor, NY (1977), pp. 1561–1577
50. Sugimura, T., et al.: Mutagenic heterocyclic amines in cooked food, in: Environmental Carcinogens Selected Methods of Analysis. Vol. 4. Some Aromatic Amines and Azo Dyes in the General and Industrial Environment (H. Egan, L. Fishbein, I.K. O'Neill, M. Castegnaro, H. Bartsch (eds.)) Internat. Agency for Res. Cancer, Lyon (1981), pp. 251–267
51. Nagao, M. et al.: Cancer Lett. *2*, 221 (1977)
52. Spingarn, N.E., Weisburger, J.H.: Cancer Lett. *7*, 259 (1979)
53. Spingarn, N.E. et al.: Cancer Lett. *9*, 177 (1980)
54. Hoffmann, D., et al.: Nature *221*, 254 (1969)
55. Masuda, Y., Hoffman, D.: Anal. Chem. *41*, 650 (1969)
56. IARC: Environmental Carcinogens, Selected Methods of Analysis. Vol. 4. Some Aromatic Amines and Azo Dyes in the General and Industrial Environment (H. Egan. L. Fishbein, M. Castegnaro, I.K. O'Neill, H. Bartsch (eds.)), Internat. Agency Res. Cancer, Lyon (1981), pp. 109–247

57. Fishbein, L.: Analysis of carcinogenic and mutagenic aromatic amines, in: Safe Handling of Chemical Carcinogens Mutagens, Teratogens and Highly Toxic Substances. Vol. 1 (D. B. Walters (ed.)), Ann Arbor Science, Ann Arbor, Michigan (1980), pp. 205–245
58. Fishbein, L.: Analysis of carcinogenic and mutagenic aromatic amines: An Overview. Toxicol. Environ. Chem. Revs. *3*, 145 (1980)
59. Parris, G.E.: Residue Revs. *76*, 2 (1980)
60. Society of Dyers & Colourists: Colour Index, revised 3rd ed. Additions and Amendments, No. 34, Perkin House, Bradford, UK (1980)
61. Anon: Chemical Profiles, Schnell Publ. Co., New York, 1 Oct. (1979)
62. Kouris, C.S., Northcott, J.: Aniline and its derivative, in: Encyclopedia of Chemical Technology (R.E. Kirk and D.F. Othmer (eds.)), 2nd. ed., Vol. 2, John Wiley, New York (1963), pp. 411–419
63. Parke, D.V.: Biochem. J. *77*, 493 (1960)
64. Williams, R.T.: Detoxification Mechanisms, 2nd ed., Chaptman & Hall, Ltd., London (1959, pp. 430–432, 464–465
65. Boyland, E., et al.: Biochem. J. *65*, 417 (1957)
66. Boyland, E., et al.: Biochem. J. *86*, 263 (1963)
67. Kiese, M.: Pharmacol. Rev. *18*, 1091 (1966)
68. Bray, H.G., et al.: Biochem. J. *47*, 483 (1950)
69. Grantham, P.H., et al.: Xenobiotica. *2*, 551 (1972)
70. Kao, J., et al.: Drug Metab. Disp. *6*, 549 (1978)
71. Piotrowski, J.K.: J. Hyg. Epidemiol. Mecrobiol. Immunol. (Praha) *1*, 23 (1957)
72. Dutkiewicz, T.: Med. Pracy. *12*, 1 (1961)
73. Dutkiewicz, T., Piotrowski, J.K.: Pure Appl. Chem. *3*, 319 (1961)
74. Piotrowski, J.K.: Pral. Lek. *24*, 94 (1972)
75. Lores, E.M., et al.: J. Chromatog. Sci. *16*, 358 (1978)
76. Hargesheimer, E.E., et al.: J. Assoc. Off. Anal. Chem. *64*, 833 (1981)
77. Crosby, D.G., Bowers, J.B.: J. Agr. Food Chem. *16*, 839 (1968)
78. Bartha, R.: J. Agr. Food Chem. *19*, 385 (1971)
79. Kaufman, D.D.: J. Agr. Food Chem. *15*, 582 (1967)
80. Moreale, A., Van Bladel, R.: J. Soil Sci *27*, 48 (1976)
81. Hsu, T.S., Bartha, R.: Soil Sci. *116*, 444 (1973)
82. Geisbühler, H., et al.: The substituted ureas, in: Herbicides: Chemistry, Degradation and Mode of Action (P.C. Kearney, D.D. Kaufman (eds.)), 2nd ed., Vol. I, Marcel Dekker, Inc., NY (1975) pp. 209–291
83. Still, G.G., Herrett, R.A.: Methyl carbamates, carbanilates and acylanilides, ibid., pp. 609–664
84. Bartha, R., Pramer, D.: Adv. Appl. Microbiol. *13*, 317 (1970)
85. Kouris, C.S., Northcott, J.: Aniline and its derivatives. in: Encyclopedia of Chemical Technology (R.E. Kirk, D.F. Othmer (eds.)), 2nd ed., Vol. 2, John Wiley, New York (1969), pp. 411–427
86. Nguyen, Y.V., Phillips, C.R.: Chemosphere *3*, 125 (1975)
87. Gitchell, A., et al.: J. Air Poll. Contr. Assoc. *24*, 357 (1974)
88. Gitchell, A., et al.: J. Air Poll. Contr. Assoc. *24*, 772 (1974)
89. Spicer, C.W., et al.: Environ. Sci. Technol. *8*, 1028 (1974)
90. Patrianakos, C., Hoffmann, D.: J. Anal. Toxicol. *3*, 150 (1979)
91. Joel, A.R., Grady, C.P., Jr.: J. Water Poll. Cont. Fed. *48*, 778 (1977)
92. Symons, J.M., et al.: Int. J. Air Water Poll. *4*, 115 (1962)
93. Tischler, L.F., Eckenfelder, W.W., Jr.: Linear substrate removal in the activated sludge process. in: Adv. Water Pollution Res., Proc. 4th Intl. Conf. Water Poll. Res. 361 (1969)
94. Neurath, G.B., et al.: Food Cosm. Toxicol. *15*, 275 (1977)
95. Meijers, A.P., VanderLeer, R.W.: Water Res. *10*, 597 (1976)
96. Greve, D.A., Wegman, R.C.C.: Schriftenr. Ver. Wasser. Buden. Lufthyg. (Berlin) *46*, 59 (1975)
97. Wegman, R.C.C., Dekorte, G.A.L.: Int. J. Environ. Anal. Chem. *9*, 1 (1981)
98. Herzel, F., Schmidt, G.: Gesund. Ing. *98*, 221 (1977)
99. Hushon, J., et al.: An Assessment of Potentially Carcinogenic, Energy-Related Contaminants in Water. Prepared for US Dept. of Energy and National Cancer Inst., Mitre Corp., McLean, VA (1980), p. 71, 72, 102
100. Shackelford, W.M., Keith, L.H.: Frequency of Organic Compounds Identified in Water. EPA Report 600/4-76-062, Environmental Research Laboratory, Athens, GA (1976), pp. 9, 58, 59, 78

101. Games, L.M., Hites, R.A.: Anal. Chem. *49*, 1433 (1977)
102. Hamaker, J.W.: Decomposition: Quantitatives aspects. in: Organic Chemicals in the Soil Environment (C.A.I. Goring, J.W. Hamaker), Marcel Dekker, New York (1972), 253
103. Zoeteman, B.C.J., et al.: Chemosphere *9*, 231 (1980)
104. Zepp, R.G., et al.: Environ. Sci. Tech. *9*, 1144 (1975)
105. Struif, B., et al.: Vom Wasser *45*, 53 (1975)
106. Kaussmaul, H.: Behavior of persistent organic compounds in bank-filtrated Rhine water. in: Aquatic Pollutants (O. Hutzinger, H. VanLelyveld, B.C.J. Zoeteman (eds.)), Pergamon Press, Oxford (1978), p. 265
107. Parris, G.E.: Environ. Sci. Technol. *14*, 1099 (1980)
108. Hauser, C.S., Krivonak, A.E.: J. Agr. Food Chem. *26*, 1164 (1978)
109. Furukawa, T., Brindley, D.D.: Clays, Clay Minerals *21*, 279 (1973)
120. Pillai, P., et al.: Chemosphere *11*, 299 (1982)
111. Helling, C.S., Krivonak, A.E.: J. Agr. Food Chem. *26*, 1164 (1978)
112. Süss, A., et al.: Z. Pflanzenernaehr. Bodenkd. *141*, 57 (1978)
113. Hsu, T.S., Bartha, R.: Soil Sci. *118*, 213 (1974)
114. Bordelleau, L.M., et al.: J. Agr. Food Chem. *20*, 573 (1972)
115. Cerniglia, C.E., et al.: Arch. Microbiol. *130*, 272 (1981)
116. Alexander, M.: Science *211*, 132 (1981)
117. Ferris, J.P. et al.: Arch. Biochem. Biophys. *156*, 97 (1973)
118. Smith, R.V., Rosazza, J.P.: Arch. Biochem. Biophys. *161*, 551 (1974)
119. Daly, J.W., et al.: Experentia *28*, 1129 (1972)
120. Booth, J., Boyland, E.: Biochem. J. *91*, 362 (1964)
121. Bordeleau, L.M., Bartha, R.: Can. J. Microbiol. *18*, 1873 (1972)
122. Kaufman, D.D., et al.: J. Agr. Food Chem. *21*, 127 (1973)
123. Kearney, P.C., Plimmer, J.R.: J. Agr. Food Chem. *20*, 584 (1972)
124. Burge, W.D.: Soil Biol. Biochem. *4*, 379 (1972)
125. Lu, A.Y.H., et al.: Arch. Biochem. Biophys. *153*, 294 (1972)
126. Archakov, A. I., et al.: Biochem. Pharm. *23*, 1053 (1974)
127. Schenkman, J.B., et al.: Biochem. Pharm. *23*, 1148 (1974)
128. Smith, R.V., Rosazza, J. P.: Arch. Biochem. Biophys. *161*, 551 (1974)
129. Bachofer, R., et al.: FEBS Letters *50*, 288 (1975)
130. Jerina, D.M., et al.: Biochemistry *9*, 147 (1970)
131. Patel, T.R., Gibson, D. T.: J. Bacteriol. *119*, 879 (1974)
132. Calamari, D., et al.: Chemosphere *9*, 753 (1980)
133. Jolley, R.L.: Chlorination effects on organic constituents in effluents from domestic sanitary sewage treatment plants. Oak Ridge National Laboratory Publ. No. ORNL-TM-4290, Oak Ridge National Laboratory, Oak Ridge, TN (1973)
134. Jenkins, R.L., et al.: Arch. Environ. Contam. Toxicol. *7*, 301 (1978)
135. Noller, C.R.: Chemistry of Organic Compounds, W.W. Saunders Co., Philadelphia (1958), pp. 475–490; 695–697
136. Geisbühler, H., et al.: Weed Res. *3*, 277 (1963)
137. Dalton, R.L., et al.: Weeds *14*, 31 (1966)
138. Kaufman, D.D., Kearney, P.G.: Appl. Microbiol. *13*, 443 (1965)
139. Bartha, R., et al.: Appl. Microbiol. *15*, 67 (1967)
140. Bartha, R., Pramer, D.: Science *156*, 1617 (1967)
141. Chisaka, H., Kearney, P.C.: J. Agr. Food Chem. *18*, 854 (1970)
142. Bartha, R.: Weed Sci. *17*, 471 (1969)
143. Walker, N. Harris, D.: J. Appl. Bact. *32*, 457 (1969)
144. Ebing, W.: Ber. Landw. *51*, 742 (1974)
145. Engelhardt, G., et al.: Chemosphere *2/3*, 85 (1977)
146. Plimmer, J.R., et al.: J. Agr. Food Chem. *18*, 859 (1970)
147. Briggs, G.G., Walker, N.: Soil Biol. Biochem. *5*, 695 (1973)
148. Tweedy, B.G., et al.: Science *168*, 482 (1970)
149. Wallnöfer, P.R., et al.: Pest. Biochem. Physiol. *7*, 481 (1977)
150. Bollag, J.M., Russel, S.: Microbiol. EC. *3*, 65 (1976)
151. Ross, J.A., Tweedy, B. G.: Bull. Environ. Contam. *10*, 234 (1973)

152. Hegeman, G.D.: J. Bacteriol. *91*, 1140 (1966)
153. Burge, W.D.: Soil Sci. Soc. Am. Proc. *37*, 392 (1973)
154. Corke, C.T., et al.: J. Agr. Food Chem. *27*, 644 (1979)
155. Fletcher, C.L., Kaufman, D.D.: J. Agr. Food Chem. *27*, 1127 (1979)
156. Jones, T.C.: TSCA Chemical Assessment Series. Preliminary Risk Assessment Phase 1. Benzidine, Its Congeners, and Their Derivative Dyes and Pigments. EPA-560/11-80-019, Office of Pesticides and Toxic Substances, Washington, DC (1980), p. 10
157. US Environmental Protection Agency. Chemicals in Commerce Information System (CICIS), Office of Pesticides and Toxic Substances, Chemical Information Division, Washington, DC (1981)
158. Boeniger, M.: The Carcinogenicity and Metabolism of Azo Dyes, Especially Those Derived from Benzidine, DHHS/NIOSH Publication No. 80–110, Robert A. Taft Laboratories, Cincinnati, Ohio (1980), p. 3
159. The Society of Dyers and Colourists: Colour Index, 3rd ed., Vol. 4, Lund Humphries, Bradford, UK (1971), p. 4742
160. The Society of Dyers and Colourists: Colour Index, 3rd ed., Vol. 6, Lund Humphries, Bradford, UK (1975), p. 6405
161. JRB Associates, Inc.: Survey of the Manufacture, Import and Uses for Benzidine, Related Substances and Related Dyes and Pigments. Prepared for the US Environmental Protection Agency. McLean, VA (1979)
162. National Toxicology Program: Second Annual Report on Carcinogens. NTP Report 81-43, US Dept. of Health and Human Services, Washington, DC, December (1981), pp. 53–55, 253–255
163. Collier, H.B.: Clin. Biochem. *7*, 3 (1974)
164. Auerbach Associates, Inc.: Benzidine Derived Dyes and/or Pigments. Rept. No. PB-284854. Prepared for US Consumer Products Safety Commission. Philadelphia, PA (1978), p. 28
165. IARC: IARC Monographs on the Evaluation of Carcinogenic Risk of Chemicals to Man. Vol. 1, Lyon (1972), pp. 80–86
166. IARC: IARC Monographs on the Evaluation of Carcinogenic Risk of Chemicals to Humans, Supplement 1, Chemicals and Industrial Processes Associated with Cancer in Humans, IARC Monographs Volumes 1–20, Lyon (1979), p. 25
167. International Labour Office: Cancer of the Bladder among Workers in Aniline Factories (Studies and Reports) Series F, No. 1, Geneva (1921)
168. Ames, B.N., et al.: Proc. Natl. Acad. Sci. USA. *70*, 2281 (1973)
169. Bos, R.P., et al.: Toxicology *16*, 113 (1980)
170. Fahmy, M.J., Fahmy, O.G.: Mutat. Res. *56*, 31 (1977)
171. Painter, R.B.: J. Environ. Pathol. Toxicol. *2*, 65 (1978)
172. Amlacher, E., Ziebarth, D.: Arch. Geschwulstforsch. *49*, 490 (1979)
173. Martin, C.N., et al.: Cancer Res. *38*, 2621 (1978)
174. Williams, G.M.: Cancer Lett. *4*, 69 (1978)
175. Pienta, R.J.: Transformation of Syrian hamster embryo cells by diverse chemicals and correlation with their reported carcinogenic and mutagenic activities. in: Chemical Mutagens (F.J. deSerres and A. Hollaender (eds.)) Vol. 6, Plenum, New York (1980), pp. 175–202
176. Bassendowska-Karska, E.: Pol. Tyg. Lek. *35*, 53 (1980)
177. Meigs, J.W., et al.: Arch. Ind. Hyg. *4*, 533 (1951)
178. Sciarini, L.J., Meigs, J.W.: Arch. Environ. Hlth. *2*, 423 (1961)
179. Kellner, H.M., et al.: Arch Toxicol. *31*, 61 (1973)
180. Bradshaw, L., Clayson, D.B.: Nature *176*, 974 (1955)
181. Clayson, D.B., et al.: Acta Uni. Int. Cancrum. *15*, 581 (1959)
182. Shriner, C.R., et al.: Rev. Environmental Effects on the Environmental Effects of Pollutants: II Benzidine (EPA-600-1-78-024 ORNL/EIS-86) US Environmental Protection Agency, Health Effects Research Laboratory and Oak Ridge Laboratory, Oak Ridge, TN (1978), pp. 41–65
183. Haley, T.J.: Clin. Toxicol. *8*, 13 (1975)
184. Zavon, M.R., et al.: Arch. Environ. Hlth. *27*, 1 (1973)
185. Radding, S.B., et al.: Rev. Environmental Fate of Selected Chemicals. EPA 566/5-75-001, US Environdmental Protection Agency, Office of Toxic Substances, Washington, DC (1975)
186. US Environmental Protection Agency: Status Assessment of Toxic Chemicals No. 5, Benzidine, Industrial Pollution Control Division, Cincinnati, OH, Sept. 6 (1977)

187. Mitre Corporation: Scoring of Organic Air Pollutants, Chemistry, Production and Toxicity of Selected Organic Chemicals, McClean, VA (1976)
188. Takemura, N., et al.: Int. J. Pollut. *9*, 665 (1965)
189. Furukawa, T., Brindley, G.M.: Clayminer *21*, 279 (1973)
190. McBride, M.B.: Clays, Clay Material *27*, 224 (1979)
191. Zierath, D. L., et al.: Soil Sci. *129*, 277 (1980)
192. Malaney, G.W., et al.: J. Water Poll. Control Fed. *39*, 2020 (1967)
193. Tabak, H.H., Barth, E.F.: J. Water Poll. Control Fed. *50*, 552 (1979)
194. Caulfield, M.J., et al.: Cancer Res. *39*, 2155 (1979)
195. Cotruvo, J.A., et al.: Ann. NY Acad. Sci. *298*, 124 (1977)
196. Simmon, V.F., et al.: Ozone methods and ozone chemistry of selected organics in water. 2. Mutagenic assays. in: Ozone/Chlorine Dioxide Oxidation Products of Organic Materials. Ozone Press International, Cleveland, OH (1978), pp. 126–133
197. Lu, P.Y., et al.: Arch. Environ. Contam. Toxicol. *6*, 129 (1977)
198. Thomas, A.W., et al.: J. Am. Ind. Hyg. Assoc. *39*, A-18 to A-24 (1978)
199. NIOSH: Spec. Occupational Hazard Rev. Benzidine-based Dyes, DHEW (NIOSH) Publication No. 80–109, Washington, DC (1980)
200. NIOSH: Techn. Rep.: Carcinogenicity and Metabolism of Azo Dyes, Especially those Derived from Benzidine. National Institute for Occupational Safety and Health, Cincinnati, OH, July (1980)
201. Anon: Chem. Reg. Reptr. *4*, 63 (1980)
202. DHEW/NIOSH/NCI, Joint Intelligence Bull. No. 24, Direct Black 38, Direct Blue 6, and Direct Brown 95, Benzidine derived dyes. NIOSH Publ. No. 78–148, Dept. of Health, Education and Welfare, Washington, DC (1978)
203. Anon: Occup. Health Safety Lett *10*, 6 (1980)
204. US Environmental Protection Agency: Benzidine: Proposed Toxic Pollutant Effluent Standards, 40 CFR Part 129. Federal Register. 41 [127] June 30 (1973)
205. The Society of Dyers and Colourists: Colour Index, 3rd ed., Vol. 2, Lund Humphries, Bradford, UK (1971), pp. 2223–2382, 2439
206. Nat. Cancer Inst. 13-Week Subchronic Toxicity Studies of Direct Blue 6, Direct Black 38 and Direct Brown 95 Dyes, NCI-66 TR-108, DHEW Publ. No. (NIH) 78-1358 Bethesda, MD (1978)
207. Walker, R.: Food Cosmet. Toxicol. *8*, 659 (1970)
208. Lazear, E.J., et al.: Toxicol. Lett. *4*, 519 (1979)
209. Nony, C.R., et al.: J. Anal. Toxicol. *4*, 132 (1980)
210. Tanaka, K.T.: Japan J. Ind. Hlth. *22*, 194 (1980)
211. Yoshida, O., Miyakawa, M.: Metabolic aspects in analytical and experimental epidemiology of cancer, in: Etiology of Bladder Cancer (W. Nakahara (ed.)), University Park Press, Baltimore, MD, (1973)
212. Hitz, H.R., et al.: J. Soc. Dyers Col. *94*, 71 (1978)
213. Cerniglia, C.E., et al.: Biochem. Biophys. Res. Communs, (1982), in press
214. Gardiner, D.K., Borne, B.J.: J. Soc. Dyers Col. *94*, 339 (1978)
215. Gerarde, H.W., Gerarde, D.F.: J. Occup. Med. *16*, 322 (1974)
216. IARC: Monogr. Evaluation of the Carcinogenic Risk of Chemicals to Man, Vol. 4, Internat. Agency for Res. Cancer, Lyon, (1974) pp. 49–55, 65–71, 97–111
217. Stanford Res. Inst.: A Study of Industrial Data on Candidate Chemicals for Testing. Prepared for US Environmental Protection Agency, EPA-560/5-77-06, Menlo Park, CA, Aug. (1977)
218. Bell, D.R.: Final Environmental Impact Statement: Proposed Regulation (administration action) Handling of Certain Carcinogens, US Occupational Safety and Health Administration, Washington, DC (1973), pp. 13–16
219. Garner, R.C., et al.: Cancer Lett. *1*, 39 (1975)
220. Hsu, R.S., Sikka, H.C.: Toxicol. Appl. Pharmacol. *64*, 306 (1982)
221. US Environmental Protection Agency: Ambient Water Quality Criteria for Dichlorobenzidine (EPA 440/5-80-040, PB 81-11717), Office of Water Regulations and Standards, Criteria and Standards Division, Washington, DC (1980), pp. C-1-C-4
222. Appleton, H.T., Sikka, H.C.: Environ. Sci. Technol. *14*, 50 (1980)
223. Sikka, H.C., et al.: Fate of 3,3′-dichlorobenzidine in aquatic environments, EPA 600/3-78-003, US Environmental Protection Agency, Washington, DC

224. Appleton, H., et al.: Fate of 3,3′-dichlorobenzidine in the aquatic environment, in: Aquatic Pollutants: Transformation and Biological Effects (O. Hutzinger, L.H. Van Lelyveld and B.C.J. Zoetman (eds.))
225. Banerjee, S., et al.: Environ. Sci, Technol. *12*, 1425 (1978)
226. Johnson, W.M., Parnes, W.D.: Ann. NY Acad. Sci. *329*, 277 (1979)
227. USDA: Quantities of Pesticides used by Farmers in 1966 Agricultural Rep. No. 179. US Dep. Agriculture, Washington, DC, April (1970)
228. Masuda, Y., Hoffmann, D.: Anal. Chem. *41*, 650 (1969)
229. Hoffmann, D., et al.: Nature *221*, 254 (1969)
230. Patrianakos, C., Hoffmann, D.: J. Anal. Toxicol. *3*, 150 (1979)
231. IARC: Monogr. Evaluation of the Carcinogenic Risk of Chemicals to Man, Vol. 16, Internat. Agency for Research on Cancer, Lyon (1978), pp. 325–336
232. Kehe, H.J., Kouris, C.S.: Diarylamines. in: Encyclopedia of Chemical Technology (R.E. Kirk and D.F. Othmer (eds.)), 2nd ed., Vol. 7, Wiley & Sons, New York (1965), pp. 40–49
233. Society of Dyers and Colourists: Colour Index, 3rd ed., Vol. 4, Yorkshire, UK (1971), pp. 4223, 4453, 4454, 4808
234. Society of Dyers and Colourists: Colouor Index, revised 3rd ed., Vol. 5, Yorkshire, UK (1975), pp. 5033, 5034, 5084
235. Veys, C.A.: A Study of the Incidence of Bladder Tumours in Rubber Workers (Thesis for Doctorate of Medicine). Faculty of Medicine, University of Liverpool, Liverpool, UK (1973)
236. NIOSH: Current Intelligence Bulletin; Metabolic Precursors of a Known Humn Carcinogen, beta-Naphthylamine, US Dept. of Health, Education and Welfare, Rockville, MD (1976), pp. 1–3
237. Scott, T.S., Williams, M.H.C.: Brit. J. Ind. Med. *14*, 150 (1957)
238. Innes, J.R.M., et al.: J. Natl. Cancer Inst. *42*, 1101 (1969)
239. Kummer, R., Tordoir, W.F., Tidjschr. Geneeskd. *53*, 415 (1975)
240. Batten, P.L., Hathway, D.E.: Brit. J. Cancer *35*, 342 (1977)
241. Stasenkova, K.P.: Toksikol. Novykh. Prom. Khim. Veshchestv. *10*, 90 (1968)
242. Ku, Y., Alvarez, G.: Chemosphere *11*, 41 (1982)
243. Pitter, P.: Water Res. *10*, 231 (1976)
244. Baird, R., et al.: J. Water Pollut. Control Fed. *49*, 1609 (1977)
245. Stula, E.F., et al.: J. Environ. Pathol. Toxicol. *1*, 31 (1977)
246. Gianatasio, P.A.: Rubber Age, July (1969) pp. 51–59
247. Wagoner, J.K.: Ann. NY Acad. Sci. *329*, 328 (1979)
248. Russfield, A.B., et al.: Toxicol. Appl. Pharmacol. *31*, 47 (1975)
249. Stula, E.F., et al.: Toxicol. Appl. Pharmacol. *31*, 159 (1975)
250. McCann, J., et al.: Proc. Natl. Acad. Sci. *72*, 5135 (1975)
251. McQueen, C.A., et al.: Toxicol. Appl. Pharmacol. *58*, 231 (1981)
252. Rappaport, S.M., Morales, R.: Anal. Chem. *51*, 19 (1979)
253. Comprehensive Cancer Center of Metropolitan Detroit: How widespread the hazard of industrial carcinogens: The Adrian Experience, in: Environmental Cancer Alert (C.M. King (ed.)) Vol. 3, Michigan Cancer Foundation, Detroit (1979), pp. 1–2
254. Zimmermann, L.: Justification for inclusion of 4,4′-methylenebis(2-chloroaniline) in the National Toxicology Program. Personal Communication, April 27 (1981)
255. Jenkins, F.P., et al.: Food Cosmet. Toxicol. *10*, 671 (1972)
256. Case, R.A.M., et al.: Brit J. Ind. Med. *11*, 75 (1974)
257. Goldwater, L.J., et al.: Arch. Environ. Hlth. *11*, 814 (1965)
258. Mancuso, T.F., El-Attar, A.A.: J. Occup. Med., *9*, 277 (1967)
259. Hueper, W.C., et al.: J. Ind. Hyg. *20*, 46 (1938)
260. Bonser, G.M., et al.: Brit. J. Cancer *10*, 353 (1956)
261. Bonser, G.M.: J. Pathol. Bacteriol. *55*, 1 (1943)
262. Conzelman, G.M., Jr., Moultox, J.E.: J. Natl. Cancer Inst. *42*, 825 (1969)
263. Purchase, I.F.H., et al.: Brit. J. Cancer *44*, 892 (1981)
264. Saffiotti, U., et al.: Induction of bladder cancer in hamsters fed aromatic amines, in: Bladder Cancer: A Symposium, Aesculapius, Birmingham, Alabama (1967), p. 129
265. Sellakamur, A.R., et al.: Proc. Am. Assoc. Cancer Res. *10*, 78 (1969)
266. Conzelman, G.M., Jr., Moultox, J.E.: J. Natl. Cancer Inst. *49*, 193 (1972)
267. Hadidian, Z., et al.: J. Natl. Cancer Inst. *41*, 985 (1968)

268. Bonser, G.M., et al.: Brit. J. Cancer *6*, 412 (1952)
269. Hicks, R.M., et al.: Experimental induction of bladder tumours by a two-stage system. in: Carcinogenesis, Vol. 2, Mechanisms of Tumour Promotion and Carcinogenesis (Slaga, T.J. (ed.)), Raven Press, New York (1978), p. 475–489, Swak, D., Bantwell, R.K.
270. Gehrmann, G.H., et al.: Occupational carcinoma of the bladder. In: Proc. IX Internat. Congr. Ind. Med., Wright, Bristol (1949), p. 427
271. Radomski, J.L., et al.: Cancer Res. *40*, 3537 (1980)
272. Radomski, J.L., et al.: Cancer Res. *31*, 1461 (1971)
273. Deichmann, W.B., Radomski, J.L., J. Natl. Cancer Inst. *43*, 263 (1969)
274. Kadlubar, F. F., et al.: Cancer Res. *38*, 3628 (1978)
275. deSerres, F.J., Ashby, E.J. (eds.): Short-term tests for carcinogens. Rep. Internat. Collaborative Program, Elsevier/North Holland, Amsterdam (1981), p. 130

Phosphate Esters

Derek C. G. Muir

Freshwater Institute
Department of Fisheries and Oceans
Winnipeg, Canada R3T 2N6

Introduction

Phosphate esters (triaryl phosphates, triaryl/alkyl phosphates, trialkyl phosphates and trihaloalkyl phosphates) have been widely used since the 1940's in industrial and consumer products as fire retardant plasticizers and as high temperature functional fluids. Use of fire retardants during the 1960's and '70's increased greatly because of the demand for improved fire safety in commercial products made from synthetic polymers [1, 2]. Triaryl phosphates (TArPs), triaryl/alkyl phosphates (TAr/AlPs), trialkyl phosphates (TAlPs) and trihaloalkyl phosphates (THAPs) are incorporated at low part per hundred concentrations in flexible plastics such as polyvinyl chloride and polyurethane foams where they provide flame retardancy as well as acting as plasticizers. The need for fire resistant hydraulic fluids and lubricants has also created a demand for various TArPs [3]. Phosphate esters have been of toxicological interest since the 1930's when outbreaks of polyneuritis in the United States were attributed to tri-ortho-cresyl phosphate, a component of tricresyl phosphate oil [4]. In the mid-1970's tris(2,3-dibromopropyl) phosphate, a fire retardant in polyester fabrics, was demonstrated to be mutagenic and was banned for use in clothing in many countries [5]. Phosphate esters have also been suggested as replacements for polychlorinated biphenyls [6] which has stimulated interest in their environmental toxicology and chemistry.

Nomenclature

The phosphate esters discussed here are all esters of orthophosphoric acid and are listed in Chemical Abstracts as the acid followed by the names of the radical substitutents. IUPAC nomenclature uses the prefix O-alkyl or O-aryl for each alkyl/aryl group on the ester oxygen followed by the name phosphate [7]. It is accepted

Table 1. Major components of commercial phosphate esters [8, 9]

Abbreviation	Common Name[a]	Structure 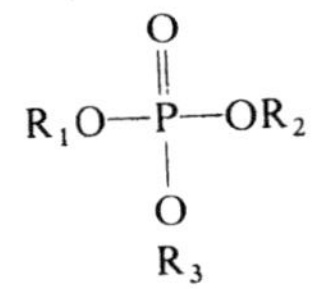 R1	R2	R3	Trade Name[b]	Manufacturer
1. Triaryl phophates (TArPs)						
BPDP	t-Butylphenyldiphenyl phosphate	$(CH_3)_3CC_6H_4$	C_6H_5	C_6H_5	Fyrquel GT Kronitex 100B Santicizer 154	Stauffer FMC Monsanto
CDP	Cresyl diphenyl phosphate	$CH_3(C_6H_4)$	C_6H_5	C_6H_5	Phosflex 112 Santicizer 140	Stauffer Monsanto
CPDPP	Cumylphenyl diphenyl phosphate	$C_6H_5C(CH_3)_2C_6H_4$	C_6H_5	C_6H_5	Pydraul 115E	Monsanto
IPDP	Isopropylphenyldiphenyl phosphate	$(CH_3)_2CHC_6H_4$	C_6H_5	C_6H_5	Reolube HYD70 Phosflex 41B Kronitex 100	Ciba-Geigy Stauffer FMC
NPDPP	Nonylphenyl diphenyl phosphate	$C_9H_{19}C_6H_5$	C_6H_5	C_6H_5	Pydraul 50E	Monsanto
TPP	Triphenyl phosphate	C_6H_5	C_6H_5	C_6H_5	TPP	Monsanto
TCP	Tricresyl phosphate (mixed isomere of o, m, p)	$CH_3(C_6H_4)$	$CH_3(C_6H_4)$	$CH_3(C_6H_4)$	TCP Phosflex 179C	Monsanto Stauffer
TXP	Trixylyl phosphate	$(CH_3)_2C_6H_3$	$(CH_3)_2C_6H_3$	$(CH_3)_2C_6H_3$	Phosflex 179A Kronitex TXP	Stauffer FMC
2. Triaryl/alkyl phosphates (TAr/AlPs)						
DBPP	Dibutyl phenyl phosphate	C_4H_9	C_4H_9	C_4H_9	Skydrol LD-4 Phosflex 4 FMC, Ashland	Monsanto Stauffer
EHDP	2-Ethylhexyl diphenyl phosphate	C_8H_{17}	C_6H_5	C_6H_5	Santicizer 141	Monsanto
IDDP	Isodecyl diphenyl phosphate	$C_{10}H_{21}$	C_6H_5	C_6H_5	Santicizer 148	Monsanto

3. Trialkyl phosphates (TAlPs)						
TBEP	Tris(2-butoxyethyl) phosphate	$C_6H_{13}O$	$C_6H_{13}O$	$C_6H_{13}O$	KP 140 TBEP	FMC Mobil, Ashland
TBP	Tributyl phosphate	C_4H_9	C_4H_9	C_4H_9	Skydrol LD-4 Phosflex 4	Monsanto, FMC Stauffer, Ashland
TEP	Triethyl phosphate	C_2H_5	C_2H_5	C_2H_5	TEP	Eastman
TEHP	Tris(2-ethylhexyl) phosphate	C_8H_{17}	C_8H_{17}	C_8H_{17}	TOF	Monsanto, Ashland
4. Trihaloalkyl phosphates (THAPs)						
TCEP	Tris(2-chloroethyl) phosphate	C_2H_4Cl	C_2H_4Cl	C_2H_4Cl	Fyrol PCF	Stauffer
TCPP	Tris(2-chloropropyl) phosphate	C_3H_6Cl	C_3H_6Cl	C_3H_6Cl	Dalt-o-Gard	ICI, Bayer, SUC Ugine Kuhlman
TDBP	Tris(2,3-dibromopropyl) phosphate	$C_3H_5Br_2$	$C_3H_5Br_2$	$C_3H_5Br_2$	TP-69	Great Lakes
TDCP	Tris(1,3-dichloropropyl) phosphate	$C_3H_5Cl_2$	$C_3H_5Cl_2$	$C_3H_5Cl_2$	Fyrol FR2	ICI

[a] Common names of major components of commercial products

[b] Trade names and manufactures in North America and Western Europe cited in recent reviews [8–10, 15]

practice to name phosphorus compounds by common names which in English are in any case similar to the IUPAC nomenclature [7]. TAr/AlPs and THAPs are thus named as phosphates of the constituent radical group. Common names of the major components of phosphate ester formulations along with abbreviations used in the text, trade names and structures are listed in Table 1.

Production and Uses

Triaryl and Triaryl/Alkyl Phosphates

Major uses of TArPs and TAr/AlPs are as fire retardant plasticizers in the manufacture of PVC plastics and as industrial hydraulic fluids (Table 2). Total quantities of TArPs and TAr/AlPs manufactured in the United States were estimated at 47×10^6 kg in 1977 and were expected to rise to 77×10^6 kg by 1986 [8]. About 3% of this production was exported to Japan, W. Europe, Canada and Mexico [8]. Production by subsidiaries of the three U.S. manufacturers (Table 1) is carried out in Europe [11]. Manufacture of phosphate esters is also carried out in the Soviet Union and Japan [12, 13].

Tricresyl phosphates (TCPs) were the dominant TAr/AlPs in hydraulic fluids and plasticizers during the 1960's and early 1970's but have recently been replaced for many uses, especially as plasticizers [14]. A yearly 5% decline in TCP production has been predicted for the 1978–82 period [14]. The major replacement for TCP is IPDP produced by two U.S. manufacturers (FMC and Stauffer) and by at least one European manufacturer (Ciba-Geigy).

Trialkyl Phosphates

TAlPs are used as fire retardant plasticizers, industrial solvents and defoaming agents. TEP, TBEP, TEHP, and TBP (Table 1) are the most widely used TAlP's

Table 2. Estimated production and uses (kg $\times 10^6$) of TArPs and TAr/AlPs in the U.S. 1977 and projections for 1986 [8]

Compound	Flame retardant plasticizer	Hydraulic fluid	Lubricant additive	Misc[a]	Total 1977	1986[b]
TCP/XCP	3.0	7.2	1.4	1.1	13.6	9.1
CDP	2.0	3.7	0.7	0.8	7.3	5.0
TPP	5.0	0	0	0.4	5.4	8.2
DBPP	0	0.8	0.1	0	0.9	1.2
EHDP	0.9	0.5	0.1	0.4	1.9	4.1
IPDP	3.5	6.4	1.1	0.7	11.7	29.5
IDDP	5.1	0	0	0.5	5.6	4.1
TBPDP	0.9	0	0	0	0.9	9.1
Total	20.4	18.6	3.4	3.9	47.3	77

[a] Miscellaneous includes exports and minor uses
[b] Estimated by MRI [8]

[15, 2] although many others have been described [3]. TEP is used as a plasticizer in cellulose acetate and as a viscosity additive in polyester resins [2]. TBP is a primary solvent and plasticizer for cellulosics and chlorinated rubber. TEHP and TBEP are used primarily as flame retardant plasticizers. U.S. production of TAlPs in the mid-1970's for use as flame-retardant plasticizers ranged from $3–5 \times 10^6$ kg/yr judging from differences in total phosphate ester (TArP and TAlP) production [2] and TArP production [8]. This estimate does not include quantities produced for other uses such as solvents and defoaming agents.

Trihaloalkylphosphates

The major uses of THAPs are as flame retardant additives in flexible and rigid polyurethane foams. Flexible foams are used in automotive and furniture industries while rigid foams are used mainly in the construction industry as insulation [9, 16]. Phosphate ester THAPs are added to the reaction mixture in the preparation of the foams but are not chemically incorporated. THAPs represented about 50% of the flame retardants used in the urethane foam market in the mid-1970's [16]. Until the withdrawal of TDBP and TDCP from use in textiles, THAPs also represented a major portion of the flame retardants used in textiles [17]. U.S. production of the THAPs was estimated to be about 10×10^6 kg/yr in 1976 rising to 20×10^6 kg/yr in 1986 [2]. THAPs are also produced in significant quantities in Japan, France and the U.K. [18]. Concern about chronic, especially carcinogenic, effects of TDBP and other THAPs may result in a decline in use of these compounds during the 1980's [16].

Chemistry

Synthesis

The manufacturing process presently in use for TArPs involves condensation of the aryl compound (cresol, phenol or p-substituted phenols) with phosphorus oxychloride in the presence of a metal catalyst [3, 8]:

$$3\,ArOH + POCl_3 \rightarrow (AlCl_3\ catalyst) \rightarrow (ArO)_3PO + 3HCl.$$

The temperature of the reaction is raised to 200 °C over 8 h. To produce TArPs with mixed substituents, materials are added in stoichiometric quantities. Unreacted phenolics are removed by vacuum distillation (50 torr) and TArP is removed by distillation at 3 torr. Further purification of the product is accomplished by washing with dilute alkali and charcoal filtration [8]. TAr/AlPs are produced by reacting stoichiometric quantities of the aryl substituents first, followed by cooling of the mixture to <20 °C before addition of the alcohol [3]:

1. $2ArOH + POCl_3 \rightarrow AlCl_3\ catalyst \rightarrow (ArO)_2POCl + 2HCl$
2. $ROH + (ArO)_2POCl \rightarrow AlCl_3 \rightarrow (ArO)_2ROPO + HCl$.

Source of the cresol for TCP is cresylic acid, a byproduct of the cracking of crude oil. During the mid-70's the price of cresylic acid rose almost 3-fold thus making

TAr/AlPs produced from synthetic phenolics price competitive [8]. TAlPs are produced by procedures similar to those for TAr/AlPs however low reaction temperatures and reduced pressures must be used to avoid formation of alkyl chlorides, a side reaction not encountered with synthesis of TArPs [3]. TEP is manufactured from diethylether and phosphorus pentoxide via a metaphosphate intermediate [2].

TCEP, TCPP and TDCP are prepared commercially by reaction of ethylene oxide, propylene oxide or epichlorhydrin, respectively, with phosphorus oxychloride. TDBP is prepared from 2,3-dibromopropanol and $POCl_3$ [9]. The reaction is carried out under nitrogen atmosphere and excess reagents are removed by vacuum distillation [3, 9].

Impurities

Industrial uses of phosphate esters require highly pure materials and most products are sold with <2.5% impurities (mainly solvent residues) [8]. However, all phosphate ester products are mixtures of various related isomers, eg. TCP, prepared from cresols and xylenols, contains a large number of structurally similar components. The presence of tri-o-cresyl phosphate (TOCP) and other o-cresyl-containing isomers in TCP and TXP formulations has been a major source of concern because of the delayed neurotoxic effects of these products. Much of the chemical methodology on TArPs involves determination of TOCP in formulations [19, 20] and in food products [21, 22]. Cresylic acids used for preparation of TCPs in recent years have had o-cresol contents of <0.1% [10]. Deo and Howard [23] could not detect o-cresol in hydrolysates of Santicizer 140 (CDP) or Fyrquel GT (BPDP). Tetra-aryl pyrophosphates and diarylchlorophosphates are also expected impurities in TArPs and TAr/AlPs [24] but free phenols were not detected in a sample of TCP analysed prior to hydrolysis [20]. Perhaps because of the difficulty of separating related TArP isomers there are no reports of determination of o-alkyl substituted contaminants in widely used products such as IPDP and BPDP.

TDBP has been reported to contain several halogenated alkanes (at 0.05% levels) and 2,3-dibromo-1-propanol, a compound subsequently shown to have mutagenic activity in the Ames/Salmonella assay [9, 17].

Thermal Stability and Flame Retardancy

A variety of volatile and non-volatile products are formed upon heating TArPs and THAPs in air at temperatures above 200 °C (Table 3). TArPs are generally much more stable than TAlPs or THAPs under these conditions [25]. Thermal oxidation of TCP produced high molecular weight compounds resulting from polymerization through the oxidized aryl moieties of the compound. Thermal stability was inversely related to the degree of alkylation of the aryl group (BPDP > TCP > TXP) [26].

TPP was found to volatilize intact during heating (210–350 °C, air or N_2) of various polymers in which it was incorporated although some orthophosphoric acid was formed at higher temperatures [27]. TBP degraded readily during combustion of polymers to yield phosphoric acids and orthophosphate. TDBP formed at least 8 major volatile products when heated (350 °C, N_2) three of which contained

Table 3. Thermal oxidation of some phosphate esters in air at 370 °C [26]

Compound	% of starting[a] material remaining	Major degradation products			
		1	2	3	4
TCP	99.7	Cresols	H_2O	Toluene	
TBP	39.6	Butene	Butanol	Water	
TDCPP	31.8	Tri and dichloropropane	HCl	Acrolein	
TDBP	36.1	Tri and dibromopropenes	HBr		
TCEP	41.0	Dichloroethane	Ethyl chloride	HCl	Acetal-dehyde

[a] Nonvolatile residue – mainly starting material

phosphorus. Bromine ions and orthophosphoric acid were also detected during combustion of TDBP [27]. TDCP and TCEP produced halogenated alkenes via substitution of halogens for the phosphoryl group during pyrolysis (260 °C, 3 mm Hg) [28].

Phosphoric acids produced during combustion of polymers, in which phosphate esters are incorporated, suppress flaming (oxidation of gaseous products) and glowing (solid phase combustion) [29]. TBP, TPP, and TDBP are thought to act primarily in the gas phase, since they volatilize undegraded from the polymer. Formation of PO is proposed to effect flame retardance by catalyzing the recombination of H atoms [27]:

$$PO + H \rightarrow HPO$$
$$HPO + H \rightarrow H_2 + PO\,.$$

In the condensed phase of the burning polymer, formation of phosphoric acids enhances char formation and may form a coating which retards oxidative degradation [27, 29]. TDBP incorporated at 3% in polystyrene has been shown to lower the temperature of the burning surface and to reduce the amount of volatile products [30]. Halogens in THAPs provide additional flame reduction by trapping free-radicals which suppresses the combustion chain reaction [17].

The fire resistance and thermal stability of TArPs is also a major reason for their use as hydraulic fluids and compressor lubricants where ruptures under high pressure onto hot surfaces could be a fire hazard.

Photolysis

Irradiation (medium pressure Hg arc lamp) of TPP and TCP (0.02 *M* in ethanol) yielded diaryl products formed by intramolecular coupling of electron-rich aryl groups. TCP was more reactive than TPP producing a dimethylbiphenyl (31–51% yield) and a mono-cresyl phosphate (2–10%) [31]. Photolysis under conditions simulating sunlight or in aqueous solutions was not studied.

Hydrolysis

Phosphate esters are generally resistant to hydrolysis in neutral or acid waters (pH 5.0–7.0) but have been found to degrade rapidly in sterile distilled water at pH 9.0–9.5 [32, 33]. The half-lives of TPP at pH 8.2, 9.0 and 9.5 (20–25 °C) have been reported to be 7.5, 3.0 and 1.3 days, respectively [32, 33]. In alkaline solution phosphate esters undergo second order nucleophilic reactions due to the hydroxide ion involving cleavage of the O-P bond. The OH ion is more effective than H_2O in causing O-P bond fission by a factor of 10^8 for TArPs and 10^4 for TAlPs [34]. It follows that alkaline and neutral hydrolysis of an aryl dialkyl phosphate may not give the same products [35]:

$$R_3OH + (R_1O)(R_2O)POO^- \leftarrow H_2O - R_3O\ PO(OR_1)(OR_2) - OH^- \rightarrow (R_3O) \cdot (R_2O)POO^- + R_3OH$$

$$R_1 = \text{aryl};\ R_2, R_3 = \text{alkyl}.$$

Alkaline hydrolysis will result in loss of R_1, while at neutral pH, R_2 and R_3 would be more susceptible to hydrolysis than the aryl group. Wolfe [35] correlated second order alkaline hydrolysis (pH 8.0, 30 °C) constants (k) with Hammett sigma constant (s) for 4 TArPs (including TPP and pTCP) and found the following relationship:

$$\log k = 1.4\, s - 0.47.$$

This equation predicts that TArPs with alkyl-substituted phenyl rings ($s < 0$) will have slower rates of hydrolysis than TArPs with electron withdrawing substituents ($s > 0$). Similar predictions were made with alkylarylphosphates [35]. Limited data on the hydrolysis of THAPs suggests a similar generalization can be made for these compounds. Thus dichloro-substituted THAPs are more labile than mono-chloro THAPs under acid conditions [16].

Diaryl and dialkyl phosphates have been found more resistant to hydrolysis than the corresponding triesters. The reduced chemical reactivity of the diesters in water is thought to be due to the fact that the primary species is an anion [36]. Mono aryl and alkyl phosphates, which exist as mono and dianions at common environmental pH's, are generally more susceptible to hydrolysis than the di- or triesters [8, 38].

Physical Properties

Phosphate ester mixtures such as TCP and TXP are liquids at 25 °C, although some of the individual compounds are solids (Table 4). Alkyl and haloalkyl-substituted esters are generally liquids in their pure form. Specific gravities range from 0.98 for TBP, 1.1–1.2 for most TArPs and TAr/AlPs and 1.4–2.2 for THAPs so that unlike hydrocarbon based oils, most phosphate esters would tend to sink when spilled into water. The viscosity and low melting points are important criteria in the selection of these compounds as plasticizers and lubricants [8]. Octanol-water partition coefficients (K_{ow}) and water solubility (S) of many phosphate esters have been measured in order to predict the environ-

Table 4. Physical properties of phosphate esters

Compound	MW	MP[a] (°C)	BP (°C/mm Hg)	VP[b] (mm Hg/°C)	WS[b] mg/l	K_{ow} [b] ($\times 10^4$)	Reference
TPP	326.3	48.5	238/10	1.3/200	1.9	4.25/4.06	[8, 37, 39]
				18.2/250			
TCP p, p, p	368.4	77–8	244/3.5	–	–	–	[40]
m, m, m		25–6	260/ 15	–	–	–	
o, o, o		11	410/760				
TCP Mixed isomers	368.4	– 28*	241–55/4	< 0.02/150	0.36	12.8/13.2	[8, 37, 39]
				0.29/200			
TXP (2,3)3	410.5	61	–	–	–	–	[10]
(2,6)3		136–8	262–4/6				
TXP Mixed isomers	410.5	– 20*	248–65/4	< 0.02/150	0.89	42.7	[8, 39]
				0.33/200			
BPDP	382.4	– 21*	261/6	1.35/200	3.2	13.2	[8, 39]
				10.2 /250			
IPDP	368.4	– 26*	220–30/1	–	2.2	20.2	[8, 39]
DBPP	286.3	–	200/20	0.5/200	96	1.88	[8, 39]
TBP	266.3	– 80	292/760	127/177	280	1.01/1.03	[39, 40, 41]
TEP	182.2	– 56.5	209–218	1/39.6			[41]
EHDP	362.4	– 80	239/1	0.5/100	1.9	53.4	[8, 39, 40]
IDDP	390.4	< –50*	–	< 0.1/150	0.75	27.3	[8, 39]
				0.5/200			
NPDPP	452.4	–	–	–	0.77	86.0	[32]
CPDPP	444.4	–	–	–	0.063	120.0	[32]
TEHP	434.6	– 74	216/5	0.23/150	≈1,000	1.68	[39, 40]
TBEP	398.5	– 70	225–8/4	0.03/150	≈1,100	0.45	[38, 37, 41]
				0.01/ 20			
TCEP	285.5	– 55	214/25	0.5/145	7,000	0.003	[40, 37, 9]
TDBP	697.7	5.5	65/0.005	–	1.6	2.43	[9, 37]
TCPP	327.6	– 40*	55–60/0.004	–	1,100	–	[9]
TDCP	430.9	26.7	236–7/5	–	100	0.554	[9, 37]

[a] *signifies pour point rather than melting point
[b] VP = vapor pressure; WS = water solubility; K_{ow} = octanol – water partition coefficient

mental behavior of these compounds. Using data for all phosphate esters that are liquids at 25 °C the following relationship is obtained:

$$\ln K_{ow} = 8.44 - 0.70 \ln S \; (r = 0.90, n = 14).$$

For TArPs which are solids at 25 °C the relationship derived by MacKay et al. [42] gives good predictions of K_{ow} from S if the melting point is known.

Analytical Methods

Determination of phosphate esters in environmental samples or technical products is best accomplished using gas chromatography (GLC) with phosphorus-specific detectors. All phosphate esters are sufficiently volatile and thermally stable for GLC analysis. However, the related isomers in such products as TCP and TXP are difficult to separate even with capillary columns. Thus for determination of specific components such as TOCP, analysis of phenolic moieties by GLC following alkaline hydrolysis of the phosphate esters is a useful technique [19, 20].

Extraction

Phosphate esters are readily extracted from water using dichloromethane [43, 44]. Hexane is less efficient than other common solvents for extraction of TEP and TBEP from water samples [44]. For large samples containing ng/L concentrations, XAD resin extraction appears to be the method of choice. Recovery efficiencies of >70% at 1, 10, and 100 ng/L concentrations of various TArPs and TAr/AlPs using XAD have been reported [45]. Hydrolysis products, diaryl and dialkyl phosphates, are also efficiently recovered from water (acidified to pH 1) using XAD resins [46, 47].

For food products and biological tissues, methods developed for the extraction of organophosphate insecticide residues are suitable for recovery of TAr/AlPs [48, 49]. Fish tissues (ground with anhydrous Na_2SO_4) have been extracted with petroleum ether [48], diethyl ether/petroleum ether [32], acetonitrile [44] or toluene-isopropanol-dichloromethane [50] to recover TArPs and TAr/AlPs. Hexane extraction recovered only unmetabolized ^{14}C-TPP from whole fish while much of the radioactivity extracted with aqueous methanol consisted of unidentified products [43].

Phosphate esters are extracted from sediments with dichloromethane [32] or acetonitrile [44]. Comparison of three extraction procedures for recovery of TArPs from sediment showed that refluxing with aqueous methanol was more efficient than more non-polar solvents [43]. Aqueous methanol will also extract diaryl phosphates from sediments [46].

Cleanup of Extracts

Water sample extracts can be analysed for phosphate esters directly by GLC following extraction when a phosphorus-specific detector is used [32, 45] but sediment and tissue extracts require extensive removal of co-extractives. Gel permeation chromatography (GPC) has been used to separate fish lipids from several TArPs and their phenolic hydrolysis products [43, 50]. Acetonitrile/petroleum ether partitioning has been used to separate TArPs from fish lipids followed by Florisil chromatography in which TArPs eluted in a 50% ether/petroleum ether eluate [48]. Activated charcoal chromatography and washes with conc. sulfuric acid and 0.5 N NaOH have also been described [44]. Silica gel and carbon-polyurethane foam columns have been used to clean up fish extracts following GPC separation [50]. Alumina column chromatography has been used to separate TArPs from sediment co-extractives [43].

GLC Separation

Applications of GLC to the analysis of phosphate esters, especially TArPs, are relatively recent in comparison to organophosphate insecticides. TCP was used as a liquid stationary phase long before its analysis by GLC was reported [51]. Hardy [52] separated several alkyl phosphates including TBP and di-n-octyl methyl phosphate on a silicone liquid phase (DC-200) while Horton [53] separated TBP and other alkyl phosphates on Carbowax 20 M. Bloom [54] separated 10 TArPs and TAr/AlPs on 1% QF-1 on Chromosorb Q (100/120 mesh) using temperature programming (70–240 °C) with flame ionization detection. Deo and Howard [23] separated and identified a large number of TArPs in commercial samples by gas chromatography-mass spectrometry (GC-MS) using a packed column (1.83 m) of 3% OV-101 on Chromosorb W (80/100 mesh). Analyses of environmental samples for phosphate esters employed similar liquid phases and packed columns with phosphorus or MS detection [32, 44, 48]. Keith et al. [55] identified TPP and TBP in drinking water extracts by MS after separation on a 30 m × 0.4 mm i.d. glass capillary column coated with methyl silicone (SP-2100). LeBel et al. [45] chromatographed 14 phosphate esters including THAPs, TAlPs and TArP mixtures on a fused silica capillary column (25 m × 0.2 mm i.d.) coated with methyl silicone. Detection limits by this technique using a phosphorus-selective detector were 50–100 pg for well resolved compounds such as TBP, TCEP and TDCP and approximately 0.5 ng for TCP mixtures [45].

Diaryl and dialkyl phosphates resulting from hydrolysis of the corresponding triesters can also be analysed by GLC following derivatization. Alkylation with various diazoalkanes [56], benzylation [57] and pentafluorobenzylation [58] of dialkyl phosphate metabolites of organophosphorus insecticides prior to GLC analysis have been reported. Other than alkylation [46] with diazoalkanes, these procedures have not been applied to the analysis of diaryl phosphates.

TDCP has been reported to yield 8 peaks when chromatographed on a 1.8 m × 2 mm i.d. column packed with 1.95% QF-1/1.5% OV-17 on Gas Chrom Q [59]. However single peak resolution of TDCP has been achieved with fused silica capilliary columns [45] and with efficient packed columns (0.28% Carbowax 20 M) [60].

Analytical Interferences

The wide use of phosphate esters as plasticizers appears to be responsible for their appearance in procedural blanks. TBEP and several TArPs have been identified in distilled water, pesticide grade solvents and o-rings used in metering valves [45, 49, 61]. TArPs have been reported in analytical grade filter paper and in airborne debris from newly replaced air-filters used in a laboratory air-handling system [49]. TDCP was identified in reagent grade water and in hexane [59]. The analysis of phosphate esters at sub-ng/g levels thus requires careful attention to possible sources of contamination.

Other Analytical Procedures

High pressure liquid chromatography (HPLC) has not been widely applied to the analysis of TArPs. Although aryl substituted phosphate esters have UV absorption maxima in the 260 nm range [20] they do not have unique chromophores which would permit specific UV or fluorescence detection. Tittarelli and Mascherpa [62] described the application of graphite furnace atomic absorption spectroscopy for detection of TArPs after HPLC separation. HPLC with size exclusion (cross-linked polystyrene particles) columns has been applied to the analysis of TCP in lubricating oils [63]. TLC has been used for qualitative identification of TArPs in extracts of plastics [64] and edible oils [21]. Bloom [54] noted that phosphate esters were difficult to separate from each other by TLC due to their similar polarity. Renberg et al. [65] obtained good resolution of some TArPs and TAr/AlPs by use of reverse-phase TLC using a solvent system of acetone: water (7:3). P 31 and proton NMR have also been used for the identification of TArPs [66] and THAPs [59].

Mass spectrometry (MS) has been used in numerous studies for the identification of phosphate esters in environmental samples. Electron impact MS of TArPs yields spectra with clearly defined molecular ions and base peaks due to the diaryl phosphate or to the aryl ring fragments [67]. TAlPs, however, yield weak molecular ions and intense peaks due to alkane and alkene fragments. Rearrangements due to the acquisition of one or more hydrogen atoms following disassociation of alkyl groups are observed [67]. Desmarcheliers et al. [68] have reviewed the mass spectra of TArPs and TAlPs. The major ions of a number of commercial phosphate esters observed following electron impact GC-MS have been tabulated [23, 45]. Negative chemical ionization MS has been used to identify TDCP [59, 60] and TDBP metabolites [69] in extracts of plastics, water and biological tissues.

Environmental Fate

Entry

Leakage of hydraulic fluid is thought to be a major mode of entry of phosphate esters into the environment (Fig. 1). Leaching from plastics that are exposed to water in piping or following disposal in landfills may also be an important mode of entry. Landfilling is the major form of removal of plastics from commercial use [8]. Monitoring studies have generally not detected those phosphate esters used as lubricants or hydraulic fluids but have observed those used as plasticizers [32, 70]. Solid wastes containing phosphate esters which are landfilled are thought to be the principal route of release from manufacturing/formulating plants. Air emissions as aerosols from lubricant seals and vapor loss from plastics, e.g. in vinyl car seats,

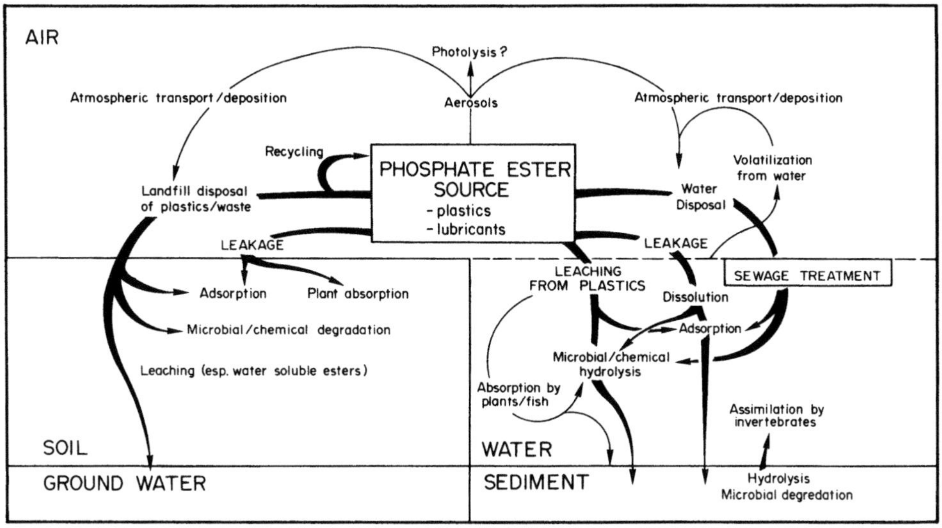

Fig. 1. Potential pathways for transport and removal of phosphate esters in the environment. (Adapted from Holiday et al. [10] with permission)

may be an important mode of human exposure to phosphate esters [8]. Burning of waste oils or plastics may also lead to release of phosphate esters since TPP, TBP and TDBP can volatilize intact from polymers heated at 200 to 350 °C [27].

Sorption to Sediments/Soils

The low water solubility and high K_{ow} of most phosphate esters suggests high soil adsorption coefficients (K_{oc}) [where K_{oc} is the soil-water partition coefficient (Kd) normalized for organic carbon content] [71]. This is confirmed by the results of Kenmotsu et al. [37] for the partitioning (Kd values) of several phosphate esters between water and a marine sediment (Table 5). In natural waters containing for example 50 mg/L suspended solids, greater than 90% of the esters would be expected to be in solution (Table 5). In soils however the high K_{oc} values suggest low leaching potential relative to many pesticides [71].

Volatilization from Water

Alkyl phosphates such as TBP and TBEP have high Henry's constants (Table 5) and could be expected to volatilize from water following entry into aquatic systems. TArPs and TAr/AlPs have Henry's constants in the 1×10^{-5} to 10^{-7} (atm m^3/mol) range and consequently have a low rate of volatilization from water [72].

Monitoring Studies

Monitoring of river water, drinking water, sediments and biota in the United States, Canada, Europe and Japan has shown that some TArPs, TAlPs and

Table 5. Sediment sorption coefficients (Koc) and Henry's Constant's (H) of some phosphate esters and predicted behaviour in aqueous solution

Compound	k[a]	Koc[b]	% in sol'n[c]	H (atm m^3/mol)[d] half-life	Volatilization (days)[e]
TPP	60	7,850	96.2	1.66×10^{-6}	60
TCP	400	27,539	93.3	2.15×10^{-7}	496
CDP	–	6,727	96.7	4.64×10^{-6}	22
EHDP	–	31,102	86.5	1.58×10^{-5}	7
BPDP	–	14,600	93.2	8.48×10^{-7}	128
IPDP	–	18,328	91.6	–	–
NPDPP	–	40,306	83.2	–	–
CPDPP	–	48,314	80.5	–	–
DBPP	–	5,037	97.5	–	–
IDDP	–	21,591	90.3	1.28×10^{-4}	1.3
TBP	3.5	3,592	98.2	2.48×10^{-2}	0.4
TBEP	500	2,311	98.9	4.70×10^{-6}	235
TEHP	–	4,738	97.7	–	–
TCEP	2.5	151	99.9	1.81×10^{-7}	516
TDBP	150	5,791	97.2	1.07×10^{-7}	1.9
TDCPP	10	2,591	98.7	–	–

[a] k values from Kenmotsu et al. [37] for a marine sediment
[b] Koc values calculated from Kow using the equation of Kenaga and Goring [71]: log Koc = 1.377 + 0.544 log Kow
[c] % in solution in water containing 50 mg/l suspended sediment with organic carbon content of 10% using equations of Wolfe et al. [78]: Fraction in solution = 1/(Koc × ratio of sediment/water × org. carbon) + 1
[d] Henry's constant = vapor pressure (25 °C) (atm)/S (mol/m^3). Values of H were obtained from references [10, 16, 8] or calculated from known vp and S (TBP, TBEP, Table 4). The results are only estimates and have not been confirmed by laboratory studies
[e] Volatilization half-life in water calculated from the equations of Mackay and Leinonen (72): $T = 0.69/K_{ol}$ for a 1 m depth system. K_{ol} = volatilization rate constant (m/hr). $K_{ol} = [1/K_L + RT/HK_G]^{-1}$ where K_L = liquid mass transfer coefficient, K_G = gas transfer coefficient, T = °K, R = Gas Constant

THAPs are present at sub- ng/g concentrations. The most frequently detected are TBP, TPP and TCEP (Table 6). TArP concentrations are lower in drinking water compared to raw water. Sheldon and Hites [73] found that TPP levels in Philadelphia water were reduced from 0.2 to 0.03 ng/ml in the water treatment plant but TBP residues were higher in drinking water (0.4 ng/L) than input water (0.3 ng/L). Williams and LeBel [70] found that cities with river water sources had 2 to 3-fold higher levels of TBEP, TBP, TCEP, and TCPP than those with lake sources although levels were < 10 ng/L except for TBEP. The presence of TBEP was attributed mainly to contamination from rubber gaskets used in sampling equipment.

Concentrations of phosphate esters in sediments and fish surveyed by Mayer et al. [32] and by the Japanese Environment Agency [74] were higher than those in water by 100–1,000 fold. These surveys suggest that with the exception of TBP and TPP, phosphate esters were not widespread contaminants of aquatic environments at ng/g levels.

Table 6. Concentrations of phosphate esters in water, sediment and fish at various locations during the 1970's

Location/Date (Reference)	Concentration of phosphate esters			
	Water (ng/l)		Sediment (ng/g)	Fish (ng/g)
1. Geneva [75]	TBP	< 2–82	n.s.	n.s.
2. River Waal [76]	TBP	< 10–10,000	n.s.	n.s.
The Netherlands	TCEP	< 10–10,000	n.s.	n.s.
3. Japan (Various [74] locations, 1975)	TBP	< 10–710 (16/100)[a]	< 3–350 (34/100)	< 5–26 (31/94)
4. Japan	TBP	10–580 (39/117)	3–240 (48/117)	2–13 (27/85)
(Various [74]	TCEP	90 (3/114)	<10 (90/114)	5–140 (9/93)
locations, 1977–78)	TCP	< 250 (0/114)	1,060–2,160 (2/114)	0.1 (0/93)
	TDCPP	< 500 (0/114)	<40 (0/1.140)	< 20 (0/93)
	TBEP	< 500 (0/114)	<10 (0/114)	< 10 (0/93)
5. U.S. Midwest	TPP	n.s.	n.s.	0.06–0.15 (31)
[48]	TCP	n.s.	n.s.	0.04 (1)
	TXP	n.s.	n.s.	0.08 (1)
	CDP	n.s.	n.s.	0.13–0.41 (3)
	NPDPP	n.s.	n.s.	0.16–0.50 (3)
6. U.S. (Delaware	TBP	60–2,000 (14/16)[b]	n.s.	n.s.
River, 1978 [73])	TBEP	300–2,000 (16/16)	n.s.	n.s.
	TPP	100–400 (13/16)	n.s.	n.s.
7. U.S. (Delaware River	TBP	300–800 (31)	n.s.	n.s.
and Philadelphia [77])	TPP	< 10–16,000 (31)	n.s.	n.s.
8. U.S. City water	TBP	< 50–500 (2/10)	n.s.	n.s.
[55]	TPP	< 10–120 (2/10)	n.s.	n.s.
9. U.S. Midwest Rivers	TPP	< 100–7,900 (32/62)	< 0.1–0.9 (16/82)	< 0.01–4 (13/40)
and Lakes and	NPDPP	< 100–20,000 (3/55)	< 0.1–0.9 (3/55)	<0.01–20 (17/40)
San Francisco [32]	CPDPP	< 100–9,500 (3/55)	< 0.1–0.1 (1/55)	<0.01–4 (15/40)
10. U.S. near TAP	IPDP	< 100–100 (7)	n.s.	0.1–8 (6)
manufacturer [8]	TCP	n.s.	n.s.	2–5 (2)
11. Canadian city	TBP	< 0.05–62 (60)	n.s.	n.s.
water [70]	TCEP	< 0.05–52		
	TDCPP	< 0.05–23		
	TPP	< 0.05–8.6		
	TBEP	1.1–560		
	TCP	< 0.05–4.3		
12. Great Lakes	TBP	0.8–29.5 (24)	n.s.	n.s.
city water [79]	TCEP	0.1–12.6		
	TDCPP	< 0.1–4.6		
	TPP	< 0.1–4.8		
	TCP	< 0.1–1.7		

[a] Indicates number of samples which were analyzed or fraction of samples in which it was detected
[b] Tri(t-butyl) phosphate not Tri(n-butyl) phosphate was identified in these samples [73, 77]
[c] n.s. = not sampled

Biodegradation in Waters and Sediments

Some TArPs undergo hydrolysis at pH levels of natural alkaline water [32, 33] so that chemical degradation may be an important mode of loss in some situations. The presence of suspended sediment and microorganisms increases the rate of hydrolysis [32, 33]. In natural waters most studies have reported a two stage process where the rate of degradation increases as the microbial population capable of degrading the compound grows. For TAlPs and TAr/AlPs the shorter the alkyl chain the more rapid the disappearance of the ester in river water die-away tests and activated sludge systems [39]. For TArPs increasing the size and number of substituent groups on the phenyl ring leads to decreasing biodegradability [39]. Among TCP isomers, m,m,m-TCP and o,o,o-TCP degraded more rapidly than the p-substituted isomer in lake water [33]. TPP and TCP degraded more rapidly in river water tests than TBP or TBEP but the most persistent esters were TDBP, TCEP and TDCP which had half-lives that were 6-, 12-, and 20-fold, respectively, greater than that of TPP [37]. Low dissolved oxygen had little effect on the rate of degradation of TPP, NPDPP and CPDPP in a simulated microcosm [32].

Degradation of phosphate esters in natural waters or activated sludge is assumed to give rise to diaryl- and dialkyl-products as well as the coresponding phenols and alcohols, however, experimental evidence for this with commercial phosphate esters is limited. Howard and Deo [33] identified diphenyl phosphate in lake water containing TPP. They also noted changes in GC chromatograms of commercial phosphate esters held in lake water due to selective degradation of some isomers. Wagemann et al. [80] monitored the release of orthophosphate from a TCP mixture (Imol 140) and reported half-lives of 96 days for this product in river water. This was much longer than observed for the disappearance of TCP itself and suggests that the diaryl and mono-aryl phosphates or other products of microbial/chemical hydrolysis degrade more slowly than the parent compound. TDBP underwent loss of HBr within 24 hrs after addition to water during toxicity testing with fish [81] while TDCP did not degrade under the same conditions [60]. In sewage sludge TDBP undergoes hydrolysis yielding dibromopropanol [82]. Oxidation of alkyl substituents may also be an important degradation pathway in natural water and activated sludge. TCP has been shown to yield p-hydroxybenzoic acid in activated sludge incubations [83].

Pickard et al. [84] found that a mixed bacterial population isolated from a lake sediment previously contaminated with fuel oils could utilize TOCP, TXP, TPP and a commercial TCP mixture as the sole carbon source. They proposed breakdown via phosphatase action to yield di- and monoaryl esters. Degradation of TPP and EHDP in river sediments under aerobic conditions yielded diphenyl phosphate as the major product. However, up to 50% of the radioactivity in sediment was unextractable after a 60 day incubation suggesting incorporation of carbon-14 into sediment organic matter [85].

TCP products incorporated into agar or soil have been shown to support growth of soil thermophilic fungi [86]; however, there are few studies on degradation of phosphate esters in soils. When incorporated as plasticizers in PVC little degradation of TXP and TCP containing plastic was observed following 8 weeks burial in soils at favorable humidity (30%) and temperature (22–25 °C) levels [87]. Hydrolysis of dialkylaryl thiophosphate insecticides (e.g. diazinon and parathion) in soil is more rapid than in water and may be catalysed by soil clay and organic matter as well as by biological systems [88]. Hydrolysis of TAlPs and TAr/AlPs in soils is likely to be comparable to structurally similar organophosphate insecticides although this has not been demonstrated experimentally.

Pharmacokinetics and Metabolism

Mammalian and Avian Species

Studies of phosphate ester uptake, elimination and metabolism have focussed largely on TOCP and more recently on TDBP with the purpose of explaining the toxic effects of these two compounds. Early studies of TCP absorption in rabbit and dog indicated that m- and p-TCP isomers were poorly absorbed relative to the o-TCP isomer by way of peritoneal and epidermal membranes [89]. Intravenous injection of TCP isomers resulted in rapid excretion in bile. Liver and spleen retained the major portion of untransformed TCP whatever the route of exposure [89]. Dermal exposure of ^{32}P-TOCP resulted in 100-fold more rapid appearance of radioactivity in plasma and urine of human subjects compared to a dog [90]. Clearance of ^{32}P from plasma was complete in humans and dog within 24 h. Liver contained the highest ^{32}P concentrations in dog tissues examined 24 h after dermal exposure [90]. In hens, up to 26.5% of ^{32}P-TOCP was excreted within 72 hrs after a single oral dose as unmetabolized product. Concentrations of ^{32}P were 25-fold higher in liver than in plasma at 72 h post-treatment [91].

Metabolism of TOCP and other TArPs proceeds via hydroxylation of alkyl substituents on the rings [92]. Eto et al. [93] isolated three metabolites of TOCP from rat liver and intestine, two of which they identified as cyclic saligenin phosphates while the third was thought to be the intermediate di(o-hydroxymethylphenyl)o-cresyl phosphate (Fig. 2). The major metabolite, 2-(o-cresyl)-4H-1:3:2-benzodioxaphosphoran-2-one (Fig. 2) was responsible for the neurotoxic activity of TOCP. In hens the cyclic saligenin phosphate was found to be the major form of radioactivity in the liver 12 to 72 hrs following oral administration of TOCP [91]. Liver of rats and hens was found to be much more active than brain, spinal cord or sciatic nerve in forming the neurotoxic metabolite of TOCP [94]. Metabolism of m-TCP in the rabbit yielded the hydrolysis products m-cresol and m-hydroxybenzyl alcohol and a neutral product tentatively identified as m-hydroxymethylphenyldi(m-cresyl) phosphate [95]. Eto et al. [95] demonstrated that tri(p-ethylphenyl)phosphate (EPP) was metabolized in rats and mice to yield 1-hydroxyethyl and 1-acetyl products. The triacetyl EPP was a potent neurotoxin. Metabolism of other widely used TArPs and TAr/AlPs by mammals and hens has not been reported, however, studies of the comparative neurotoxicity of some commercial products [96] give insight to the type of oxidative metabolism that must occur. At least 1 hydrogen must be present on the alpha-carbon of ring substituents for a neurotoxic metabolite, i.e. with a hydroxylated alkyl group, to be produced. BPDP is not neurotoxic because oxidation of the t-butyl group cannot produce a cyclic saligenin phosphate or acetyl derivative, while o-isopropylphenyl diphenyl phosphate (a possible impurity in IPDP) showed neurotoxic activity because of oxidation of the α-carbon. The relative importance of O-dearylation (frequently observed with OP insecticides [92]) to yield polar diaryl or diarylalkyl phosphates and oxidation of alkyl substituents, to produce neurotoxic products, has not been systematically investigated.

TDBP and TDCP are readily absorbed from skin and gastrointestinal tract of rabbits and rats [97, 98, 99]. Absorption of TDBP from garments by children and

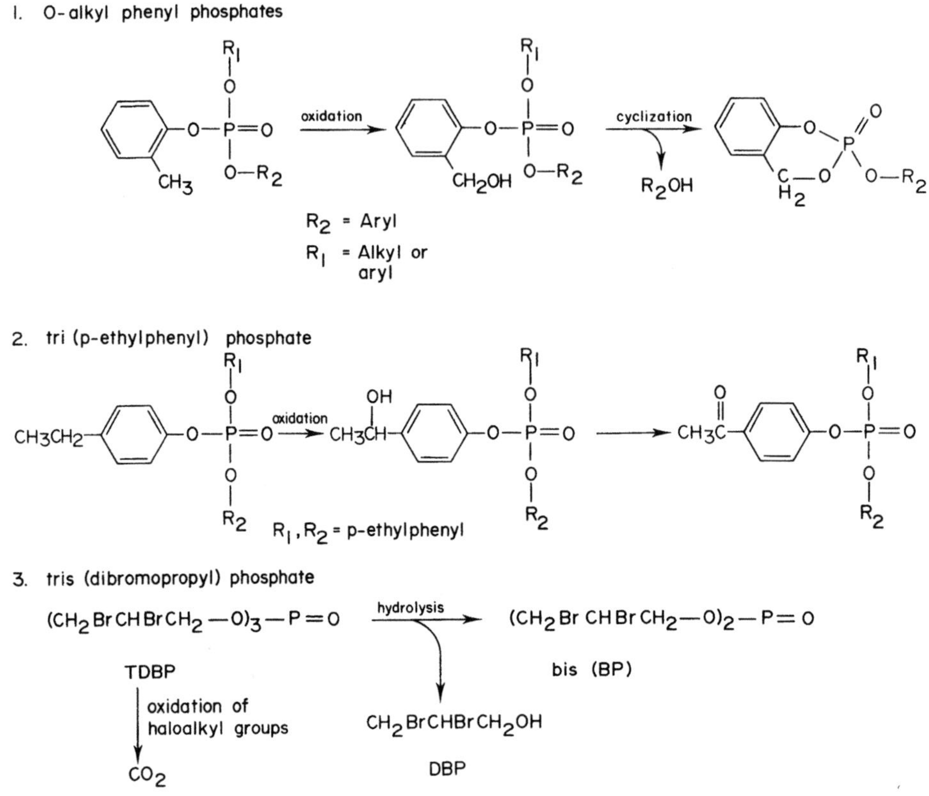

Fig. 2. Metabolism of 1. o-alkyl phosphates, 2. tri(p-ethyl phenyl) phosphate, 3. tris (dibromopropyl) phosphate

excretion of the metabolite dibromopropanol (DBP) was also demonstrated [69]. TDBP and TDCP were rapidly eliminated and metabolized within 24 hrs after intravenous (iv) administration to rats, 44–54% in urine and 14–20% as CO_2 [97, 100]. Elimination of TDCP was more rapid than TDBP. Kidney, lung and liver tissues contained the highest concentrations of TDBP and TDCP following i.v. administration [97, 100]. Bis(dihaloalkyl) phosphates were the major metabolites of TDBP and TDCP following i.v. administration to rats. Bis(dichloropropyl) phosphate accounted for 67% of total radioactivity in urine while bis(dibromopropyl) phosphates were the major form of radioactivity in plasma following i.v. dosing [97, 100]. The corresponding propanol derivatives were also detected. Microsomal oxidation of TDBP to reactive electrophiles, 2-keto and α-bromoaldehydes was proposed as the metabolic pathway which explained the mutagenicity of TDBP [101]. TDBP is also oxidized in visible light in the presence of riboflavin (i.e. by singlet oxygen) to produce similar mutagenic metabolites [102]. Chloropropandiol was observed in microsomal degradation of TDCP [97] however the chlorohydroxypropyl-containing phosphate ester which could be a reactive electrophile similar to those proposed with TDBP was not detected [97]. Bis(dichloropropyl) phosphate was found not be metabolized further by rat liver microsomes perhaps because the polarity of the diester would favor partitioning out of the microsomes into the aqueous phase [97].

TAlPs, including TBP, also undergo dealkylation to corresponding dialkyl phosphates following oral or intraperitoneal administration to rats [103]. S-alkyl cysteine metabolites of the alkyl moiety were also observed [103]. The dialkyl- and cysteine metabolites were rapidly excreted in urine within hours of administration of the trialkyl derivative.

Insects and Fish

Uptake from water and tissue distribution of phosphate esters in aquatic animals has been studied with a greater number of compounds than with mammals or birds (Table 7). Accumulation via aquatic food chains is not thought to be a major route of exposure to phosphate esters because most compounds are readily eliminated by aquatic animals. Depuration half-lives of phosphate esters in fish range from < 12 hrs for TCEP, TBP and TDCP to 4–6 days for some TArPs (TPP, IPDP, BPDP, and TCP isomers) [104, 105]. Bioconcentration of phosphate esters by fish in flow-through and static tests is generally less than predicted from the relatively high K_{ow} values for these compunds (Table 7). TAlPs and THAPs have low BCFs in static and flow-through exposures as predicted from their generally low K_{ow} values [104]. Species differences are evident in the rates of elimination of TArPs and TAr/AlPs suggesting that metabolism is more rapid in some species. TCP isomers differ in observed BCFs; mTCP is more rapidly eliminated by fathead minnows and rainbow trout fry than pTCP [105].

During exposure of rainbow trout to ^{14}C-EHDP in water, highest levels of radioactivity were observed in liver, kidney and intestine. When fish were transferred to clean water highest levels were observed in the intestine [110]. The major route of metabolism of TArPs and TAr/AlPs in fish appears to be O-dearylation or O-dealkylation to yield diaryl esters [50, 110]. Diphenyl phosphate (DPP) was identified as a major metabolite of EHDP and IPDP in rainbow trout. DPP was eliminated intact or as a conjugate [50]. Fish liver supernatants (10,000 g) will degrade pTCP, mTCP, EHDP, and BPDP to yield hydroxylated products similar to those observed in mammals and birds [111]. P-hydroxymethylphenyldicresyl phosphate and p-carboxyphenyldicresyl phosphate were identified following incubation of pTCP with fish liver supernatants [111]. However, hydrolysis to phenolic or alcohol components may be the predominate pathway in fish in vivo [50]. Metabolism of TPP by the housefly yielded p-hydroxyphenyldiphenyl phosphate [112]. A similar product was produced using the Udenfriend chemical oxidation system [112].

Fish can also accumulate phosphate esters from their diet. Hodson [113] fed rainbow trout diets containing 1–10,000 µg/g TPP for 16–32 weeks. Although uptake of chemical was not measured, fish showed reductions in growth rate at 1,000 µg/g and 50% mortality at 10,000 µg/g exposure. *Chironomus* larvae exposed to several sediments containing TPP accumulated more of the ester than those exposed to TPP in water only [114]. Assimilation of TPP from ingested sediment explained the higher levels in sediment-exposed larvae [114].

Table 7. Bioconcentration factors (BCF) of phosphate esters in aquatic animals using flow-through or static tests

Compound	Species	System[a]	Exposure conc.[b]	BCF[c,d]		Reference
				Observed	Predicted	
TPP	Rainbow trout	F	0.8–4.5(90 d)	217 (ed)	1,714	[48]
TPP (^{14}C)	Rainbow trout	S	5–50(24 h)	573		[105]
	Fathead minnow	S	5–50(24 h)	561		[105]
TPP	Rainbow trout	F	0.22–1.4 (90 d)	271		[32]
	Killifish	S	250(70 h)	250		[106]
	Killifish	F	10–30(18 d)	155		[106]
	Goldfish	S	250(100 h)	110		[104]
TCP (I-140)	Rainbow trout	F	900(120 d)	169 (g, a)	4,377	[107]
TCP	Fathead minnow	F	31.6(32 d)	165		[108]
pTCP (^{14}C)	Bluegill	F	0.124(28 d)	1,589		[109]
	Daphnids	F	0.124(24 h)	722		[109]
	Rainbow trout	S	5–50 (24 h)	1,420		[105]
	Fathead minnow	S	5–50 (24 h)	928		[105]
mTCP (^{14}C)	Rainbow trout	S	5–50 (24 h)	784		[105]
	Fathead minnow	S	5–50 (24 h)	596		[105]
BPDP (^{14}C)	Rainbow trout	S	5–50 (24 h)	1,096	4,521	[105]
	Fathead minnow	S	5–50 (24 h)	785		[105]
EHDP (^{14}C)	Rainbow trout	S	5–50 (24 h)	1,314	14,470	[110]
IPDP (^{14}C)	Fathead minnow	F	2.5 (28 d)	495	6,450	[50]
CPDPP	Rainbow trout	F	0.22–2.0 (90 d)	2,807	29,330	[32]
NPDPP	Rainbow trout	F	0.36–4.2 (90 d)	691	22,097	[32]
TDBP (^{14}C)	Bluegill	F	0.187 (28 d)	622	1,066	[109]
	Daphnids	F	0.187 (24 h)	50		[109]
TBP	Killifish	S	4,000 (100 h)	30	505	[106]
	Goldfish	S	4,000 (100 h)	7		[106]
	Killifish	F	100–840 (38 d)	21		[104]
TCEP	Killifish	S	1,000–3,000 (100 h)	2	4	[106]
	Goldfish	S	1,000–3,000 (100 h)	1		[106]
	Killifish	F	2.3–12.7 (11 d)	1		[104]
TDCP	Killifish	S	1,000 (100 h)	77	303	[106]
	Goldfish	S	1,000 (90 h)	3		[106]
	Killifish	F	40–400 (30 d)	43		[104]

[a] F = flow-through; S = static exposure
[b] Time of exposure in parentheses
[c] On a whole fish basis except where indicated. ed = edible portion; g, a = gut and adipose tissue
[d] Predicted from the equation log BCF = 0.851 log Kow − 0.7 [108]

Aquatic Plants

Duckweed (*Lemna minor*) rapidly accumulated EHDP and TPP from water in laboratory and field studies [115, 116]. A concentration factor of 283 in duckweed was observed for EHDP following 120 h exposure. In field studies maximum concentrations of TPP and EHDP (^{14}C labelled) were observed in duckweed within 24 h of the addition of the compounds to a small pond. Cattails (*Typha sp*) exposed in the same pond accumulated 100 fold lower concentrations of TPP and EHDP [116].

Biological Effects

The biological effects of phosphate ester flame retardants and lubricants have been the subject of much investigation over the past 50 years. The delayed neurotoxic effects and mechanisms of action of TOCP and related isomers have generated the most interest. Recently the mutagenicity and mechanisms of activation of TDBP and TDCP have been the subject of intense scrutiny. The reviews of Sandmeyer and Kerwin [39], Holiday and Engelhardt [16], Holiday et al. [10], Midwest Research Institute [8], Eto [92] and Johnson [117, 118] should be consulted for detailed information on the toxicology of TArPs, TAr/AlPs and THAPs.

Acute Toxicity

In general commercial phosphate esters exhibit low acute lethality in standard tests with mammals or birds [39]. TCP isomers, TPP, TCEP, TBP and DBPP all have LD50's (oral) to rats in the 3–12 g/kg range [40, 96]. TEHP, EHDP, and IDDP are among the least toxic to rats or chickens (LD50 >10 g/kg) [8, 40]. In aquatic tests, TPP has been found to be the most acutely toxic to all species with 96 h LC50's of 0.3–0.4 mg/L for rainbow trout in static systems [32]. By contrast LC50's for TCP mixed isomers, IDDP, NPDPP, and CPDPP range from 6.7–>100 mg/L for rainbow trout and bluegill [32, 109, 119]. LC50's for daphnia and midge larvae for TPP, NPDPP, CPDPP and Pydraul 50E were similar to those for rainbow trout [32]. Mysid Shrimp were the most sensitive aquatic species studied with 96 h LC50's with TPP of >0.18–<0.32 mg/L [32].

Delayed Neurotoxicity

Johannsen et al. [96] investigated the delayed neurotoxicity of many of the TArPs and TAr/AlPs in commercial use during the 1970's. TOCP and CDP (containing o-cresyl isomers) were the most neurotoxic in hens, exhibiting effects at cumulative dosages (administered by gastric intubation) of 1–2 g/kg. TBP and DBPP exhibited neurotoxic effects at levels similar to their LD50's in hens (2–4 g/kg) while cumulative doses of 120 g/kg of BPDP, IPDP, NPDPP, CPDPP, EHDP, and IDDP failed to produce delayed neurotoxicity. TCEP and TDCP have also been found not to induce delayed neuropathy at doses of 10 g/kg [120]. The neurotoxic responses are characterized in hens by a 8–14 day delay following initial administration of the ester, ataxia and finally paralysis of legs. The minimum dose of TOCP required to produce these effects in hens is 250 mg/kg [92]. In rodents subacute responses are characterized by hyperexcitability, tremors and general placid paralysis with onset of signs of poisoning varying from 5 h to several days following initial dosing [121]. These responses to TOCP are atypical of that noted in humans or hens and are similar to responses to cholinesterase inhibitors rather than delayed neurotoxins [121].

Inhibition of a specific protein "neurotoxic esterase" (NTE) which is phosphorylated by the neurotoxic ester, is associated with delayed neurotoxicity [117]. NTE hydrolyses phenyl phenylacetate or phenylvalerate and can be distinguished analytically from other esterases by specific inhibitors [118]. The genesis of neuropathy requires a high degree of NTE inhibition (>70%) during or shortly after the dosing regimen [122]. Measurement of the degree of inhibition of NTE has been used to screen phosphate esters for neurotoxic effects [118]. The actual biochemical mechanisms underlying development of delayed neurotoxic response are not completely understood. Baron [121] in summarizing work prior to 1981 concluded that there were no substantive hypotheses that explained all aspects of induction of delayed neurotoxicity. Inhibition of a protein (phospholipase A2) capable of degrading a lytic component (lysolecithin), prolonging the existence of a cell-damaging agent, has been proposed to explain the degradation of the axon and of the mylein sheath which are associated with delayed neuropathy [121].

Mutagenic Effects

Several phosphate esters and their metabolites have undergone mutagenicity testing following the observation that TDBP and its metabolite DBP were mutagenic in salmonella strains TA100, TA1535 and TA1538 [123]. Results with various mutation assays are summarized in Table 8. TDCP has been judged a weak mutagen having been found to induce weak but repeatable mutagenic responses with Salmonella strain TA100 when activated by induced rat or mouse liver, while TCEP is not considered mutagenic [17, 127]. TDBP has been identified as an animal carcinogen in rats and mice by oral or dermal exposure [129].

Structure-activity relationships for the mutagenicity of 9 THAPs and 9 haloalkanols were studied by Nakamura et al. [130] using the Ames/Salmonella microsomal assay. The presence of halogen was necessary for mutagenicity since TEP was inactive in any strains of *Salmonella typhimurium* with and without rat liver activation. Bromoalkyl derivatives were more mutagenic than chloroalkyl derivatives. Compounds such as TDBP and tris(1,2-dichloropropyl) phosphate, having adjacent β- and γ-halogens were active mutagens while TDCP (the 1,3-dichloropropyl isomer) showed positive but weak mutagenicities [130]. IPDP has shown some evidence of mutagenicity in tests for transformation of mouse BALB/3T3 cells and for positive forward mutations in activated mouse lymphoma L5178Y Tk+/−cells (Table 8) [10].

Table 8. Results of mutagenicity testing of some commercial phosphate esters

Compound	1. Ames/Salmonella plate assay, TA100, TA1535m, TA1538	2. Transformation of Mouse BALB/3T3 cells	3. Mouse Lymphoma L5178YTK± forward mutation	4. Chromosome aberation in L5178Y cells
TDBP	+ (act./non-act.)	+	+	−
TDCP	± (act. only)	−	−	+ (with act.)
TCEP	−	+	−	−
IPDP	−	+	+ (with act.)	−
BPDP	−	−	−	−

[a] Act. = activation by phenobarbitol or Arochlor 1254 induced mouse and rat liver S-9 fractions
References: TDBP [123–126]; TDCP [123, 127, 128, 130]; TCEP [128, 17]; IPDP and BPDP [10]

Chronic Effects in Fish

Laboratory studies of the effects of TPP, NPDPP, and CPDPP on fish have demonstrated growth reduction, reduction in vertebral collagen, spinal abnormalities, enlarged livers and eye disorders [32]. Sub-μg/L concentrations of Pydraul 50E, NPDPP, and CPDPP as well as TPP at > 1.4 μg/L were found to reduce vertebral collagen and to produce eye cataracts in rainbow trout after 90 day exposures [32]. Similar effects were observed at higher levels of Pydraul 50E (90 or 120 day exposures) with lake trout (at 2.6 μg/L) and with fathead minnows at > 138 μg/L [32]. The chronic effects of Pydraul 50E were attributed to the NPDPP and CPDPP components of this lubricant. Lockhart et al. [107] reported elevated activities of liver enzymes LDH and GOT and blue-grey dicoloration of internal tissues following chronic exposure of rainbow trout to 900 μg/L of a TCP lubricant. Sasaki et al. [106] reported deformation of spines of killifish with relatively high water concentrations of TCEP (200 mg/L), TDCP (3.5 mg/L), and TPP (1.1 mg/L). Survival and reproduction of *Daphnia magna* were significantly reduced at 100 μg/L concentrations of BPDP [131]. Survival at higher concentrations was observed with caged animals in outdoor ponds [131]. MATCs of BPDP for fathead minnows were estimated to be > 140– < 250 μg/L in laboratory studies (30 days, flow-through) and > 453– < 826 μg/L in field studies in small ponds containing sediment [131].

Effects on Plants

Of 6 TArPs studies, TPP and CDP were the most toxic to green algae with IC50's of 0.26 and 0.70 mg/L respectively [132]. A lake phytoplankton was more sensitive with IC50's of 0.20 and 0.50 for TPP and CDP, while TXP and TCP isomers were about 10-fold less toxic [132]. NPDPP, CPDPP, and Pydraul 115E were non-toxic to algae with LC50s (96 h) of > 1,000 mg/L [32]. TOCP was found to be more toxic than m-TCP while p-TCP was non-toxic to algae [132]. TBP applied to vegetation was found to reduce growth of the roots of rice plants and of radish and soybean plants at concentrations of 10–100 μg/g (soil) [74].

References

1. Hutzinger, O., Sundström, G., Safe, S.: Chemosphere *1*, 3 (1976)
2. Weil, E.D., in: Encyclopedia of Chemical Technology. Kirk, R.E., Othmer, D.F. (Eds.) Vol. 10, 3rd Ed. pp. 348–419, 1981. Wiley-Interscience, N.Y.
3. Chadwick, D.H., Watt, R.S., in: Phosphorus and its compounds. Vol. 2. (Ed.) Van Wazer, J.R., Interscience, N.Y. 1961, pp. 1221–1279
4. Smith, M.I., Elvove, E., Frazier, W.H.: Public Health Rep. *45*, 2509 (1930)
5. Abelson, P.H.: Science *197*, 113 (1977)
6. Broadhurst, M.G.: Environ. Health Perspect. *2*, 81 (1972)
7. Fest, C., Schmidt, K.-J.: The Chemistry of Organophosphorus Pesticides. Springer-Verlag, N.Y., Tokyo 1982
8. Mid-West Research Institute: "Assessment of the need for limitation on triaryl and trialkyl/aryl phosphates". Report prepared for the United States Environ. Protect. Agency. Contract No. 68-01-4313, 1979
9. Lande, S.S., Santodonato, J., Howard, P.E., Greninger, D., Christopher, D.H., Saxena, J.: "Investigation of selected potential environmental contaminants: Haloalkyl phosphates". Report prepared for the United States Environ. Protect. Agency, Contract No. 560/2-76-0007. NTIS No. BP257910 1976

10. Holiday, M.G., Engelhardt, F.R., Holmes, A.: "A Review of the environmental and health aspects of triaryl/alkyl phosphates". Report prepared for the Dept. of Nat'l Health & Welfare, Ottawa, Canada. 1980
11. European Chemical News.: May 7, 13 (1976)
12. Druyan, E.A.: "Quantitative determination of tricresylphosphate isomers". Patent No. 449296. U.S.S.R. 1974. Cited in CA 83,022082u
13. Mukai, S., Yamaguchi, O.: "Purification of phosphoric acid ester oils for electric insulation". Patent Nr. 78 15596. Japan 1977. Assignee: Nisshin Denki Co. Ltd. Cited in CA 89, 092169a
14. Anon.: Chemical Marketing Reporter. *213*, 9 (1978)
15. Chemcyclopedia. Vol. 1, Amer. Chem. Soc. (1982)
16. Holiday, M.G., Engelhardt, F.R.: "The environmental and health aspects of tris(haloalkyl)phosphates: A review". Report prepared for Dept. of Nat'l Health & Welfare Ottawa Canada. 1980
17. Ulsamer, A.G., Osterberg, R.E., McLaughlin, J. Jr.: Clin. Toxicol. *17*, 101 (1980)
18. Kuryla, W.C., Papa, A.J.: Flame retardancy of polymeric materials. Vol. 1, Marcel Dekker, N.Y. 1973
19. Murray, D.A.J.: J. Fish. Res. Board. Can. *32*, 457 (1975)
20. Sugden, E.A., Greenhalgh, R., Pettit, J.R.: Environ. Sci. Technol. *14*, 1498 (1980)
21. Sengupta, P., Sen, A.R. and Mathew, T.V.: Res. Ind. *18*, 56 (1973)
22. Rapaport, L.I., Lebedinskaya, N.N.: Vopr. Gig. Tr., Profpatol. Toksikol. Proizvod. Ispol'z. Fosfororg. Plastif., 103 (1973). Cited in CA 84, 85006
23. Deo, P.G., Howard, P.H.: J. Assoc. Offic. Anal. Chem. *61*, 266 (1978)
24. Boyden, J.W., Deacon, M.T.: Analyst *95*, 935 (1970)
25. Paciorek, K.J.L., Kratzer, R.H., Kaufman, J., Nakahara, J.H., Christos, T., Hartstein, A.M.: Amer. Indust. Hyg. Assoc. J. *39*, 633 (1978)
26. Cho, L., Klaus, E.E.: Amer. Soc. Lubric. Eng. Trans. *24*, 119 (1980)
27. Brauman, S.K.: J. Fire Retard. Chem. *4*, 18 (1977)
28. Okamoto, Y., Kimura, N., Sakurai, H.: Bull. Chem. Soc. Jpn. *47*, 1299 (1974)
29. Van Wazer, J.R., in: Phosphorus and its compounds. Vol. 2. Van Wazer, J.R. (Ed.) p. 1937, Interscience, N.Y. 1961
30. Inagaki, N., Sakurai, S., Katsuura, K.: J. Appl. Polymer Sci. *23*, 2023 (1979)
31. Finnegan, R.A., Matson, J.A.: J. Amer. Chem. Soc. *94*, 4780 (1972)
32. Mayer, F.L., Adams, W.J., Finley, M.T., Michael, P.R., Mehrle, P.M., Saeger, V.M., in: Aquatic Toxicology and Hazard Assessment: Fourth Conference, Branson, D.R., Dickson, K.L. (Eds.) Amer. Soc. Test. Mater., pp. 103–123, 1981
33. Howard, P.H., Deo, P.G.: Bull. Environ. Contam. Toxicol. *22*, 337 (1979)
34. Barnard, P.W.C., Bunton, C.A., Llewellyn, D.R., Vernon, C.A., Welch, V.A.: J. Chem. Soc. 2670 (1961)
35. Wolfe, N.L.: Chemosphere *9*, 571 (1980)
36. Bunton, C.A.: Acc. Chem. Res. *3*, 257 (1970)
37. Kenmotsu, K., Matsunaga, K., Ishida, T.: Okayama-ken Kankyo Hoken Senta Nempo *4*, 103 (1980). (Can. Trans. Fish Aquat. Sci. 4907, Dept. Fisheries & Oceans, Winnipeg, Canada)
38. Mellan, I.: Industrial Plasticizers. Pergamon Press. N.Y. 1963
39. Saeger, V.W., Hicks, O., Kaley, R.G., Michael, P.R., Mieure, J.P., Tucker, E.S.: Environ. Sci. Technol. *13*, 840 (1979)
40. Sandmeyer, E.E., Kirwin, C.J. Jr., in: Patty's Industrial Hygiene and Toxicology. Vol. 2A. Clayton, G.D., Clayton, F.E. (Eds.) Wiley-Interscience, N.Y. 1981
41. Sax, N.I.: Dangerous Properties of Industrial Materials. 5th Ed. Van Nostrand Reinhold, N.Y. 1979
42. Mackay, D., Bobra, A., Shiu, W.Y., Yalkowsky, S.H.: Chemosphere *9*, 701 (1980)
43. Muir, D.C.G., Grift, N.P., Solomon, J.: J. Assoc. Offic. Anal. Chem. *64*, 79 (1981)
44. Kenmotsu, K., Matsunaga, K., Ishida, T.: Shokuhin Eiseigaku Zasshi *21*, 18 (1980)
45. LeBel, G.L., Williams, D.T., Benoit, F.M.: J. Assoc. Offic. Anal. Chem. *64*, 991 (1981)
46. Muir, D.C.G., Grift, N.P.: J. Assoc. Offic. Anal. Chem., *66*, 684 (1983)
47. Daughton, C.G., Crosby, D.G., Garnas, R.L., Hsieh, D.P.H.: J. Agric. Food Chem. *24*, 236 (1976)
48. Lombardo, P., Egry, I.J.: J. Assoc. Offic. Anal. Chem. *62*, 47 (1979)
49. Daft, J.L.: Bull. Environ. Contam. Toxicol. *29*, 221 (1982)

50. Huckins, J.N., Petty, J.D.: Presented at the 184th Amer. Chem. Soc. Nat'l Meeting, Kansas City 1982. Environ. Chemistry Div. Paper No. 16
51. Keulmans, A.I.M.: Gas Chromatography, 2nd Ed., Reinhold, N.Y. 1959
52. Hardy, C.J.: J. Chromatog. *13*, 372 (1964)
53. Horton, A.D.: J. Chromatog. Sci. *10*, 125 (1972)
54. Bloom, P.J.: J. Chromatog. *75*, 261 (1973)
55. Keith, L.H., Garrison, A.W., Allen, F.R., Carter, M.H., Floyd, T.L., Pope, J.D., Thruston, A.D. Jr., in: Identification and analysis of organic pollutants in water. Keith, L.H. (Ed.) Ann Arbor Sci., Ann Arbor, MI. 1976
56. Shafik, M.T., Bradway, D.E., Enos, H.F., Yobs, A.R.: J. Agric. Food Chem. *21*, 625 (1973)
57. Daughton, C.G., Cook, A.M., Alexander, M.: Anal. Chem. *51*, 1949 (1979)
58. Reid, S.J., Watts, R.R.: J. Anal. Toxicol. *5*, 126 (1981)
59. Hudec, T., Thean, J., Kuehl, D., Dougherty, R.C.: Science *211*, 951 (1981)
60. Ahrens, V.D., Maylin, G.A., Henion, J.D., St. John, L.E., Jr., Lisk, D.J.: Bull. Environ. Contam. Toxicol. *21*, 409 (1979)
61. Williams, D.T., LeBel, G.L., Benoit, F.M.: J. Assoc. Offic. Anal. Chem. *64*, 635 (1981)
62. Tittarelli, P., Mascherpa, A.: Anal. Chem. *53*, 1466 (1981)
63. Majors, R.E., Johnson, E.L.: J. Chromatog. *167*, 17 (1978)
64. Campbell, G.B., Foxton, A.A., Worsdall, R.L.: Lab. Pract. *19*, 369 (1970)
65. Renberg, L., Sundström, G., Sundh-Nygard, K.: Chemosphere *9*, 683 (1980)
66. Gurley, T.W., Ritchey, W.M.: Anal. Chem. *48*, 1137 (1976)
67. Quayle, A., in: Advances in Mass Spectrometry. Waldon, J.D. (Ed.) Pergamon, N.Y. 1959
68. Desmarcheliers, J.M., Wustner, D.A., Fukuto, T.R.: Res. Rev. *63*, 77 (1976)
69. Blum, A., Gold, M.D., Ames, B.N., Kenyon, C., Jones, F.R., Hett, E.A., Dougherty, R.C., Horning, E.C., Dzidic, I., Carroll, D.I., Stillwell, R.N., Thenot, J.-P.: Science *201*, 1020 (1978)
70. Williams, D.T., LeBel, G.L.: Bull. Environ. Contam. Toxicol. *27*, 450 (1981)
71. Kenaga, E.E., Goring, C.A.I., in: Aquatic Toxicology. Eaton, J.G., Parrish, P.R., Hendricks, A.C. (Eds.) Amer. Soc. Test. Mater. p. 78 1980
72. Mackay, D., Leinonen, P.J.: Environ. Sci. Technol. *9*, 1178 (1975)
73. Sheldon, L.S., Hites, R.A.: Environ. Sci. Technol. *13*, 574 (1979)
74. Japan Environment Agency, Office of Health Studies, Dept. Environ. Health, Environmental Survey Reports 1977, 1978, 1979
75. Grob, K., Grob, G.: J. Chromatog. *90*, 303 (1974)
76. Meijers, A.P., Van Der Leer, R.Chr.: Water Res. *10*, 597 (1976)
77. Sheldon, L.S., Hites, R.A.: Environ. Sci. Technol. *12*, 1188 (1978)
78. Wolfe, N.L., Zepp, R.L., Paris, D.F., Baughman, G.L., Hollis, R.C.: Environ. Sci. Technol. *11*, 1077 (1977)
79. Williams, D.T., Nestmann, E.R., LeBel, G.L., Benoit, F.M., Otson, R., Lee, E.G.H.: Chemosphere *11*, 263 (1982)
80. Wagemann, R., Graham, B., Lockhart, W.L.: Can. Fish. Mar. Ser., Tech. Rpt. 486 (1974)
81. Maylin, G.A., Henion, J.D., Hicks, L.J., Leibovitz, L., Ahrens, V.D., Gilbert, M., Lisk, J.D.: Bull. Environ. Contam. Toxicol. *17*, 499 (1977)
82. Kerst, A.F.: J. Fire Retard. Chem., Suppl. *1*, 205 (1974)
83. Ku, Y., Alvarez, G.H.: Appl. Environ. Microbiol. *43*, 619 (1982)
84. Pickard, M.A., Whelihan, J.A., Westlake, D.W.S.: Can. J. Microbiol. *21*, 140 (1975)
85. Muir, D.C.G., Grift, N.P. Unpublished data. Dept. Fisheries & Oceans, Winnipeg, Canada 1982
86. Mills, J., Eggins, H.O.W.: Int. Biodetect. Bull. *10*, 39 (1974)
87. Booth, G.H., Robb, J.A.: J. Appl. Chem. *18*, 194 (1968)
88. Khan, S.U.: Pesticides in the Soil Environment. Elsevier. N.Y. 1980
89. Gross, E., Grosse, A.: Arch. Exper. Pathol. Pharmacol. *168*, 473 (1932)
90. Hodge, H.C., Sterner, J.H.: J. Pharmacol. Exp. Therap. *79*, 225 (1943)
91. Sharma, R.P., Watanabe, P.G.: Pharmacol. Res. Commun. *6*, 475 (1974)
92. Eto, M.: Organophosphorus Pesticides Organic Chemistry and Biological Chemistry. C.R.C. Press Inc. 1974
93. Eto, M., Casida, J.E., Eto, T.: Biochem. Pharmacol. *11*, 337 (1962)
94. Morazain, R., Rosenberg, P.: Toxicol. Appl. Pharmacol. *16*, 461 (1970)
95. Eto, M., Abe, M., Takahara, H.: Agr. Biol. Chem. *35*, 929 (1971)

96. Johannsen, F.R., Wright, P.L., Gordon, D.E., Levinskas, G.J., Radue, R.W., Graham, P.R.: Toxicol. Appl. Pharmacol. *41*, 291 (1977)
97. Nomeir, A.A., Kato, S., Matthews, H.B.: Toxicol. Appl. Pharmacol. *57*, 401 (1981)
98. St. John, L.E. Jr., Eldefrawi, M.E., Lisk, D.J.: Bull. Environ. Contam. Toxicol. *15*, 192 (1976)
99. Ulsamer, A.G., Porter, W.K., Osterberg, R.E.: Toxicol. Appl. Pharmacol. *45*, 251 (1978)
100. Lynn, R.K., Garvie-Gould, C., Wong, K., Kennish, J.M.: Toxicol. Appl. Pharmacol. *63*, 105 (1982)
101. Soderlund, E.J., Nelson, S.D., Dybing, E.: Acta Pharmacol. Toxicol. *45*, 112 (1979)
102. McCoy, E., Hyman, J., Rosenkranz, H.S.: Mutat. Res. *77*, 209 (1980)
103. Jones, A.R.: Experientia *26*, 492 (1970)
104. Sasaki, K., Suzuki, T., Takeda, M., Uchiyama, M.: Bull. Environ. Contam. Toxicol. *28*, 752 (1982)
105. Muir, D.C.G., Yarechewski, A.L., Grift, N.P.: Chemosphere, *12*, 145 (1983)
106. Sasaki, K., Takeda, M., Uchiyama, M.: Bull. Environ. Contam. Toxicol. *27*, 775 (1981)
107. Lockhart, W.L., Wagemann, R., Clayton, J.W., Graham, B., Murray, D.: Environ. Physiol. Biochem. *5*, 361 (1975)
108. Veith, G.D., DeFoe, D.L., Bergstedt, B.V.: J. Fish. Res. Board Can. *36*, 1040 (1979)
109. Sitthichaikasem, S.: Some Toxicological Effects of Phosphate Esters on Rainbow Trout and Bluegill. Ph.D. Thesis, Iowa State University, 1978
110. Muir, D.C.G., Grift, N.P.: Chemosphere *10*, 847 (1981)
111. Muir, D.C.G., Grift, N.P.: Manuscript in preparation 1983
112. Eto, M., Miyamoto, H., Hashimoto, Y.: Botyu-Kagaku *40*, 106 (1975)
113. Hodson, P.: Great Lakes Biolimnology Lab., Dept. Fisheries and Oceans, Can. Centre Inland Waters, Burlington, Ont. Personal Communication
114. Muir, D.C.G., Townsend, B.E., Lockhart, W.L.: Environ. Toxicol. Chem., in press 1983
115. Lockhart, W.L., Billeck, B.N., deMarch, D.G.E., Muir, D.C.G., in: Aquatic Toxicology, Sixth Conference. Bishop, W.E., Cardwell, R.D. & Heidolph, B.B. (Eds). Amer. Soc. Test Mater., p. 460, 1983
116. Muir, D.C.G., Grift, N.P., Lockhart, W.L.: Environ. Toxicol. Chem. *1*, 113 (1982)
117. Johnson, M.K.: C.R.C. Critical Rev. Toxicol. *3*, 289 (1975)
118. Johnson, M.K.: Arch. Toxicol. *34*, 259 (1975)
119. Dawson, G.W., Jennings, A.L., Drozowski, D., Rider, E.: J. Hazard. Mater. *1*, 303 (1975)
120. Sprague, G.L., Sandvik, L.L., Brookins-Hendricks, M.J., Bickford, A.A.: J. Toxicol. Environ. Health *8*, 507 (1981)
121. Baron, R.L.: Ann. Rev. Entomol. *26*, 29 (1981)
122. Johnson, M.K., Lotti, M.: Toxicol. Lett. *5*, 99 (1980)
123. Prival, M.J., McCoy, E.C., Gutter, B., Rosenkranz, H.S.: Science *195*, 76, (1977)
124. Carr, H.S., Rosenkranz, H.S.: Mutat. Res. *57*, 381 (1978)
125. Brusick, D., Matheson, D., Bakshi, K.: Mutat. Res. *53*, 31 (1978)
126. Furukawa, M., Sirianni, S.R., Tan, J.C., Huang, C.C.: J. Nat'l Cancer Inst. 60, 1179 (1978)
127. Gold, M.D., Blum, A., Ames, B.N.: Science *200*, 785 (1978)
128. Matsuoka, A., Hayashi, M., Ishidate, M., Jr.: Mutat. Res. *66*, 277 (1979)
129. Van Duuren, B.L., Loewengart, G., Seidman, I., Smith, A.C., Melchionne, S.: Cancer Res. *38*, 3236 (1978)
130. Nakamura, A., Tateno, N., Kojima, S., Kaniwa, M., Kawamura, T.: Mutat. Res. *66*, 373 (1979)
131. Adams, W.J., Kimerle, R.A., Heidolph, B.B., Michael, P.R., in: Aquatic Toxicology, Sixth Conference. Bishop, W.E., Cardwell, R.D. & Heidolph, B.B. (Eds). Amer. Soc. Test. Mater., p. 367, 1983
132. Wong, P.T.S., Chau, Y.K.: Sci. Total Environ. In press 1983

Phthalic Acid Esters

C. S. Giam, E. Atlas, M. A. Powers, Jr., J. E. Leonard
Department of Chemistry, Texas A & M University
College Station, TX 77843, USA

Introduction

The twentieth century can as properly be called the age of organic chemistry as it can the atomic age. The dramatic increase in the production of synthetic fibers, processed foods, pesticides, plastics, pharmaceuticals, and a thousand other materials has completely altered our immediate human environment and provided a wealth of new materials. But it has also transformed our natural environment both intentionally and unintentionally.

The phthalate esters are among these new high volume synthetic materials. Phthalic acid esters (structure *1*, also called phthalates, phthalate esters, or PAEs) are derivatives of phthalic acid (*2*) whose systematic name is 1,2-benzenedicarboxylic acid. There are two other benzenedicarboxylic acids, viz., the 1,3-isomer (isophthalic acid, *3*) and the 1,4-isomer (terephthalic acid, *4*). Terephthalic acid is important in the manufacturer of certain types of polyester fabrics. Monomeric esters of this acid are not common in commerce, and the polymeric esters are extremely insoluble and nonvolatile. Hence, terephthalates have a limited environmental distribution. The derivatives of isophthalic ester are of commercial importance only as specialty chemicals. There is little data concerning the environmental significance of the terephthalates and isophthalates, so the present review will be largely confined to the phthalates themselves.

PAEs are widely distributed because of their high production and use. In 1979 (the last year for which official figures are available) the United States produced approximately 2 kg of phthalate esters per person, most of which was used for making polyvinyl chloride into a flexible material. In such materials the phthalate plasticizer may provide up to 67% by weight of the final product. Thus, the phthalate esters are very widely distributed in the human and natural environment.

Concern for the possible health effects of these materials has been reported for a number of years, primarily because of their use in items such as blood transfusion bags and other medical materials from which they could easily be transferred in-

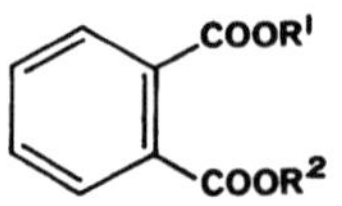

1

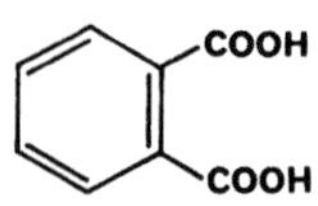

2

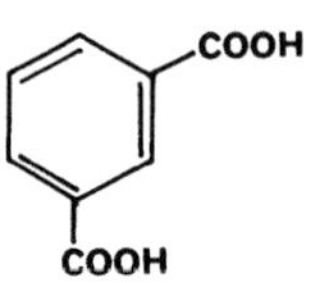

3

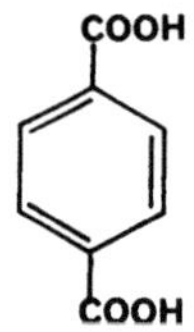

4

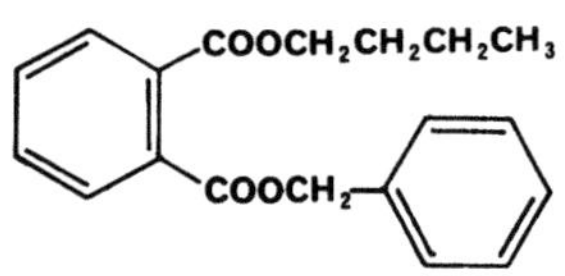

5

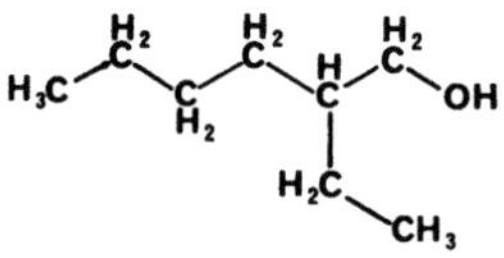

6

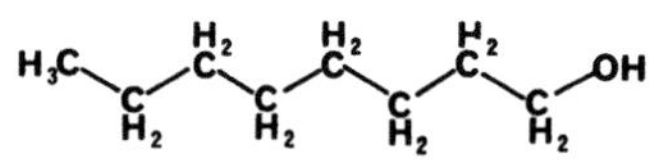

7

travenously to human patients [15, 100, 117, 150–152]. The relatively low acute toxicity of most phthalate esters, though, quieted most of these fears ([15] and Table 22 and 24). Recently the concern has shifted from their short-term, acute effects to their long-term, chronic effects. There are reports that these materials are potential human carcinogens. Recently the National Toxicology Program of the US Public Health Service reported the deliberations of a special conference on the mutagenicity and carcinogenicity of phthalates esters (Environmental Health Perspectives, volume 45, 1982).

Chemical and Physical Properties

The general structure of a dialkyl phthalate is a rigid planar aromatic ring with two flexible, often nonlinear, aliphatic side chains. The polar carboxyl group is buried in the middle of the molecule (except for small R groups such as methyl and ethyl) and exerts only a small effect on the physical properties of the compound.

The alcohol moieties in the phthalate esters (ROH, related to R^1 and R^2 in *1*) are usually aliphatic rather than aromatic, and a wide variety of alcohols have been used. Indeed, some commercial plasticizers have two different alkyl groups, as in butyl benzyl phthalate (*5*). Table 1 lists some common phthalate esters and the abbreviations that will be used in this paper for the compounds.

Table 1. Common phthalic acid esters (PAEs)

R groups[a]	Abbreviation[b]	CA[c]	RTECS[c]
Bis(2-butoxyethyl)	BBEP	*[117-83-9]*	TI0175000
Bis(2,3-epoxypropyl)	BEPP	*[7195-45-1]*	TI0200000
Bis (2-ethoxyethyl)	BEEP	*[67329-05-9]*	TI2050000
Bis(2-ethylhexyl)[d] (di-2-ethylhexyl)	DEHP[c, e]	*[117-81-7]*	TI0350000
Bis(2-methoxyethyl)	BMEP	*[117-82-8]*	TI1400000
Bis(1-methylheptyl)[d]	BMHpP	*[131-15-7]*	TI0360000
Bis(2-methylpropyl) (diisobutyl)	DiBP[b]	*[84-69-5]*	TI1225000
Bis(phenylmethyl) (dibenzyl)	DBzP[b]	*[523-31-9]*	
2-butoxyethyl butyl (Butylglycol butyl)	BGBP[b]	*[33374-28-6]*	
Butyl decyl	BDcP	*[89-19-0]*	TI0527000
Butyl 2-methylpropyl (butyl isobutyl)	BiBP[b]	*[17859-53-5]*	
Butyl phenylmethyl (butyl benzyl)	BBP[b]	*[85-68-7]*	TH9990000
Decyl hexyl	DcHP	*[25724-58-7]*	TI0545000
Decyl octyl	DcOP	*[119-07-3]*	TI0550000
Dibutyl	DBP	*[84-74-2]*	TI0875000
Dicyclohexyl	DCHP	*[84-61-7]*	
Didecyl	DDcP	*[84-77-5]*	TI0900000
Diethyl	DEP	*[84-66-2]*	TI1050000
Dihexyl	DHP	*[84-75-3]*	TI1100000
Diisodecyl	DiDcP[f]		
Diisononyl	DiNP[f]		
Dimethyl	DMP	*[131-11-3]*	TI1575000
Dinonyl	DNP	*[84-76-4]*	TI1800000
Dioctyl	DOP	*[117-84-0]*	
Dipentyl	DPeP	*[131-18-0]*	
Diphenyl	DPhP	*[84-62-8]*	
Di(2-propenyl) (diallyl)	DAP[b]	*[131-17-9]*	CZ4200000
Dipropyl	DPP	*[131-16-8]*	
Ditridecyl	DTDP	*[119-06-2]*	TI1950000
Diundecyl	DUDP	*[3648-20-2]*	TI1980000
Mono(2-ethylhexyl)	MEHP	*[4376-20-9]*	TI2500000
Mono(2-methoxyethyl)	MMEP	*[16501-01-2]*	TI2275000

[a] The first name given is the Chemical Abstracts name (1,2-benzenedicarboxylic acid,... ester); parenthetic names are commonly employed in trade and the chemical literature

[b] Abbreviations are derived from the Chemical Abstracts name *unless* there is an established abbreviation which is widely used in the literature (e.g., DEHP, not BEHP)

[c] CA = Chemical Abstracts Registry Number; RTECS = National Institute for Occupational Safety and Healthy "Registry of Toxic Effects of Chemical Substances" Accession Number

[d] Also called dioctyl; a confusing nomenclature

[e] In the older literature DEHP is sometimes abbreviated DOP, but the practice is mercifully dying out

[f] DiDcP and DiNP are not pure compounds but mixtures of materials with average side chains of length 10 and 9, respectively

The carboxyl group is central to the compounds' chemical properties. Like other esters, the principal chemical reaction of phthalates is hydrolysis (Eq. 1). This reaction is catalyzed by both acid and base. It is relatively slow for phthalate esters because of their low water solubility, but the solubility and the rate of reaction rise with increasing temperature. The rate of ester hydrolysis is strongly influenced by steric hindrance in the alcohol moiety; hence, esters derived from hindered alcohols such as 2-ethyl-1-hexanol (*6*) would be expected to be much more inert toward hydrolysis than those from straight-chain alcohols such as 1-octanol (*7*). In biological media this reaction is also enhanced by esterases, and other biochemical reactions become important. The photochemical, oxidative and metabolic reactions of phthalates are detailed in the section below on their environmental fate.

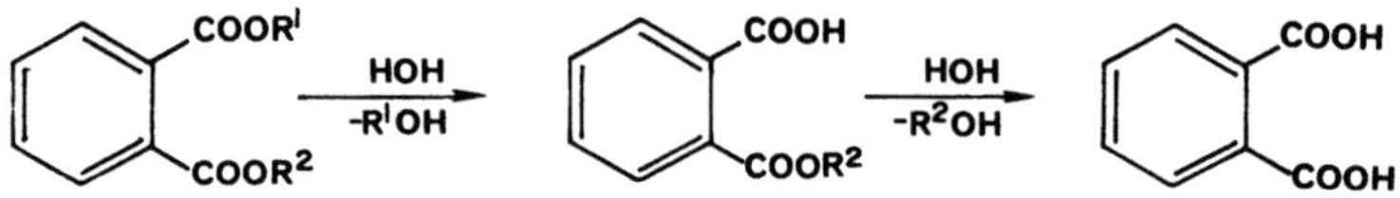

(1)

The phthalates are more reactive than many other classes of poorly soluble, nonvolatile organics, such as polynuclear aromatic hydrocarbons, polyhalogenated biphenyls, and chlorinated pesticides. Thus, they are generally more easily degraded than many of these other environmental contaminants. However, they are ubiquitous, and their environmental and metabolic reactions can lead to toxic as well as nontoxic compounds.

A plasticizer confers flexibility on a polymeric material by filling and lubricating the polymer. Thus, the most important properties of plasticizers are chemical inertness, high fluidity, high solubility in the polymer (since up to 67% by weight may be added [115]), low solubility in water (to prevent loss by leaching and be hydrolysis), and low volatility (to prevent loss by volatilization). Table 2 lists the chemical/physical properties of several common phthalates.

The phthalates have a remarkably large liquid range; for DEHP it extends from −50 °C to 384 °C. This property gives them the fluidity and low volatility required for plasticizer use. Their characteristically low water solubility causes several problems in handling and analyzing them accurately (as discussed in the analysis section below), and it also means that it is difficult to obtain high quality solubility data for them. Schwarz and Miller [210] used both elution chromatography and uv spectroscopy to determine the solubilities of a number of hydrophobic compounds; they find good agreement between the two methods. For DEP the difference was only 7%, and it was one of the worst-case compounds. The vapor pressures of most PAEs are extremely low at room temperature. The data listed in Table 2 have been extrapolated from higher temperatures. They are useful for comparison of the different phthalates, but the absolute values may be systematically skewed. The difficulty of measuring such low vapor pressures is indicated by the divergence of the experimental values reported.

Production and Use

Phthalic acid esters are synthesized from phthalic anhydride and the corresponding alcohol by the Fischer esterification shown in Eq. 2. Phthalic anhydride is pro-

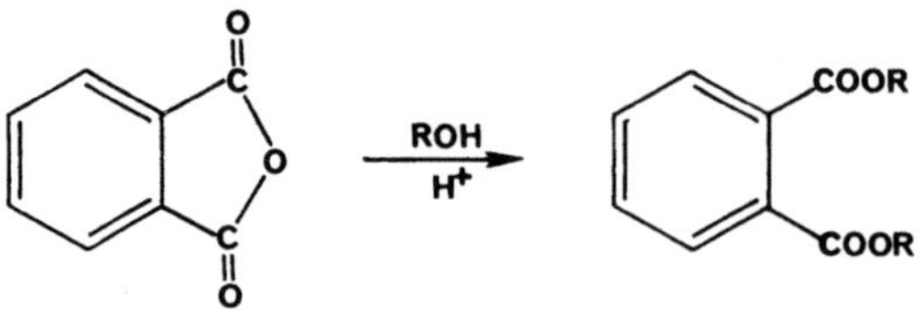

(2)

Table 2. Chemical and physical properties of common phthalates[a]

Compound	MW	mp (°C)	bp (°C)	Water solubility (ppm)	Vapor pressure (mTorr at 25 °C)	log K_{ow}^{h}
DMP	194.19	2	284	4,290[h] 4,320[i]		1.53
DEP	222.24	− 3	298	928[h] 896[i]	2.8[c]	2.35
DPP	250.30		304-5	108[h]		3.27
DBP	278.35	−35	340	11[b] 10[h] 13[i] (3)[j]	0.028[f]	4.57
DiBP	278.35		295-8	20[h]		4.11
BMEP	282.0		$230^{10\ Torr}$	110,000		
BEEP	310.35	34	345			
BBP	312.39			3[h] 3[d]	1.5×10^{-2}[d]	4.91
DPhP	318.33	73	$257^{14\ Torr}$			
DCHP	330.43	66			7×10^{-4}[f]	
DBzP	346.39	42-3	$277^{15\ Torr}$			
DHP	356.50				3.5×10^{-6}[e]	
DEHP	390.56	−50	384	0.04[h] 0.4[i] (1)[j]	3×10^{-4}[f] 1.4×10^{-3}[g] 3.3×10^{-7}[e]	9.64
DOP	390.56		$220^{5\ Torr}$	3[i]	4×10^{-8}[e] 9×10^{-7}[f]	
DcOP	424.62				1×10^{-7}[e]	
MBP	222.24	73-4				
MEHP	194.19	2	decomposes			

[a] Most data from 5, 15, 263. K_{ow} is the octanol-water partition coefficient
[b] Ref. 210
[c] Ref. 263
[d] Ref. 93
[e] Ref. 99
[f] Ref. 268
[g] Ref. 127
[h] Ref. 145
[i] Ref. 278
[j] Solubility in 35‰ instant ocean, Ref. 81

duced by the gas phase oxidation of naphthalene (Eq. 3) or o-xylene (Eq. 4) over vanadium pentoxide [267, 273] or the liquid phase oxidation of o-xylene with soluble salt catalysis. The naphthalene route is older (since it uses a common coal tar derivative), it does not require separation of isomers in either the product or starting material, and it proceeds in high yields. The xylene route uses a product separated from the methylation of benzene or toluene, so it can more easily exploit the currently available petrochemical feedstocks, and (unlike the naphthalene pro-

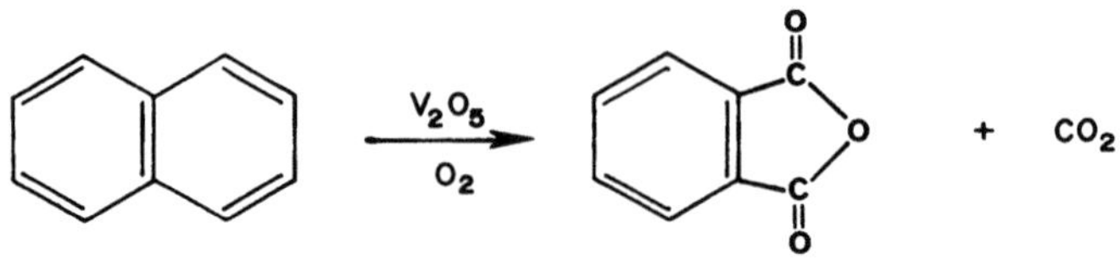

(3)

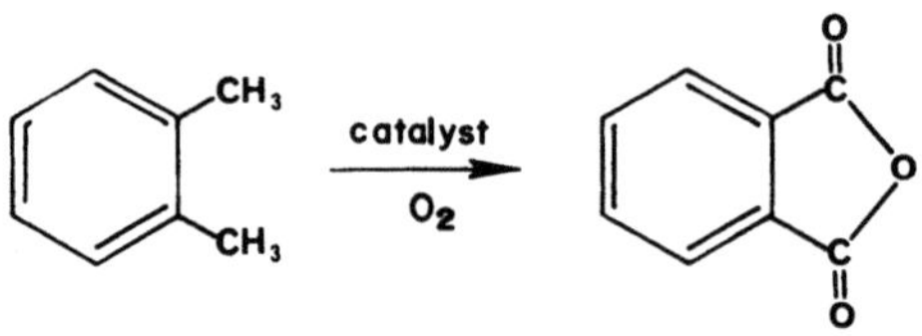

(4)

cess) all carbons appear in the product. Despite these advantages for the xylene process, Wittcoff and Reuben [274] estimate that the naphthalene route accounts for about a third of the US phthalic anhydride production and a quarter of the production in Japan and Western Europe.

The PAEs are principally used as plasticizers for poly(vinyl chloride) products. In his 1973 review of their use and production, Graham briefly outlines the history of this use and its relationship to the plastics industry as a whole [97].

"The first commercially significant plasticizers were discovered shortly after the development of cellulose nitrate in 1846. Castor oil was patented for this use in 1856. In 1870 camphor became the plasticizer of choice for cellulose nitrate. Celluloid remained the major thermoplastic material until shortly before World War II."

"The introduction of phthalate esters in the 1920's overcame the excessive volatility and undesirable odor of camphor. However the commercial availability of poly(vinyl chloride) and the synthesis of di-2-ethylhexyl phthalate in 1933 quickly shifted emphasis away from cellulose nitrate and started the rapid growth of the flexible poly(vinyl chloride) industry."

Until 1979 the United States International Trade Commission regularly tabulated the US production of a wide variety of synthetic organic chemicals for use in tariff regulation. However, a new international agreement on tariffs ended the need for such detailed and precise information, and only rough estimates are currently available in this area. Table 3 gives the data on plasticizers from the 1979 ITC report on domestic production.

Table 3. United States production of phthalate plasticizers, 1979[a]

Plasticizer	Production (millions of pounds)			
All Plasticizers				2,133
Phthalates			1,291	
"Dioctyl"		336		
DEHP	301			
DiDP		175		
DBP+DiBP		17		
DTDP		27		
Others		725		
Benzenoids, Imported[b]			17	

[a] Source: Ref. 248
[b] Source: Ref. 249

Table 4. End uses of plasticizers in the United States (1972)[a]

Use	Millions of pounds
Building and construction	387
Home furnishing	203
Transportation	114
Apparel	72
Food containers & medical	46
Other	200
Total plasticizer usage	1,022

[a] Source: Ref. 97

The most striking features of these data are the great importance of phthalates compared to all other types of plasticizers; the importance of DEHP compared to the other phthalates, and the large variety of phthalates which are produced. The ITC does not separate out compounds which have a single production source, since this would require divulging proprietary information. Hence, not all of the individual phthalates produced are shown in this table. It is instructive to note, though, that the production for the "other" category is greater than that for any individual compound. According to Schouten and coworkers [208], in Europe DEHP is even more important relative to other PAEs. It accounts for 88% of European plasticizer use, and DBP accounts for an additional 7%. In his 1973 review Graham also tabulated plasticizer consumption by end use. These data are shown in Table 4; current data would probably differ in detail from these, but this is likely to be a fairly accurate representation of the general use.

Plasticizers constitute the principal market for phthalates, but there are dozens of minor uses which together consume several tens of millions of kg annually. These uses are quite diverse, but they exploit the same properties of stability, fluidity and low volatility which make PAEs effective plasticizers. For example, the low molecular weight phthalates are used in insect repellents, usually in combination with other compounds, since they are not necessarily the most potent available single ingredient [34, 120, 121]. However, in addition to repellency, they provide shelf stability, low odor, and little or no additional color. These same properties also are useful in the cosmetics industry where they are used as alcohol denaturants, perfume fixatives, and ingredients in paint-on fingernails [63, 69, 94]. A more extreme use for this low reactivity is exhibited by their use in munitions productions [257].

The plasticizer and nonplasticizer uses of PAEs introduce large quantities of phthalates into the immediate environment of almost all human activities. This ubiquity can be a severe handicap in ultraclean environments (see, for example, [99], for its effect on the US space program), and a constant source of concern around the ill, the young, and the aging whose health may not be as tolerant toward these materials as that of the general population. The size of the problem can be seen simply by noting that in medical implements alone, phthalates are contained in such medical and paramedical devices as heart valves, vascular grafting material, intrauterine devices, catheters, dialyzing units, blood sets, and disposable syringes, to name but a few [187]. Furthermore, phthalates are used as textile finishes in synthetic fabrics, which means that we are constantly exposed to them by the clothes that we wear [96]. Thus, the high production and ubiquitous use of phthalates make them one of the most important environmental organic contaminants.

Analysis

The analytical data in this review cover a wide spectrum of concentrations, from % (g/100 g) to ppt (parts per trillion, pg/g or ng/L). Clearly such a broad range of analytical regimes requires modifications in technique and alterations in the method of choice. For the most part, however, this review will be concerned with the ppm (parts per million, μg/g or mg/L) to ppb (parts per billion, ng/g or μg/L) range of analysis, so there will be a central coherence to the methodologies described. It is worth noting that throughout this review we will use mass-based units such as ppt, ppb, and ppm in the strict sense of a dimensionless constant (pg/g, etc.) for biota and solid samples and in the somewhat looser sense (for aqueous solutions) which equates a liter of water with a mass of 1,000 g. Like most literature reports, the results of air analyses will be in mass per unit volume rather than mass ratios. The conversion from mass to volume units can be carried out (if the molecular weight of the material is known); Eq. 5 shows the conversion between these units for air (20% oxygen, 80% nitrogen) at 25 °C.

$$1\ \mathrm{ppm_{mass}} = \frac{\mathrm{MW}}{28.8}(1\ \mathrm{ppm_{volume}}) = 849\ (\mathrm{l\,g/m^3})\,. \qquad (5)$$

The general analytical procedure for phthalic acid esters is similar to that for other organic contaminants and, indeed, owes much to workers who developed these techniques for use with other substances and hence are not cited directly in this review. First, an environmental sample is taken. The *sampling* procedure will depend upon the particular sample medium (air, water, soil or sediment, or biota) which is under consideration, but there are common features. In the next step the organic fraction is isolated from the medium; this *isolation* simply separates the organics from an aqueous or solid phase, and it results in a very heterogeneous mixture of organic compounds. The next step in the analysis is *fractionation* or cleanup of the organic isolate to separate the moderately polar organic fraction (which contains the PAEs) from the other organics. After the fractionation step, the sample is then characterized to identify the particular PAEs present and to quantitate them. *Characterization* is almost invariably done by gas chromatography (GC) with any of a number of detection systems (electron capture = ECD; flame ionization = FID; mass spectrometry = MS) or high performance liquid chromatography (HPLC) with ultraviolet (UV) or MS detection. The material in this section is divided along these general lines to facilitate comparison of the different methodologies.

Sampling and Sample Handling: Variation with Sample Type

Accurate phthalate analysis requires extreme care at the start of the process. The final accuracy achieved will depend very greatly on the planning and precautions used in sample gathering and handling. An important initial consideration is the appropriate sample size. The proper sample size is determined by estimating the concentration expected and the final instrumental detection limit. Data presented later in this chapter can assist with both of these estimates. For analyses at the ppb and ppt levels, sample concentration is usually required. This concentration process may be simply a matter of extracting a large volume of sample with a small volume of solvent, using an adsorbent through which a large volume of water or air is passed, or the actual evaporation of sample solution to reduce its volume. The choice of concentration (and hence, sampling) method, then, will be influenced by the stage at which concentration must be carried out and the environmental medium in which the PAEs are found.

Air Analysis. For air analysis calibrated pumps are used to determine volume of air sampled; the throughput must be matched to the expected contaminant level. Samples of the particulate-adsorbed and aerosol phthalates can be obtained on carefully cleaned glass fiber filters. These filters are then extracted with a low-boiling organic solvent in a Soxhlet extractor [25]. A more common method of air sampling involves trapping the phthalates with an adsorbent placed at the air intake. This method retains both vapor phase and particulate phthalates. If a glass filter is placed upstream of the adsorbent trap, the two quantities can be measured separately [14, 82].

A variety of adsorbents have been used. Ethylene glycol was used in an early study [232] and over 90% recovery was reported. However, most groups have found it more convenient to employ a solid adsorbent. Table 5 lists a number of adsorbents which have been used.

One early approach was to seek an adsorbent which would trap all of the organic pollutants present in one pass. However, the state of the art is essentially the same as in 1973 when Mieure and Dietrich [164] stated that they knew "of no universal column packing which is capable of quantitatively retaining and later releasing all volatile organic pollutants." They worked around this problem by using three traps (Chromosorb 101, 105 and Tenax) for the different classes of organic contaminants. The difficulty of achieving one-pass trapping has directed most efforts in this area toward finding selective adsorbents which will guarantee high recovery of the desired compounds, even though they may allow other compounds to pass through easily.

Giam and his collaborators [79, 84] have used several different solid adsorbents for trace and ultratrace level analyses. They stress the importance of considering humidity as a factor when determining the recovery of phthalates on solid adsorbents. Thus, the frequently employed Amberlite XAD-2 resin is quite satisfactory for phthalate adsorption (for DMP, DEP, DBP, and DEHP) in dry air, but it does not perform well in high humidity. Florisil and polyurethane foam both performed better than XAD-2 under wet conditions, but they are satisfactory only for the larger phthalates tested (DBP and DEHP). For example, with DEHP the recovery measured on polyurethane foam was 100% in dry air and 99% in moist. The comparable figures for DBP, DEP, and DMP were (99%, 97%), (67%, 57%), and (59%, 9%), respectively. Based on field tests of their procedure for the ultratrace analysis of DBP and DEHP, they state that polyurethane foam is the adsorbent

Table 5. Adsorbents employed for phthalate analysis

Adsorbent	Medium	Comments	Ref.
Amberlite XAD-2	Air	DMP, DEP, DBP, DEHP analysis. Efficient when air is dry	[79]
	Water	Acetonitrile extracted	[76]
	Water	Ether extracted	[224, 241]
Amberlites XAD-2/XE-347	Water	Broad applicability	[103, 104]
Charcoal	Air	Headspace analysis of perfumed detergents	[63]
Cellulose esters	Air	DHP analysis. Pentane extracted. Recovery >92%. Sulfuric acid a problem	[213]
	Air	NIOSH method for DBP, DEHP	[230]
Ethylene glycol	Air	>90% recovery	[232]
Florisil	Air	Good for DBP and DEHP, but not for DMP, DEP	[79]
	Air	Quick monitoring method for heavier phthalates	[84]
Poly(tetrafluoroethylene) (PTFE)	Air	DiNP, DiDcP analysis. Methanol extracted. Recovery >92%	[213]
Polyurethane foam	Air	Very good for DBP and DEHP, but not for DMP, DEP. Best adsorbent for ultratrace air analysis	[79, 278]
Reversed-phase beads	Water	5% acetone extracted. *n*-Octadecyltrichlorosilane treated porous glass beads. >80% adsorption	[60]
Tenax	Water	For basic, neutral and high boiling compounds	[164]
	Air		[164]

of choice for air analysis [79]. For more contaminated environments (with smaller volumes required), Florisil is a better alternative [13].

Selover and his collaborators [213] investigated cellulose esters and poly(tetrafluoroethylene) (PTFE) as adsorbents for the higher concentrations of PAEs which are found in the workplace. They found that PTFE worked quite well for the larger phthalates (DiNP and DiDcP in their studies), but that mixed cellulose esters were preferable for the lighter DHP. Sulfuric acid mist in the air was a serious problem for adsorption on cellulose esters; it interfered with adsorption directly and also enhanced hydrolysis during sample storage.

Biota Analysis. In contrast to the air analyses, the actual sampling of biota is not an experimentally complex process *if* care is taken to avoid contamination (see below). Because of the high probability of biodegradation of any PAEs in the sample, it is important that biota samples be frozen as soon after collection as possible and analyzed quickly. Giam [76] and Giam and Chan [82] have discussed the problems of sampling biota; the latter reference contains a flow chart of the analytical procedure.

Soil, Sediment, and Solid Analysis. For soil, sediment, and solid waste samples, all that is usually required is the sample be collected with scrupulously cleaned tools and be stored *cold* in carefully cleaned glass vessels. Cold storage is especially important in wet or high-organic samples, since biodegradation is more likely to be a problem. Giam and coworkers [174, 175] have reported satisfactory results by storing samples in commercial freezers. Sediment analyses can be performed on either wet or dry materials, but there has been no careful study on the effects of sample drying on the concentration found. Low molecular weight phthalates could be susceptible to volatilization or degradation under these conditions, and certain soil types could catalyze hydrolysis during heating. Schwartz and collaborators [209] avoid most of these problems by freeze-drying the sediments.

Water Analysis. Water analysis is frequently carried out simply by collecting a water sample in the field and then returning it to the laboratory for extraction or other isolation procedures. As noted, extreme care must be taken to avoid contamination and degradation, but otherwise this procedure is straightforward. In many cases, though, it would be more convenient to obtain the field sample in such a way that a smaller volume of material could be transported to the laboratory. In such cases it is customary to use a solid adsorbent similar to those used in air analysis studies. By far the most common adsorbent for this purpose is Amberlite XAD-2 [76, 103, 104, 224], although other materials have been used (see Table 5). Giam and coworkers [90] have described the use of XAD-2 for concentrating up to 96 L of seawater for analysis of DBP and DEHP of ppt levels. Harris and coworkers [104] suggested the use of a mixed resin (XAD-2/XE-347) for wastewater analysis, but the choice was primarily based on the recovery of phenols. Plasticizer phthalates were recovered well by both resins.

An interesting reversed phase adsorbent, *n*-octadecyltrichlorosilane-treated porous glass beads, has been used by Ehrhardt and Derenbach and their collaborators [50, 60]. This material adsorbs a wide range of organics with better than 80% recovery; 5% aqueous acetone is used for removing the adsorbates. In their analyses they distinguish particulate-adsorbed and dissolved PAEs by removing particulates by filtration on glass fiber filters in a stainless steel holder. Since humic materials [39, 155, 181] and sediments [225, 226] adsorb phthalates strongly, this separation of the liquid and particulate phases is important for accurate solution phase data.

Sampling and Sample Handling: Contamination, Adsorption, and Degradation

In their review of analytical procedures, Giam and Chan [82] point out three principal obstacles to accurate identification and quantitation of phthalates:
1) sample size limitations,
2) contamination from solvents, glassware, and other laboratory materials, and
3) adsorption to sample vessels and other materials.

To this we should add 4) biological and chemical degradation following sample collection [169, 201, 208, 209, 213]. As mentioned above, estimation of the appropriate sample size depends on information which is contained in later sections. However, it is appropriate to consider the other three potential obstacles here.

Contamination. We have described above the ubiquity of phthalates and the wide diversity of their uses. This point must be kept clearly in mind when attemp-

Table 6. DEHP content of some common laboratory items[a]

Material	ppm DEHP
Tygon tubing	20,000
Neoprene stopper	1,600
Black rubber tubing	250
Black rubber stopper	30
Cork	6
Amber latex tubing	4
Glass wool	1
Teflon sheet	0.8
Polyethylene tubing	0.8
Aluminum foil	0.3
Tap water	0.0015
Sodium chloride	0.0015
Sodium sulfate	0.0002
Florisil	0.00005
Distilled water	<0.00005

[a] Source: Ref. 86; reprinted with permission. Copyright 1975 American Chemical Society

ting to carry out trace and ultratrace analyses, since large blanks may otherwise result. Tepper [231] pointed out the serious contamination problem associated with the analysis of these substances. In that same year Zitko [282] reported 4.25 μg/g phthalates in Wheaton extraction thimbles, and Stalling and coworkers [222] noted problems with controlling laboratory-produced contamination. Mathur [153] described the contamination problem in some detail and provided some representative data on phthalates in laboratory materials. Schwartz and coworkers [209] have discussed the problems which they had with extraction thimbles, filter paper and Florisil, and Schouten and coworkers [208] state that the lower limit on their HPLC analysis of phthalates in water (10 ng) is set by the contamination level of the hexane used for extraction.

Table 6 (adapted from [86]) provides a glimpse of the magnitude of the problem; it is a tabulation of the phthalate levels found in our laboratory for a variety of common substances. Many other items could also be listed and must be considered as potential contamination sources. Ishida [115] and Webster [264] have both published similar lists. Contamination problems are particularly severe if the contaminants are introduced before the organics are concentrated.

Giam and his co-workers have described the precautions necessary to minimize or avoid contamination [76, 85, 92]. Glassware, scoops, and other heat-stable items should be carefully cleaned by the procedures described and then heated above 300 °C with aluminum foil closures. Distilled water should be purified by continuous extraction with high purity hydrocarbon solvent (hexane, petroleum ether, etc.). Some chromatographic solvents are pure enough to use as received, but each lot should be checked by concentrating 500–1,000 fold and then examining the concentrate by gas chromatography. As mentioned, the avoidance of contamination in the analytical laboratory requires control even over the maintenance and cleaning procedures followed by outside personnel. Other workers have also described procedures for reducing the contamination problem [115, 208, 209, 264]. In our experience phthalate contamination is the most difficult analytical problem to overcome in obtaining reliable, reproducible results at very low concentrations.

Adsorption. Sullivan, Atlas, and Giam [225, 226] have published two studies describing the problems resulting from adsorption of phthalates on various materials. They found that upon standing, phthalate samples reversibly adsorbed to glass, leading to anomalously low analytical results. This adsorption also cause a synergistic adsorption of polychlorinated biphenyls (PCBs) to the surface. Furthermore, they used radioactive tracer techniques to show that phthalates were very strongly adsorbed to clay particles. This phenomenon is important in accounting for the environmental distribution of phthalates, and it can lead to erroneous estimates of the amount of dissolved (as opposed to suspended) PAEs present in an aqueous system. They note that the amount of adsorption increases with increasing salinity and decreasing water solubility of the PAE. Thus, DBP is much less strongly adsorbed than DEHP.

In their study of resins for phthalate analysis in water samples, Harris and collaborators [104] found that adsorption to the vessels was worse with the lower molecular weight phthalates. Over 20% of ^{14}C-DMP used in the study was recovered from the wall of the reservoir. Takeuchi and Ishii [229] report that addition of 1 ppm ethylene glycol to the solvent reduces the losses of phthalates by adsorption to glass sample vessels.

Overcoming the adsorption problem is important in sediment analysis, since phthalates must be completely stripped from the sediment if an accurate result is to be obtained. Payne and Benner [189] compared the two most common methods for sediment analysis, extraction with ultrasonication and Soxhlet extraction. They found little difference in the percent recovery of DEP by the two methods. The ultrasonication method is faster than the Soxhlet method (which requires 4-h reflux), but the stainless steel ultrasonicator probe must be replaced after every 150–200 extractions, and this adds significantly to the cost of the analysis. They recommend the Soxhlet method as being more cost-effective. Peterson [194], however, states that the extraction efficiency of the ultrasonic method is equal to or better than the Soxhlet method for sediment analysis. The simplicity and efficiency of the ultrasonication method has been utilized by other groups as well [172].

The importance of adsorption by humic substance is a much-debated topic. Schnitzer and his coworkers pointed out some time ago [155, 181] that phthalates were found associated with suspended humic acid substances in aqueous systems. Carlberg and Martinsen [39] found that two widely used methods for phthalate analysis failed to give accurate results in humus-containing water. On the other hand, Tomson and his coworkers [241] used spiked samples to show that humic acid did not significantly interfere in determining the PAE concentration in Phoenix, Arizona, ground water and waste water.

Yamasaki and Kuwata [280] found in their study of Osaka (Japan) urban air that the amount of DEHP found was correlated with amount of dust present, but that was not true of DBP. They attribute this to DEHP being particle-related and DBP being vapor-phase. Thus, adsorption problems are not of concern only for aqueous and sediment systems.

Biological and Chemical Degradation. As mentioned previously, phthalates are more chemically and biologically reactive than are many other common organic contaminants. Thus, the possibility must always be kept in view that storage of the material will increase the rate of sample loss through chemical or biological degradation.

This can be a serious problem. Payne and Benner [189] state that the "need to minimize the lag time from sampling to extraction is demonstrated when a water sample from the ecosystem was analyzed, held at ambient temperature, and re-analyzed in 24 and 48 h. These re-analyses revealed DEP losses of 64 and 89%, respectively, from the initial concentration. When it is not possible to extract the samples promptly, the need for immediate refrigeration cannot be overemphasized."

Similarly, Schouten and coworkers [208] found rapid loss of DEHP and DBP from river water samples which could not be attributed to adsorption. They observed a 50% loss of DEHP within 5 days and a 90% loss of DBP within 3 days. They also observed microbial contamination of their laboratory standards. They were able to overcome these problems by refrigerating or extracting the river water samples immediately and by autoclaving their standards and flaming the lip of the standards jar after removing aliquots. They acknowledge that these results differ from those of other authors, but they attribute the variation to different microbial populations and to a higher concentration of PAEs in the standard. Antimicrobials did not completely remove the problem. They emphasize that the addition of the extractant to the samples at the field site and the immediate treatment in the laboratory must be followed for reproducible results. By contrast, the same group of workers, in a related study of sediment samples from the same Dutch rivers, found that there was little or no biodegradation observed in their samples [209]. Thus, the site, concentration and sample variability of this effect is quite high. Sullivan and coworkers [225] advocate the use of sodium azide as an antimicrobial for standard solutions.

Chemical degradation can also be a problem. We have already mentioned the interference of sulfuric acid with the workplace analysis of DHP [213]. Rhoades and coworkers [201] reported on the effects of a variety of parameters (pH, temperature, residual chlorine) on the stability of phthalates (DMP, DEP, BBP, DEHP, DOP) in waste water samples; on the basis of these studies they provide recommended preservation techniques for such samples. The principal variable is pH; they recommend acidification if samples are to be stored prior to analysis.

Isolation and Fractionation

Air Samples. The isolation of organics from air is usually accomplished by the adsorption method described above. Frequently the solvent extract of the adsorbent, or the vapors released by heating the adsorbent, can be analyzed directly by gas or liquid chromatography. When interferences such as PCBs are present, additional cleanup (like those described for biota or water sediments) may be required.

Biota Samples. Samples of biota present a special analytical problem because of the presence of large quantities of lipid materials which coextract with the phthalates. Burns et al. [35] point out that if these lipids are not removed, they cause serious problems with lack of repeatability and column and detector degradation upon subsequent GC/ECD analysis.

Separation from the lipids is customarily done by column or thin-layer chromatography. Burns et al. [35] extracted fish lipids and chromatographed them on a gel permeation column. Following this treatment they used sulfuric acid impregnated alumina layers to remove traces of lipids. They report a recovery of 79%–86% for DEHP using this system. A thin-layer system for separating lipids in phthalate analysis was reported several years ago [222].

In his discussion of procedures for PAE analysis, Giam [76] recommends that biota samples be prepared for analysis by maceration in acetonitrile. The wet acetonitrile solution is then extracted by adding 5% NaCl solution and 1:5 (v/v) dichloromethane: petroleum ether. The hydrophobic layer is washed with water, dried over anydrous sodium sulfate and then concentrated to less than 10 mL in a Kuderna-Danish concentrator. This solution is then chromatographed on Florisil (3% water deactivated) using 6%, 15%, and 20% ether in petroleum ether as successive eluants. Under these conditions, DEHP comes out cleanly in the second-fraction; DEP and DMP come out in the third. DBP elutes in both of these fractions, but it is resolved well in the subsequent GC analysis. This procedure allows the simultaneous determination of PCBs and chlorinated pesticides. Mayer and coworkers [157] reported the use of Florisil for the cleanup of fish samples for PAE analysis, but the procedure in Giam [76] is preferable.

In connection with a study of the liquid-liquid extraction of 63 acidic pharmaceuticals from tissue samples, Müller [173] reported the partition coefficients and their pH dependence for these compounds, including some phthalates. Zitko [282] extracted phthalates from biological samples with hexane and obtained partial separation from lipids by chromatography on alumina. This method requires further cleanup which is achieved by the extraction of phthalates from hexane into *N,N*-dimethylformamide (DMF). He reports distribution coefficients for various phthalates between hexane and other organic solvents (DMF, dimethylsulfoxide and acetonitrile).

Soil, Sediment, and Solid Samples. We have discussed the comparative advantage of ultrasonication and Soxhlet extraction. The isolation of PAEs from solid samples generally involves solvent extraction (hexane, acetonitrile, acetone: hexane, etc.). The preferred solvent depends on whether the solids are wet or dry. Müller and his coworkers [172] use a particularly thorough cleanup procedure in their sediment studies. Following extraction with acetone: hexane (1:1) they passed the extract through an alumina/Florisil/copper column with dichloromethane: hexane and then further purified by sequential treatment with Celite, sulfuric acid and mercury. Eglinton [59] used purification on TLC (silica gel) after extraction with heptane: 2-propanol (1:4).

Water Samples. We have described above the methods by which water samples can be treated with adsorbents. However, in many cases it is simpler (and *much* easier to avoid contamination) to use solvent extraction for the isolation of the organics. Harris and coworkers [103, 104] found that extraction with dichloromethane was 101% efficient for DEP, while a mixed XAD-2/XAD-347 resin was 88% efficient. This difference would probably be less for higher molecular weight phthalates.

To increase efficiency and speed, ultrasonication [189] or continuous extraction methods [193] are often used. Peters [193] recently tested five different types of extraction methods for their ability to strip organics from aqueous solutions. The ten compounds tested included two chlorobenzenes, six phenols, one hydrocarbon and DMP. The apparatus tried were a liquid-liquid extractor (LLE) in which the heavy solvent is dripped through the water; a flow-under and a flow-over extractor (FUE and FOE) in which only the interfaces of the two solvents are in contact, and a steam-distillation and a steam co-distillation extractor (SCE and SCDE) in the aqueous solution is steam distilled either into (SCE) or whith (SCDE) an organic solvent. He found that the LLE and the FUE gave similar high recoveries and the FOE was more variable (but quite good for DMP). The SCE and SCDE both gave much lower recoveries. In actual practice the FUE or FOE has the advantage that emulsions, which are common with the LLE, seldom form.

The procedure of Giam [76] for water analysis is typical. The sample is extracted three times with high purity petroleum ether (or dichloromethane) and the organic layers are combined and dried over sodium sulfate. The sample is then concentrated under nitrogen and chromatographed through Florisil as described above for biota samples. EPA method 625 uses dichloromethane as extractant and an alumina column [245, 247]. Variations on this work up include separating the solution of organic compounds into acid, base, and neutral fractions by liquid/liquid extraction prior to further cleanup [14, 211, 245] and using thin-layer chromatography on silica gel rather than column chromatography [60, 211]. Mori [168] has reported a method for the direct analysis of hexane extracts by HPLC without further cleanup or concentration. The lower detection limit is 20 ppb in hexane by this method; since he used a 1 : 10 (v/v) ratio of hexane : water in sampling, this corresponds to 2 ppb in the original sample.

Characterization

Gas Chromatography. Gas chromatography has been the method of choice for phthalate analysis in most laboratories for years. The low volatility of these materials require that they be chromatographed at high temperatures, but they are sufficiently stable and polar to chromatograph at or above 200 °C. For a number of years such analyses were performed on packed columns. A list of the packing materials and the temperatures at which they were run is given in Table 7. As better equipment and more reliable supports have become available, the gas chromatographic conditions have come to employ temperature programming and a very small set of columns packings. It is important to use a high quality support, since these materials tend to tail badly on active supports. A second column has often been used to confirm the identification of PAEs in environmental samples [76]. The development of glass capillary columns has provided a major enhancement in the resolution possible for PAEs. This enhancement allows environmental samples to be run with less prior fractionation than had previously been possible. Common coatings for such columns are listed in Table 7.

Table 7. GC columns for phthalate analysis

Stationary phase	Support	Temperature (°C)	Ref.
Packed Columns: Single packings			
Carbowax 20M (aue packing)	Chromosorb W	90–250	[46]
DC200 5%	Chromosorb W-HP	200	[92]
DEGS 3%	Gas Chrom Q	180	[264]
Dexsil 3%	Gas Chrom Q	120–210	[59]
Dexsil 300 GC 5%	Chromosorb W-HP		[164]
6%	Chromosorb W-HP	255	[25]
EGSS-X 10%	Gas Chrom P	204	[152]
Igepal 8%	Chromosorb W-DMCS	200/220	[94]
Liquid Crystal	Chromosorb W-HP		[135]
OV-1 0.3%	Supelcoport	160/225	[201]
OV-7 0.3%	Corning GLC-110	155	[157]
0.3%		182	[222]
OV-11 0.3%	Corning GLC-110	170–235	[222]
OV-17 4%	Chromosorb W-HP	230–270	[203]
3%	Chromosorb W-HMDS	100–270	[95]
1.7%	Shimalite W	240	[115]
1.5%	Shimalite W	210	[170]
OV-101 5%	Chromosorb W-HP	225	[270]
5%	Supelcoport	210/230	[230]
4%	Chromosorb W-HP	240	[208]
3%	Chromosorb W-HP	150–220	[266]
3%	Gas Chrom Q	230	[93]
3%	Supelcoport	210	[35]
OV-210 3%	Chromosorb W-HP	225	[202]
Poly-I 110 3%	Gas Chrom Q	205	[264]
QF-1 5%	Gas Chrom Q	210	[47]
SE-30 5%	Shimalite W	230	[170]
3% (Ultraphase)	Chromosorb W AW-DMCS	210	[232]
3%	Anakrom ABS	170/240	[282]
3%	Gas Chrom Q	220	[76]
SE-30 3%	Chromosorb W-HP	210	[76]
3%	Gas Chrom Q	200	[189]
			[131]
SE-52 1.5%	Chromosorb W	230	[115]
SP-2100 3%	Supelcoport	200	[176]
SP-2250 3%	Supelcoport	75–325	[245]
1%	Chromosorb W-HP	60–220	[266]
SP-2310 5%	Chrom W-AW	220	[213]
UCW-98 10%	Diataport S	100–330	[136]
XE-60 3%	Chromosorb W-HP	225	[270]
Packed Columns: Mixed packings			
SP-2250 (1.5%) + SP-2401 (1.95%)	Supelco AW-DMCS ·	195	[76]
	Supelcoport	160–225	[201]
		180/220	[245]

Table 7 (continued)

Stationary phase	Support	Temperature (°C)	Ref.
OV-17 (3%) + SP-2100 (3%) (precolumn) with SP-2100 (1%) + SP-2401 (2%)	Supelcoport	200	[176] [35]
OV-17 (1.5%) + QF-1 (1.95%)	Gas Chrom Q	200 215	[47] [264]
Capillary Columns			
Liquid crystal	WCOT		[135]
OV-101	WCOT	80–200	[60]
OV-17	SCOT	100–250	[99]
SE-52	WCOT	150–210	[11]
SE-54	WCOT	110–290	[44]
SP-2100 fused silica			[174]
SP-2100	WCOT	50–250	[241]
UCON 50 HB5100 + SE-30	WCOT		[172]

Krishen [136] investigated the behavior of plasticizers in programmed temperature GC using FID detection and separation with a 10% UCW-98 on 60/80 mesh Diatoport S column. Both phthalate and nonphthalate plasticizers were included; PAEs examined include DBP, BzBP, DEHP, and DDcP. He was able to show that there was a linear relationship between side-chain carbon number and retention time relationship PAEs.

For environmental applications the choice of detectors is also critical. Some groups have employed flame ionization detectors [2, 46]. In connection with their studies of PAE metabolites, Albro and his coworkers [4] developed a GC method for many of these compounds and tabulated FID response factors for them. For many trace and ultratrace applications FID is simply too insensitive. Giam and Chan [82], for example, found that the electron capture detector was 100 times more sensitive than the FID on the same samples. The reported sensitivities are in the range of 1–10 ng/injection for FID and 20–100 pg/injection for ECD [60, 201]. In developing Level 2 analysis procedures for the EPA, Thrun and coworkers [236] reported that GC/FID and GC/MS had a detection limit for esters of about 50 ng injected, while GC/ECD detected as little as 2–13 pg (depending on the particular ester). The improved sensitivity of the ECD is not without problems, however. Several groups (e.g. [35]) have noted that impurities in PAE samples, particularly lipids, can seriously degrade the performance of the ECD over time. Also, the high sensitivity of ECD toward chlorine compounds means that polyhalogenated biphenyls, which exhibit similar chromatographic behavior, can interfere with detecting PAEs. Both lipid and PCB/PBB interferences can be removed by chromatography on Florisil [76, 84]. Veith and Kuehl [256] have described an automated gel-permeation method for lipid removal. Gurka and coworkers [101] have described the application of the Fourier transform infrared (FT-IR) detector to environmental problems. They used computer searching to identify the components and this severely limits the method's sensitivity. The minimum injected quantity required for the peak locator to function is 1.1 μg for DMP, 2.0 for DEP and 2.1 for DBP.

Gas Chromatography/Mass Spectrometry. GC/MS has been used for phthalate analysis for over a decade [99, 232]. However, in the past five years it has dominated work in this area. One important reason is that mass spectroscopy is self-confirming. The identification of particular peaks can be carried out by matching to a known standard without running a second column, or it can be carried out by comparison to computer-stored spectra libraries which are readily available on modern instruments. Phthalate esters (except for DMP) all show a large *m/z* 149 peak, so that they can be readily detected in the presence of large amounts of other contaminants. Modern instruments usually allow scanning at one or more fixed values of *m/z*, for selection of particular compound classes, *plus* monitoring of total ion current for quantitative comparison. Quantitation on GC/MS has progressed rapidly in recent years, and sensitivities as low as 10 pg/injection have been reported for PAEs [12]. This sensitivity allows the examination of extracts of water samples at the ppb level without further purification [60]. Occasional problems have been reported with some compounds; Ehrhardt and Derenbach [60] reported problems with DEHP decomposition in their system. The problem was idiosyncratic in that DBP, DMP, and DEP could all be determined with good accuracy and repeatability.

The immense power of computer-controlled GC/MS has been exploited in a variety of studies for determining the level of environmental contaminants; example of such studies are listed in Table 8. Two representative studies of the flexibility of this method are the water analysis studies of Ellis and co-workers [61] who found 243 compounds in the secondary effluent of wastewater treatment plants. Thirty of these compounds had not been previously identified as water pollutants and twenty were on the US EPA priority pollutant list. Similarly, Waggot [259] was able to identify 150 different organic contaminants in the River Lee, England. Dymerski and his collaborators [55] describe the use of a mobile GC/MS laboratory for water analysis problems.

High Performance Liquid Chromatography. HPLC has received less attention than GC as an analytical method for phthalate esters. However, a number of groups have used it in recent years and have found it comparable to GC for most purposes. A variety of columns have been used in HPLC analysis. A brief list is provided in Table 9. Takeuchi and Ishii [229] used a precolumn consisting of 0.1–0.2 mm i.d. PTFE tubing packed with various materials (silica gel and porous polymer) to improve the analysis of PAEs. They were able to achieve a detection limit below 1 ng with this system.

One major problem which had to be overcome for successful HPLC was the lack of suitable detectors. The principle detection method has been UV absorbance. The selection of a suitable wavelength for monitoring PAEs requires a knowledge of their ultraviolet-visible absorption spectra. Leyder and Boulanger [145] recently reported absorbance data for several different PAEs. They found no significant difference among the different dialkyl phthalates investigated (on a molar basis). Their data are for absorbance in ethanol; they note that the spectrum

Table 8. GC/MS in environmental studies of phthalates

Sample	Ref.
Air Analysis	
Aerospace	[99]
Atmospheric contaminants	[232]
DBP, DEHP on atmospheric particulates	[24]
Diesel emission particulates	[44]
Municipal incinerator fly-ash particles	[46]
Oceanic air	[13]
Organics in size-fractionated indoor aerosols	[265]
Organics in Arctic aerosol	[266]
Solid waste leachates	[261]
U.S. EPA method 625	[245]
Biota Analysis	
Fish	[270]
Marine biota	[13]
Metabolites in fish	[222]
Sediment and Solids Analysis	
Chester River and Chesapeake Bay	[194, 195]
Composted municipal rubbish	[95]
Marine sediment	[13]
Survey analysis of solid wastes	[166]
Usk River (Severn Estuary, England)	[59]
Water Analysis	
Charles and Merrimack Rivers	[108]
Groundwater and sewage	[241]
Groundwater, groundwater treatment plants, and the Elizabeth River NJ (Mobile laboratory)	[55]
Kiel Bight, Baltic Sea	[60]
Marine dissolved lipids	[124]
Oceanic water	[13]
Pharmaceutical plant wastewater	[122]
River Lee, England	[259]
Tokyo Bay Seawater	[224]
Wastewater treatment plants (Secondary effluent)	[61]
Water treatment influent and effluent	[200]

is solvent dependent. They observed the following adsorption peaks: $\lambda_{max}(\log \varepsilon) =$ 198.5 nm (4.54), 229.5 (3.90), 275.5 (3.14), 281 (sh, 3.1).

Fixed wavelength mercury lamps at 254 nm can be used for detection, but the low absorbance of phthalates at that wavelength limits studies to high concentrations of phthalates, such as mixtures of plasticizers [6, 213]. Selover and coworkers [213] adsorbed DHP onto mixed cellulose esters and then used HPLC analysis with 254 nm detection to quantitate down to 50 $\mu g/m^3$ DHP; the upper limit was above 20 mg/m^3, the highest concentration tested. However, variable ultraviolet detectors now provide easy access to the 220–240 nm range where PAEs have a large absorption band suitable for sensitive detection.

Table 9. HPLC columns for phthalate analysis[a]

Stationary phase	Solvent	Ref.
Normal Phase Chromatography		
Alkynitrile-bonded silica	2% 2-propanol/benzene	[189]
Crosslinked polystyrene	hexane	[168]
Silica gel (5 μm)	1:2 hexane: DCM with 0.1–0.2% EtOH	[208]
	1:2 hexane: DCM with 0.1–0.2% EtOH	[209]
Reversed Phase Chromatography		
Alkyl-bonded silica	MeOH (DiDcP)	[213]
	water/MeOH (2:98) (DiNP)	[213]
	water/MeOH (50:50)	[114]
(5 μm)	water/MeOH gradient	[253]
	water/MeCN (35:65)	[229]
	water/MeCN (50:50)	[114]
	water/MeCN/THF	[6]
	water/THF (50:50)	[114]
Crosslinked polystyrene	MeOH	[168]
Gel Permeation Chromatography (styrene-divinylbenzene copolymer)		
(1000 dalton exclusion)	chloroform	[168]
(1500 dalton exclusion)	THF	[116]
(4–8 μm)		[116]

[a] Solvents: DCM = dichloromethane; EtOH = ethanol; MeCN = acetonitrile; MeOH = methanol; THF = tetrahydrofuran

Schouten [208] and Schwartz [209] and their collaborators used both GC and HPLC methods in their studies of the water and sediments for a series of Dutch rivers. They report that they had an HPLC detection limit of 5–10 ng (comparable to GC/FID) and that they could reduce their solvent and procedural blanks to very low levels. Thus, they were able to detect 0.5 ppm without difficulty. They note that there is a variation in the ethanol content of commercial dichloromethane solvents which can produce variations in the HPLC results. Addition of 0.1% ethanol or 0.4% 1-butanol to the solvent overcame this problem; column life was 4–6 months.

Mori [168] studied the separation of phthalates using three different separation modes: normal phase, reversed phase and gel permeation. He used detection at 224 nm and found an absolute detection limit of 2 ng for DEHP. Using this system he was able to analyze phthalate concentrations in water as low as 2 ppb without any concentration of the hexane extract. The gel permeation column provided the highest resolution, but the sensitivity was higher in normal phase mode because of difference in absorbance of the solvents required. Ishiguro and coworkers [116] obtained high speed separations and 19,000 to 23,000 theoretical plates on a 30-cm using 4–8 μm styrene-divinylbenzene copolymer gel permeation column packing.

The application of mass spectrometric detection to HPLC is still being developed, and there has been little application of it to phthalates. Schauenberg and collaborators [207] used a moving belt device to connect an MS to an HPLC. They found that chemical contamination of a new belt provides high chemical background which declines with time. They used their system for the investigation of minimal detectable levels of a number of pollutants.

Miscellaneous Analytical Methods. Thin-layer chromatography (TLC) has been used by some groups for the analysis of phthalates. Isaaq and coworkers [114] describe the use of alkyl silica gel bonded phases for separations both by TLC and HPLC; phthalates are one of the classes they examined. One early study [179] combined TLC and IR in an investigation of the toxicological importance of plasticizers in medical devices. They provide tables of R_f values (silica gel/ethyl acetate : hexane 1 : 9) and of IR absorption maxima.

Isaaq and collaborators [114] reported recently on a comparative study of HPLC and TLC reversed-phase separation of several organic compounds. They used three different chain lengths of alkyl bonded silicas and three different solvent systems (water plus acetonitrile, methanol and tetrahydrofuran). The results were satisfactory for all six methods on both HPLC and TLC except for TLC on the shortest side chain (C2) silica. They recommended the use of C18 HPLC columns and TLC plates.

Sdika and collaborators [211] developed a semiquantitative method for concentration and identification of the main organic micropollutant classes in waters. This method separates these micropollutants into 14 main classes using sequential extraction, separation into acid/base/neutral fractions and TLC. The method is inexpensive and rapid and is designed to allow the assessment of the potential toxicological risk associated with the repurification of municipal water.

A variety of methods have been used to confirm the identification of organic pollutants as phthalates. We have already mentioned the use of second-column GC methods and of GC/MS and GC/FT-IR for this purpose. However, simple, inexpensive chemical means have also been exploited. Giam and his coworkers [83] describe the derivatization of phthalic acid (produced by alkaline hydrolysis of the PAE) by fusion with 2-chloroethylamine hydrochloride to produce *N*-(2-chloroethyl)phthalimide. The product can then be extracted from the reaction mixture and analyzed by GC in isooctane. The background is low and the phthalimide route provides a direct method for measuring total phthalates. The reported yield is 90% based on total esters. Burns and coworkers [35] used the saponification of the sample and disappearance of the PAE GC peaks for confirmation.

Nuclear magnetic resonance has been used for confirmation [228]. Giam, Goodwin, and Yano [91] have reported on the ^{13}C-NMR magnetic resonance of unsymmetrically esterified phthalic acids, including monomethyl, monoethyl, mono-1-propyl, mono-2-propyl, mono-1-butyl, mono-2-butyl, mono-t-butyl, and mono-2-ethylhexyl phthalates.

Zitko [282] reported that total PAE levels can be confirmed and even quantitated by their fluorescence in concentrated sulfuric acid. The alcohol moiety does not affect the fluorescence and as little as 200 ng DEHP per mL of solution can be measured. Terephthalates and phthalic acid also fluoresce (but with different characteristics), but isophthalates are nonfluorescent. Optimal excitation is at 270 nm and 308 nm for phthalates and terephthalates, respectively, with corresponding emission maxima of 360 and 380 nm.

Mass spectrometry, with or without gas chromatography, is a powerful tool for the confirmation of structure. Atlas and Giam [12] have recently provided a very extensive review of the mass spectrometry of phthalates. Galyashin [71, 72] has reported on the use of metastable ions for determining the positional isomers of benzenepolycarboxylic acids; such data is of particular importance for naturally occurring compounds of this type.

In a continuing study of DEHP metabolites, Harvan and coworkers [105] have reported detailed mass spectrometric data for nine metabolites. They describe the electron impact, methane positive chemical ionization (CI), oxygen negative CI and collision induced dissociation mass analyzed kinetic energy spectra (CID-MIKES) of these compounds. They report that the CID-MIKES of selected fragment ions provides particularly useful structural information.

Distribution and Transport

Overview and Environmental Data

In this section we will review the data in the literature concerning the concentrations of phthalates in various environmental compartments and describe briefly some of the transport processes which occur between compartments. To simplify the presentation we will not consider the loss of phthalates during this transport, but we will remove this restriction in the following section. Scheme 1 presents a simplified model showing the five major environmental compartments and the principal directions of PAE transport. In the first part of this section we will look at the observed concentrations of phthalates within these compartments, and in the second part we will discuss the experimental and theoretical approaches to understanding the flows between these compartments.

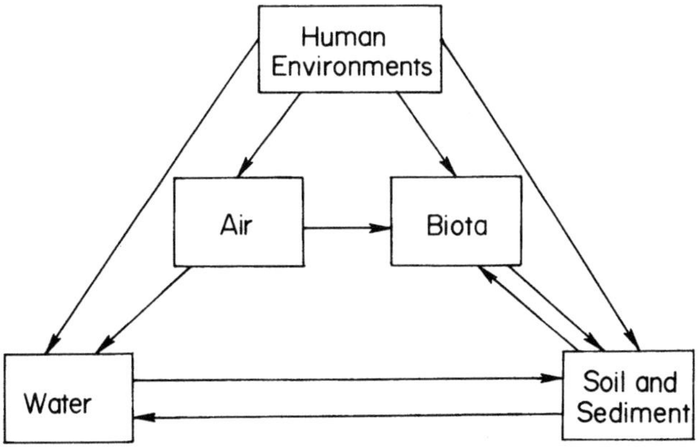

Scheme 1. Environmental compartments and phthalate transport

Human and Natural Phthalates. The human environment is the principal – perhaps the sole – source of PAEs. This phrase refers to all areas of human activity from which phthalates are released to the other environmental compartments, including the industrial areas in which they are produced, the myriad locations in which PAE-containing products are in use and subject to loss by volatilization and leaching, and to waste water and solid waste disposal and treatment areas. It has been a matter of some controversy whether or not there are naturally occurring phthalates. Graham, in his 1973 review [97], gives a brief list of natural occurrence of phthalates, but in 1974 Mathur argued that all such occurrences were suspect because of the prevalence of phthalate contamination [153]. Although one would hesitate to argue strenuously for natural biogenic phthalates as a major source, there are examples of biologically related PAEs which are not easily accounted for by anthropogenic sources.

Several observers have reported phthalates in humus and other plant remains, both recent and fossil. Balabanova-Radonova and coworkers [19] reported the isolation of branched C4–C8 alkyl phthalates from lignite humus substances from the Marits-Ivtok deposit. IR and MS were used to characterize these materials. Klindukhov and coworkers [126] found PAEs in bitumenous coals of different degrees of metamorphism, and Phillips and Breger [27, 197] found a branched dioctyl phthalate (identified by IR of the ester and the saponification products) in a Utah crude oil. In a later discussion of this finding, Breger [28] reported on the care taken to avoid contamination. Liao and collaborators [146] reported a variety of compounds, including phthalates, as the products of the hydrolytic and oxidative degradation of aquatic humic substances. These reports are ironic, since some of the early reports on the relationship of humic substances and phthalates assumed that the phthalates were contaminants [153, 155, 181].

Unusually high concentrations of some phthalates have been reported in other biota. Musial and his collaborators [176] reported that North Atlantic herring and mackerel contain unexpectedly high concentrations of the commercially minor compound DHP. The amounts reported were about six times the levels of DEHP found in these same species. Similarly, Ehrhardt and Derenbach [60] report elevated concentrations of BiBP in the waters of the Kiel Bight of the Baltic Sea. This compound is not sold in commerce, but its concentration in these waters is comparable to those for the important plasticizers DBP and DiBP. They argue for a biogenic origin on the basis that it is unlikely to be formed by an abiotic ester exchange mechanism. It is, of course, possible that anthropogenic PAEs are reworked biologically to produce DHP and BiBP. However, such biological ester exchange in PAEs has not been reported to occur in laboratory studies.

Two surprising observations of "natural" phthalates are the identification of high concentrations of DEHP in the heart mitochondria of cows, rats, rabbits, and dogs by Nazir and coworkers [177] and the isolation of DOP from bovine pineal glands by Taborsky [228]. Nazir and collaborators describe in some detail the precautions taken against contamination, and they note that these concentrations were not found in all tissue fractions. The concentrations found in heart mitochondria are in the parts per million range and are comparable to the highest levels reported in other biological systems (cf. Table 13). In the pineal gland study, Taborsky isolated and characterized (NMR, IR, elemental analysis) almost half a gram of DOP, corresponding to an organ concentration of 700 ppb. It was the third most abundant material in the lipid obtained by chloroform extraction of the gland. Although these phthalates could conceivably be endogenous, the fact that they are common plasticizers makes it likely that they are due to localized bioaccumulation of environmental PAEs. The highest levels found are similar to those found in seal pup blubber by Zitko [282] in which the author assumed that the PAEs found were exogenous, and the particular phthalates found in the bovine studies are common plasticizers.

Phthalate plasticizers can be a nuisance in some cases. Carmignagni and Bennett [40] reported that new fish tanks (for closed aquaculture systems) have DEHP concentrations of 440 ppb initially and must be flushed for 10 days with warm water before the concentration drops to the tap water background of 24 ppb. They describe the use of a polyurethane foam adsorbent for the removal of this background contaminant [41].

Environmental phthalate concentrations are highly variable in space and time. PAE concentrations associated with human activities are included in Tables 10 to 13 along with the concentrations found in "natural" areas. Some of the highest reported levels of PAEs are found associated with human activity; specific cases include the remarkably high 14,000,000 ng/m^3 DBP in a Russian factory [165], the elevated levels found in air and solids associated with waste incineration [46, 232], and the water concentrations in regions of industrial activity [60].

Atmospheric Phthalates. From the human environment PAEs move slowly into the rest of the world. Air and water are the principal vehicles for this transport, and volatilization and solubilization are the principal processes involved (Scheme 1). Table 10 lists the levels of phthalates found in a variety of atmospheric studies. The studies of oceanic air [9, 11, 78, 84] indicate that there is now a Northern Hemisphere atmospheric background of about 1–2 ng/m^3 for the major plasticizer phthalates. Without additional inputs this would eventually be washed out and removed from the atmosphere. Concentrations of phthalates in municipal areas are considerably above the levels reported in air of the remote Pacific, and the atmospheric concentrations in some areas with human activities are even greater.

The highest atmospheric levels are indoor air from industrial environments. Milkov and coworkers [165] reported levels of phthalates which were up to 66,000,000 ng/m^3 in a company manufacturing artificial leather and PVC films. In closed rooms which are tiled with PVC, PAE concentrations may reach 150,000 to 260,000 ng/m^3 [190], and Men'shikova [162] found that the concentrations of DBP in a ship's living quarters were as high as 1,220,000 ng/m^3 *three years* after the quarters had been redecorated with various plastic materials. It is little wonder, then, that indoor air pollution is a matter of some concern [52, 265]. Probably more typical of interior spaces are the values reported for laboratory air of about 20–150 ng/m^3, but this can be altered considerably by decorations and climate. Mieure and Dietrich found that the "new car smell" contained as much as 300,000 ng/m^3 PAE on a sunny day [164].

Outside air levels can also be high in industrial area. In their study of the air in Osaka, Japan, Yamasaki, and Kuwata [280] found average concentrations of 192 ng/m^3 DBP and 99 ng/m^3 DEHP during the period from October 1977 to

Table 10. Environmental concentrations of phthalates in air (ng/m³)

Location	DBP	DEHP	ΣPAEs	Other	Ref.
Oceanic Air					
Enewetak Atoll	0.9	1.4			[11][d]
Gulf of Mexico		0.5–1.7			[79][d]
Gulf of Mexico		0.2–1.0			[79][e]
Gulf of Mexico	0.3	0.4			[87][f]
Gulf of Mexico	1.3	1.2			[81][f]
North Atlantic	1.0	2.9			[90][d]
Pigeon Key, Florida	18.5	16.6			[11][f]
Municipal Air					
Antwerp (Belgium)	55–79	21–44			[43][e]
Barrow, Alaska	1.0	20		0.1[a]	[266]
College Station, Texas	3.8	2.4			[11][d]
Hamilton (Canada) Municipal incinerator	700	300		750[b]	[232]
Houston, Texas	6.2	23.5			[13]
LaPaz (Bolivia)	19–36	17–20			[43][e]
New York City	4	14			[24][e]
Osaka (Japan)	25–60	27–36		5–6[c]	[279][d]
Osaka (Japan)	192	99			[280][f]
Sterling Forest, New York	1	3			[24][e]
Laboratory Air					
Bermuda Biol. Sta.	50–145	28–124			[77][d]
Texas A&M	17–21	35–38			[77]
Industrial and Indoor Air					
Coal preheater effluent	4×10^6			9×10^6	[236]
Meatpacking plant (Colorado)				to 420,000[g]	[65][e]
"New car smell"			300,000		[164]
Russian PVC plant			$10–66\times10^6$		[165]

[a] DEP
[b] Butylphthalyl butyl glycolate
[c] DHP
[d] Vapor phase
[e] Particulates
[f] Vapor phase + particulates
[g] DCHP
[h] DOP

October 1978. The concentrations varied seasonally (proportional to temperature), and the less volatile DEHP was directly related to particulate concentrations, but DBP was not.

The distinction between particulate and vapor phase concentrations is probably an important variable in all atmospheric measurements, and Table 10 indicates whether the data are for one or both of these phases. A recent example of the information obtainable from particulate studies is the report by Clement and Karasek [46] of their investigation of PAEs and other organic contaminants in fly ash obtained from an electrostatic precipitator from an Ontario municipal refuse incinerator. The ash was sieved and the organic contaminants were then determined on the size-fractionated ash. The highest PAE concentrations were associated with a fraction which they identified as being due to the burning of newsprint. Only alkanes were a more abundant contaminant of the ash than phthalates. In his report on size-fractionated aerosols sampled from inside a New Jersey office building, Wechsler [265] states that phthalates were the most abundant compound found and DEHP appeared to be one of the most uniformly distributed substances across the various size fractions. Besides DEHP he observed DBP, DHP, DNP, and DDcP.

Aqueous Phthalates. Water is the principal medium of phthalate transport. Indeed, it is probably much more important than atmospheric transport for higher phthalates. Branson [26] estimated that the half-life for the volatilization of DEHP from water is fifteen years. There is little solid evidence regarding the relative importance of these two transport routes, but Giam and coworkers [9, 80] estimated the flux of DEHP to the Gulf of Mexico based on information in the literature and their own extensive analyses of marine samples [87, 88]. They estimate that in 1975 2.3×10^6 g DEHP entered the Gulf from rivers annually. An additional 0.5×10^6 g was dumped (both as waste and as components of other materials) and between 0.8 and 1.6×10^6 g entered by way of the atmosphere. Table 11 lists the concentrations of phthalates found in a number of types of water samples.

Surface and ground water can be contaminated by anthropogenic PAEs either directly – by dissolving phthalate materials from sources such as industrial wastes, solid waste dump sites, and exterior plasticized PVC objects – or indirectly by the washing out of atmospheric PAEs. In their study of organic contaminants in remote area atmospheres, Atlas and Giam [11] reported that rainfall at Enewetak Atoll had a concentration of about 50 parts per *trillion* for the major plasticizers, DBP and DEHP. This is related to the atmospheric concentrations of 1–2 ng/m^3; a higher atmospheric concentration would lead to a higher concentration in rainfall, other conditions being equal. However, even the highest levels of atmospheric contamination would be unlikely to produce the high levels of phthalates observed in some surface water samples, so direct contamination is important.

Surface waters, both salty and fresh, regularly contain phthalates. The concentrations range from less than 0.1 ppb to almost 1,000 ppb. The higher concentrations are usually located near industrial activities, although it is not necessarily easy to locate the actual source of the elevated phthalates level. In addition, some of the earlier high values are possibly due to contamination of the samples during the analytical procedures. Laboratory tap water, distilled and deionized water show a wide variability in PAE concentration. This variability is probably due both to the source and to contamination after water treatment. Takeuchi and Ishii [229] note that the distilled water in their laboratory contained over 4 ppb phthalates; we have normally found concentrations below our detection limit of about 0.05 ppb. The difference is almost surely due to the distribution system and subsequent handling, since the low volatility of the phthalates makes it unlikely that significant quantities are distilled over.

Phthalates in Soils, Sediments, Biota, and Wastes. Phthalates have a low solubility in water; thus, they tend to move from the aqueous phase to the contiguous solid and biological phases. Concentrations of phthalates found associated with

Table 11. Environmental concentrations of phthalates in water (μg/L or ppb)

Location	DBP	DEHP	ΣPAEs	Other	Ref.
Saltwater					
Baltic Sea	80			20	[60]
Gulf of Mexico Region					
Various sites				0.6–3.0	[47, 48]
Gulf Coast	0.07	0.13			[90]
Mississippi Delta		600			[47]
	0.09	0.07			[90]
Open Gulf	0.09	0.08			[90]
Galveston Bay, Texas		0.25			[174]
Rivers and Lakes					
Charles and Merrimack Rivers (Massachusetts)		1.5			[108]
Ijsell River (Netherlands)	1.9	1.2			[208]
Lake Huron	0.04	5.0			[222]
Lake Superior		7.0			[33]
(Black Bay, Ont.)		300			[222]
Lower Saginaw River, Bay City, Michigan		0.43		400	[93]
Meuse River (Netherlands)	0.33	1.2			[208]
Missouri River	0.09	4.9			[157]
St. Louis (Missouri)				0.2[a]	[93]
Weldon Springs, Mo. (Missouri)				0.2[a]	[93]
Monatiquot River (Massachusetts)			1–30		[109]
Rhine (Netherlands)	0.41	1.0			[208]
River (unnamed, Japan)	45	10			[168]
Tama River (Japan)	2.0	2.5			[170]
Water Supplies and Wastewaters					
Tohuku University (Japan)					
City Water	2.04				[115]
Well Water	2.49	4.82			[115]
Deionized water	0.83	1.31			[115]
Phoenix, Arizona					[241]
Sewage		0.25	0.38		
Ground water		0.37	0.38		
Tokyo city tap water	2.3	1.3			[170]
Tokyo city raw water	4.5	2.7			[170]
Tap water					
Chikusa-Ku (Japan)	14	4.0			[229]
Amsterdam (Netherlands)	0.04	0.54			[253]
Distilled water					
Texas A&M	<0.05[b]	<0.05[b]			[13]
Amsterdam	0.5	1.6			[253]
Chikusa-Ku (Japan)	1.5	2.8			[229]
Mineral water					
Amsterdam C	<0.05[b]	0.2			[253]
Amsterdam A	0.08	1.2			[253]
Rainfall					
Enewetak Atoll	0.03	0.06			[11]

[a] BBP [b] Below detection limit

Table 12. Environmental concentrations of phthalates in soils, sediment and solid wastes (ng/g = ppb)

Location	DBP	DEHP	ΣPAEs	Other	Ref.
Saltwater Sediments					
Baltic Sea		100			[172]
Gulf of Mexico Region					
Mississippi Delta	13	69	82		[90]
Nearshore	7.6	6.6	14		[90]
Offshore	3.4	2.0			[90]
Galveston Bay, Texas		22			[174]
San Luis Pass, Texas[a]	60	94	159	5[b]	[175]
Freshwater Sediments					
American lakes/rivers (several)				428[c]	[93]
Chesapeake Bay (1975–1979 core)	42	180		42[b]	[195]
Lake Constance (Switzerland)	200	500			[78]
Lake Superior (Canada)	100	200			[222]
Ijssel River[a] (Netherlands)	1,100	20,100			[209]
Lower Saginaw River, Bay City, Mich.				400[c]	[93]
Meuse River[a] (Netherlands)	500	8,200			[209]
Ohio River (Pennsylvania)			80,000		[45]
Rhine River[a] (Netherlands)	2,100	20,900			[209]
Usk River (England)	8,000	30,000	60,000	14,000[c]	[59]
Solid Wastes					
Incinerator Fly Ash (Ontario, Canada)[a]	6,400			6,300[b]	[46]
Ink pigment wastes	100,000	2,000		3,700[b]	[166]
Wastwater sludge	2,000	42,800		2,500[f]	[166]
Soil Humus					
Prince Edward Island	52	54	360		[181]

[a] Dry weight basis
[b] DEP
[c] BBP
[d] DOP

solids are listed in Table 12. There is an enormous variation in the levels reported. Lake Constance has several hundred ppb, while the levels reported in the Ohio River reach 80,000 ppb. Müller and his collaborators [172] reported that they could correlate PAE concentrations in Baltic Sea sediment cores with the level of plasticizer production in Europe. Peterson and Freeman [195] reported a similar correlation between American production and Chesapeake Bay sediments. Thus, sediments and soils are an important sink for phthalates. Organisms also serve to store environmental phthalates; typical analyses are listed in Table 13. Again, there is a wide variation in the level of phthalate contamination reported, ranging from a few ppb to tens of thousands. The ultimate limit on environmental phthalate accumulation is the rate of their degradation by chemical and biological process which we will consider below.

Comparison of these data with comparable environmental concentrations found for other organic contaminants will help put them into proper perspective.

Table 13. Environmental concentrations of phthalates in biota (ng/g = ppb)

Location	DBP	DEHP	ΣPAEs	Other	Ref.
Amphibians, Fish, and Invertebrates					
Gulf of St. Lawrence					
Herring		4,710	17,000[c, a]		[35, 176]
Mackerel		6,500	27,000[c, a]		[35, 176]
Cod		5,900	< 10[a]		[35, 176]
Plaice		< 10	< 10[a]		[35, 176]
Redfish		< 10	< 10[a]		[35, 176]
Iowa					
Channel catfish	200	400			[157]
Dragonfly naiads	200	200			[157]
Tadpoles	500	300			[157]
Lake Superior, Black Bay, Ontario					
Walleye		800			[157]
Mississippi/Arkansas					
Channel catfish	tr	3,200			[157]
Commercial fish (Canada)					
Eel		104			[270]
Tuna	78	110			[270]
Salmon	37	76			[270]
Gulf of Mexico[b]	< 0.1	4.5			[90]
Atlantic salmon *(Salmo salar)*	14	700[c]			[282]
Birds					
Cormorant eggs *(Phalacrocorax auritus)*	14,100[c]				[282]
Herring gull eggs *(Larus argentatus)*	15,700[c]				[282]
Mammals					
Heart muscle					
Rat		12,900[c]			[177]
Rabbit		11,800[c]			[177]
Dog		3,600[c]			[177]
Seal pup blubber *(Phoca vitulina)*		10,600[c]			[282]

[a] DHP
[b] Based on twenty samples including fish, crabs, shrimp, eel, rays, jellyfish, and starfish
[c] Based on weight of lipids or organ; not whole organism

Murray, Ray, and Giam [174, 175] analyzed samples of water, sediment and biota from the Galveston Bay area of the Gulf of Mexico for a variety of pollutants, including DDT and its degradation products, PCBs, phthalates, hexachlorobenzene, and pentachlorophenol. They found that phthalate concentrations were generally higher than concentrations of other contaminants, in some cases they were much higher. DEHP concentrations ranged widely over the sites studied (<2 to 12,000 ppb in water and <1 to 100 ppb in sediment). Ehrhardt and Derenbach [60] reported that in the Baltic Sea total PAEs were in two orders of magnitude greater in abundance than chlorinated compounds. Thus, phthalates are one of the most important trace organic contaminants in aquatic systems.

Phthalate contamination can be reduced either by increasing their degradation rate or by decreasing the rate at which they are transferred from the human to the natural environment. The second involves fewer sites and hence is generally more practical. The removal of phthalates from solid and aqueous wastes is a matter of some concern, particularly since six phthalates (DMP, DEP, DBP, BBP, DEHP, and DOP) are listed by the US Environmental Protection Agency (EPA) among the 129 priority pollutants [245]. Furthermore, the increasing demand for water reuse requires that efficient means be developed for reducing the general level of trace contamination. The municipal water data in Table 11 illustrate the fact that this may be more of a problem in some areas than in others. Lue-Hing and coworkers [149] examined the fate of priority pollutants at four of the Metropolitan Sanitary District of Greater Chicago water treatment plants and found that relatively few organics are currently present in detectable quantities; DEHP was found in low concentrations in influent and at 67 ppb in the primary sludge. The removal of PAEs from wastewater is not well understood. In a report prepared for the Environmental Protection Agency, E. C. Jordan, Inc. [119], reported that while DEP was reduced during treatment at a publicly owned water treatment plant, DBP, DOP, and DEHP have poorly correlated influent/effluent concentrations. Indeed, DOP was actually higher in the effluent than in the influent during the 36-day study.

In the next major section we will consider the biodegradation of PAEs and hence their removal by activated sludge and similar treatment methods. However, abiotic means are also being used. Shuckrow and Pajak [218, 219] assessed various concentration technologies for the treatment of hazardous aqueous wastes. The methods tried included adsorption on resins or activated carbon, aerobic, and anaerobic biological treatment and air stripping. While most treatment methods are at least moderately effective in reducing organic contamination, a process train consisting of adsorption on granular activated carbon followed by activated sludge treatment achieved high levels of removal for short periods of time. They applied this technology to an actual hazardous waste site. Hwang [112] discusses the suitability of treating PAE-containing waste water by solvent extraction. He presents a review of solvent extraction as a water treatment process and calculates treatabilities based on distribution coefficients and extraction column models for the six priority pollutant phthalates and several other compound classes.

Phthalate contaminated soils and sediments pose a different and less well studied problem than wastewater treatment. Miller and coworkers [166] reported on the evaluation of methodology for the survey analysis of solid wastes, including five of the six priority pollutant PAEs, (DMP, DEP, DBP, DOP, and DEHP) and they report values for the residual concentration of these compounds in a variety of solid wastes as a basis for planning the treatment of such materials. Warner and collaborators [261] present a simple, low-cost, reproducible method for assessing the potential leachability of purgeable and semivolatile organic priority pollutants (including DEHP, DMP, and DOP) from various solid waste materials based on a tumbling device designed by the US National Bureau of Standards. They found that distilled water was as effective as other media for leaching most organic compounds, but DOP and DMP were poorly leached under all conditions. The quality of this method for predicting leaching of components under environmental conditions was not tested. Neumayr [180] reviewed the systematic cleanup procedures required for soil and grandwater contaminations. He proposes the use of soil-air analysis for location of contaminants, and he cautions against producing secondary pollution during soil exchange at contaminated sites.

Models and Data for Environmental Transport

Models for Distribution and Persistence. The preceding discussion used a very simplc scheme to describe the major environmental compartments. More sophisticated versions of this scheme can be used to quantitatively predict the movement and localization of environmental pollutants. These theoretical models are designed to predict the impact of contaminants on the environment by considering the transport processes which move them between these various compartment. There is a wide variation in the sophistication of these models, and most are concerned principally with physico-chemical processes (i.e., movement between water, air, and sediment) rather than with bioaccumulation or biodegradation. PAEs have been used as compounds for testing many of these models because they are priority pollutants and there is a wealth of both physico-chemical and environmental data available for them, such as those tabulated in Table 2.

The purpose of such modelling is to allow the estimation of risk from various organic contaminants. The three principal compound characteristics involved in risk estimation are toxicity, bioaccumulation and persistence. Zoeteman and collaborators [283] provided a general discussion of the persistence of organic pollutants in river and ground water in the Netherlands. They define persistence as the ability of an aqueous pollutant to resist a reduction of the original concentration after a certain period of time while undergoing a variety of physical, chemical, and biological processes. They divide pollutants into five categories on the basis of their estimated half-lives (in days) in water. On this basis DMP is in Category I ($t_{1/2} < 0.3$ da), DBP is in Category II (0.3–3.3), and Category V (> 300 da) includes such chemically inert pollutants as Dieldrin and DDE. They further subdivide their categories to take into account the differences in persistence expected in river water (least persistent), lake waters, and ground waters (most persistent). They estimate the half-life of DMP and DBP in surface waters as 0.2 and 0.4 days, respectively.

Klöpffer, Rippen, and Frische [127] describe the use of an elaborated three-compartment (water, air, and soil/sediment) model for predicting the distribution and persistence of organic contaminants. Like Zoeteman, they subdivide the compartments, so they have a total of twelve sub-compartments. The transport of materials between these twelve compartments is then defined by 27 different processes. They distinguish between processes which are physico-chemically limited (15 processes) and those which are environmentally, e.g. meteorologically, limited (12 processes). In practice this scheme is much too large to be practical, and they revert to a three-compartment scheme to provide a numerical model for the distribution of pollutants. In order to compute this distribution they use data for the vapor pressure, water solubility and octanol-water partition coefficient. On the basis of their simplified model they predict that DEHP will be concentrated almost entirely in the soil/sediment phase of the system (air 0.1%, water 1.9%, soil/sediment 98%).

They also compute distributions for the refrigerant trichlorofluoromethane (98.98%, 0.02%, 0% [sic]), the industrial chlorocarbon hexachlorobenzene (53.10%, 0.05%, 46.85%), and the herbicide 2,4,5-trichlorophenoxyacetic acid (0%, 70%, 30%); the precision roughly reflects the precision of the parameters available. This model is closely related to their view that persistence and accumulation are the most important criteria for assessing environmental pollution potential of organic compounds [68].

Neely [178] has described a four-compartment model in which he distinguishes soils from bottom sediments. He uses the vapor pressure, water solubility and molecular weight as physicochemical parameters. With these three he calculates the environmental distribution of 84 of the 129 priority pollutants and their potential as water pollutants. Five of the six priority pollutant PAEs are included in his model. His predictions of the percentage of each that will be found in water and bottom sediments are: DMP 97%; DEP 63%; DEHP 52%; DOP 51%; DBP 14%. These predictions do not include the degradation rates of these compounds. It is not clear what the relative distribution between water and sediment is, nor how this somewhat counterintuitive result was obtained. (Why, for example, are the two "dioctyl" phthalates bracketed by smaller compounds? Are they retained by soil – as opposed to sediment – in his model?)

Wolfe, Burns, and Steen [276–278] enlarged their available set of parameters by using linear free energy relationships to estimate the rate constants for hydrolysis and biolysis, the sediment-water partition coefficients, and biosorption required for modelling the fate and transport of DMP, DEP, DBP, DOP, and DEHP. They then applied the Exposure Analysis Modeling System (EXAMS) to four defined environments: a pond, an oligotrophic lake, a eutrophic lake, and a river using environmental parameters representative of the Southeastern United States. Their results suggest that the higher molecular weight PAEs are not significant transformed and that export from the ecosystem is the dominant process.

The development of laboratory and field data bases on which such modelling efforts can be based is important to refining and testing these models. Two types of data bases are required. The first type is an accurate and self-consistent set of physico-chemical and biological parameters for use in modelling, while the second is a set of results from monitoring of field or laboratory systems for the purpose of comparing them to existing models. The two data base types are not easily separated in practice, but it is important to note that studies of this kind may have different objectives.

Assessment of Modelling Parameters. A number of groups have studied various aspects of the physico-chemical properties of PAEs. Leyder and Boulanger [145] recently published new water solubility and octanol-water partition coefficient data. Atlas, Foster, and Giam [10] reported the air-water partition coefficients and mass transfer rates of several important organic pollutants. They found that in general the data could be interpreted in terms of the Whitman resistance model for mass transfer across the interface, but that high values for DBP and dieldrin were probably due to interactions at the bubble/water interface. Dobbs and Cull [53] developed a method for estimating the vapor pressure of high molecular weight PAEs by determining their rate of volatilization.

Table 14. Bioaccumulation factors (BF) for phthalates[a]

Organism	PAE	Concentration (ppt)	Exposure time (da)	BF	Ref.
Plants					
Algae *(Oedogonium cardiacum)*	DOP	64	33	28,500	[205]
	DOP	3,450	3	6,800	[205]
Mentha aquatica	DEHP	1,000	27	18,300	[221]
Chara chara	DEHP	1,000	27	18,300	[221]
Invertebrates					
Amphipod *(Gammarus pulex)*	DEHP	1,000	27	24,500	[221]
Scud *(Gammarus pseudolimnaeus)*	DBP	100	1	1,700	[206]
	DBP	100	7	6,500	[206]
	DEHP	100	1	2,800	[206]
	DEHP	100	7	13,600	[206]
	DEHP	62,800	1	30	[206]
	DEHP	62,800	7	120	[206]
Brine shrimp	DOP	64	33	2,600	[205]
	DOP	64	3	9,300	[205]
Damselfly *(Ischnura verticalis)*	DBP	100	1	1,000	[206]
	DBP	100	7	2,700	[206]
Dendrocoelum lacteum	DEHP	1,000	27	4,100	[221]
Glass shrimp *(Palaemonetes kadiakensis)*	DBP	80	1	1,500	[206]
Helobdella sp.	DEHP	1,000	27	2,000	[221]
Limnephilus sp. (larvae)	DEHP	1,000	27	19,200	[221]
Mayfly *(Hexagenia bileanata)*	DBP	80	1	500	[206]
	DBP	80	7	1,900	[206]
	DEHP	100	1	850	[206]
	DEHP	100	7	2,300	[206]
Midge larvae *(Chironomus pulmonus)*	DBP	180	1	3,500	[206]
	DBP	180	7	6,600	[206]
	DEHP	30	1	2,400	[206]
	DEHP	30	7	3,100	[206]
	DEHP	1,000	27	1,200	[221]
Mosquito larvae *(Culex pipens quinqefasciatus)*	DOP	64	33	9,400	[205]
	DOP	3,450	3	5,500	[205]
Mussels *(Mytilus edulis)*	DEHP	4,100–42,100	28	2,500	[32]
	DiDcP	4,400–41,700	28	3,500	[32]
Oyster *(Crassostrea virginica)*	DBP	100,000	1	21.1	[275]
	DBP	500,000	1	41.6	[275]
	DEHP	100,000	1	11.2	[275]
	DEHP	500,000	1	6.9	[275]
Shrimp *(Peneaus aztecus)*	DMP	100,000	1	3.1	[275]
	DMP	500,000	1	6.3	[275]
	DBP	100,000	1	2.9	[275]
	DBP	500,000	1	30.6	[275]
	DEHP	100,000	1	10.2	[275]
	DEHP	500,000	1	16.6	[275]
Sialis sp. (larvae)	DEHP	1,000	27	2,300	[221]

Table 14 (continued)

Organism	PAE	Concentration (ppt)	Exposure time (da)	BF	Ref.
Snails *(Physa)*	DOP	64	33	13,600	[205]
	DOP	3,450	3	400	[205]
Snails *(Planorbis corneus)*	DEHP	1,000	27	17,500	[221]
Sowbug *(Asellus*	DEHP	1,900	7	80	[206]
brevicaudus)	DEHP	62,300	7	20	[206]
Water fleas *(Daphnia magna)*	DBP	80	1	2,200	[206]
	DBP	80	7	5,000	[206]
	DEHP	300	1	1,200	[206]
	DEHP	300	7	5,200	[206]
	DEHP	3,200–100,000	21	210	[31]
	DiDcP	3,200–100,000	21	120	[31]
	DOP	3,450	3	9,400	[205]
	DOP	64	33	2,600	[205]
Vertebrates					
Bluegill	BBP		21	663	[93]
(Lepomis macrochirus)					[243]
	DMP		21	57	[244]
	DEP		21	117	[244]
Fathead minnows	DEHP			886	[159]
(Pimophales promelas)					
Sheepshead minnow	DMP	100,000	1	6.0	[275]
(Cyprinodon variegatus)	DMP	500,000	1	4.7	[275]
	DBP	100,000	1	11.7	[275]
	DEHP	100,000	1	10.7	[275]
	DEHP	500,000	1	13.5	[275]
Minnow *(Phoxinus phoxinus)*	DEHP	1,000,000	27	178	[221]
Mosquito fish	DOP	3,450	3	1.2	[205]
(Gambusia affinis)		64	33	9,400	[205]
River lamprey	DEHP	1,000	27	10,600	[221]
(Lampetra planeri)					
Stickleback *(Pugitus pugitus)*	DEHP	1,000	27	300	[221]

[a] BF is the concentration of PAE in the organism divided by the final concentration of PAE in the medium

The two most important biological parameters are the rate of biodegradation (below) and the bioaccumulation factor (BF). Concentration of lipophilic organic compounds from the environment by organisms is widely observed. The extent of concentration is expressed quantitatively by BF, the ratio between the final concentration in the organism and the final concentration in the medium. The relatively high concentration of phthalates in organisms (compared to the corresponding water concentrations) in Tables 11 and 13 are evidence for the importance of phthalate bioaccumulation. The bioconcentration of phthalates has been studied directly in some organisms; typical data are listed in Table 14. The reported BFs vary widely and they also depend on the exposure time. Increasing the length of the exposure to an elevated PAE concentration can either increase or decrease their ability to accumulate or degrade the compounds. Wofford and his collaborators [275] found that the rate of accumulation depended both on the organism (oysters > shrimp > fish), and on the alcohol moiety (DBP > DEHP > DMP); however the differences were not large.

The depuration rate, the rate at which the compound is removed from an organism's tissues in a clean environment, is a related parameter, which has been occasionally investigated. In a follow up to their study of bioaccumulation of PAEs in mussels (*Mytilus edulis*), Brown and coworkers [32] found that the half-life for loss of DEHP after transferring the fish to clean water was 3.5 da. They monitored the organismic PAE concentration for a period of four half-lives. The reported half-lives for depuration of DMP, DEP, and BBP in bluegills are between 1 and 2 days [244].

There is usually little bioaccumulation of PAEs observed in mammals. Daniel and Bratt [49] administered large dietary doses of DEHP to rats. The compound accumulated primarily in the liver and the fat and a steady state concentration was reached within a week in the liver and two weeks in the fat. The steady state concentrations attained at 1,000 ppm were 36 and 8 ppm in the liver and fat, respectively, and at 5,000 ppm the comparable numbers were 115 and 80. The concentrations dropped 80% in the liver within one week after removing DEHP from the diet; in the fat only about one-third was gone after three weeks. The dose-dependent increase in the steady state concentrations indicates that the metabolic system was not saturated even at these high concentrations. Thomas [233] points out that the localization appears to depend on the route of administration, so that dietary DEHP is accumulated in the fat and liver, while intravenous administration does not appear to involve high fat concentrations. Some workers [258, 271] have reported high kidney concentrations.

Williams and Blanchfield [271] studied the retention, excretion, and metabolism of DEHP in male rats. They found that there is only limited retention and accumulation of 7-^{14}C-DEHP by male rats whether given as a single dose or fed in the diet. They reported that the adipose tissue was the only tissue or organ which consistently contained radioactivity. In a fifteen day feeding study with a diet containing 1,000 ppm radiolabelled DEHP, the adipose tissue accumulated 6–9 ppm DEHP. Activity was also detected in the kidney, liver, testes, skeletal muscle, lungs, and heart.

Similar results on the distribution and elimination of phthalates have been observed in male and female mice. Gaunt and Butterworth [74] report an autoradiographic study of orally administered DEHP in male mice. A single dose (6.72 mg/animal) of ^{14}C-DEHP was administered and its excretion and organ distribution were monitored. The radioactivity was excreted in urine and feces mainly in the first 24 h and completely within 3–5 days. No radioactivity was found in the central nervous system, skeleton or thymus, and little activity was detected in the testes. There was no evidence of tissue storage. Lindgren and coworkers [147] described a similar study on the distribution of ^{14}C-DEHP in pregnant and nonpregnant female mice. High activity was observed initially in brown fat, liver, gall bladder, intestinal contents, kidney, and urinary bladder. After administration of $^{14}C{=}O$ labelled (but not alkyl-labelled) DEHP, marked retention was observed in the skin, cartilage, and tendons. In the early embryo a high concentration was observed in the neuroepithelium. They suggested that this accumulation may be related to the DEHP-induced malformations exencephaly and spina bifida which are observed in mice.

Södergren [221] found that aquatic plants in a laboratory microcosm have very large bioaccumulation factors. He examined the possibility that biomagnification might occur through the food chain, that is, organisms which eat plants might have larger concentrations than plants, and the predators might have the largest concentrations of all. In fact this was not the case. The predators appeared to have the most efficient degradative mechanisms for PAEs. The largest BFs in the system were for plants and bottom feeders (who could ingest sediment-adsorbed PAEs).

The observed bioaccumulation rate is a composite of the rates for uptake, degradation, and excretion of a compound, and this net rate determines the steady state concentration of contaminant at a given environmental concentration. In long term studies the chemical stability of the compound must also be considered. Thus, the rates of chemical and biological degradation cannot be completely ignored when predicting the rate of passage of phthalates through the ecosystem, since they will affect the chemical nature of the material in passage and ultimately lead to its decomposition. We will provide move discussion of these rates later.

Plants have lower metabolic rates relative to animals, so biodegradation rates should be lower. Thus, we might expect that the steady state bioaccumulation factor would be related primarily to the uptake and excretion rates alone, and there should be a correlation between the bioaccumulation factors and the lipophilicity of the compounds. Geyer and coworkers [75] found that for the alga *Chlorella fusca* there was a correlation between log BF and log (water solubility) for 34 organic compounds from a variety of functional group classes. The relation is shown in Eq. 6 and the water solubility (WS) is in ppb. The correlation coefficient was 0.875 for the 34 compounds.

$$\log \mathrm{BF} = 4.549 - 0.46 \log \mathrm{WS} \tag{6}$$

Interestingly, the bioaccumulation factor found for DEHP (5,400) was too high using the value of 600 ppb for the solubility of DEHP. But, a closer fit is obtained using the value of 40 ppb reported recently by Leyder and Boulanger [145]. A better correlation might be expected with the octanol-water partition coefficient K_{ow}, but it was not used in this study. K_{ow} is difficult to measure for compounds with low solubility, but Leyder and Boulanger report that for most dialkyl phthalates it is highly correlated with aqueous solubility. They were hesitant, though, to extrapolate their correlation into the range expected for the K_{ow} of DEHP.

Veith and his coworkers [254, 255] have described a method for obtaining the value of K_{ow} from the HPLC retention time on a C18 reversed phase column; they have correlated BF with $\log K_{ow}$ for a variety of compounds.

Some workers have suggested that fat solubility might also be a good predictor of lipophilicity, and hence of bioaccumulation. Recently, however, Dobbs and Williams [54] found that fat solubility alone is not a good predictor of either BF or K_{ow}, and they caution against its use as a measure of either.

The available BF data are probably a weak foundation on which to build any substantial theoretical edifices, though. The data in Table 14 show that there is a sizeable time dependence for the value obtained, which probably indicates that often we are not dealing with a true steady state phenomenon. Thus, experiments whose time scale is days to weeks may not accurate predict the effect of months or years of exposure.

Freitag and coworkers [67] have combined bioaccumulation studies with other methods to provide an ecotoxicological profile analysis. They use measurements in five different systems to determine the several environmental parameters. *Bioaccumulation* is determined in algae (*Chlorella fusca*) and fish (golden orfes, *Leuciscus idus melanotus*); *retention* and *elimination* are investigated in rats; *biodegradability, metabolism, accumulation* and *bound residues* are measured in the activated sludge test, while the possibility of *photodegradation* (photomineralization) is studied with UV light ($\lambda > 290$ nm). They report results for 35 different environmental organic compounds including DEHP. Their methods are similar to those used by others, but they have attempted to provide a consistent rationale and methodology for such studies and their work is clear evidence of the consensus which is developing regarding the relevant parameters to measure.

Model Ecosystem Studies. Södergren [221] investigated the importance of considering interfacial interactions in ecosystem modelling, a point emphasized in the results of Atlas, Foster, and Giam [10] above. He exposed fish, invertebrate and plants to ^{14}C-DEHP in a static model ecosystem for 27 days. After 5 days only 2% of the original DEHP was in the aqueous phase, but 62% of the original DEHP could be recovered from the various surfaces: glass walls, sediment and surface microlayer. The other 36% could not be directly accounted for, but is believed to be due to extraction losses and to biodegradation to water-soluble metabolites and CO_2. Less than 1% of the material was in the organisms, even though bioaccumulation factors as high as 18,000 were observed. The relationship between these microcosm values and the natural world is complex, since the volume to interfacial areas of the two systems are so different.

Overcash, Shea, Weber, and Miles [184–186, 215] made an extensive study of the potential impact of DBP on an agricultural ecosystem. They found that DBP applied at the rate of 1,000 to 10,000 ppm showed a half-life of greater than 6 months. The rate of volatilization was below 5%, so degradation was the principal loss mechanism.

As our knowledge of the physicochemical and biological parameters of the phthalates is broadened and deepened, and our understanding of the important interactions in natural systems increases, even more complex studies than these will become possible.

Environmental Fate: Chemical and Biological Degradation

Chemical Degradation

Phthalate esters eventually disappear from the environment because they are subjected to a variety of degradative processes. Table 15 gives the relative rates for these various processes for BBP as reported in Gledhill and coworkers [93]. Hydrolysis (Eq. 1) is an important means for the primary degradation of PAEs, but a

Table 15. Environmental persistence of BBP[a]

Process	% degradation		Time (days)	Halflife (days)
	Primary	Ultimate		
Biodegradation				
Activated sludge	93–99		1	
CO_2 evolution (aerobic)		96	28	
Gas production (anaerobic)		<10	28	
River water	100		9	2
Lake water microcosm	> 95		7	< 4
Lake water microcosm		51–65	28	
Photodegradation	< 5		28	
Hydrolysis	< 5		28	>100

[a] Source: Ref. 93; reprinted with permission. Copyright 1980 American Chemical Society

purely chemical reaction it quite slow. Indeed, under natural and simulated natural condiditons the only rapid processes for removing PAEs are biological. As we shall see, BBP is more easily degraded than many larger phthalates (DEHP, DOP. etc.), but the relative importance of chemical and biological degradation probably holds for almost all phthalate plasticizers.

The results of Callahan and coworkers [36] give an idea of how slow this process is. They estimate that the half-life for hydrolysis at pH 7 and 25 °C range from 3.2 years for DMP to 2,000 years for DEHP. Wolfe, Steen, and Burns [277, 278) have measured the base-catalyzed hydrolysis rates for DBP, DMP, DiBP, and DEHP and used a linear free energy relationship (Taft σ^*, E_s) to apply these to other compounds. They found that the diester was hydrolyzed about twelve times faster than the monoester for these compounds.

Because of the slowness of the chemical hydrolysis reaction, PAEs deposited on sediments appear to be relatively stable, as indicated by studies of the relationship between the date of sediment deposition and the relative production of PAEs at that date [172, 194, 195]. Similarly, atmospheric photodegradation is only a minor degradative pathway because of the low volatility of these compounds and because their principal absorptions are well into the UV region of the spectrum. Aqueous photodegradation can and does occur when phthalates are exposed to UV light, primarily due to oxidation by the hydroxyl radical. Kotzias, Parlar, and Korte [132] studied the aqueous photochemical oxidation of several organic chemicals in water in the presence of nitrate and nitrite. They found that the rate of photochemical hydroxylation of PAEs by HO· to give phenols was greatly enhanced by the presence of nitrites. This reaction is probably not a significant degradative route for environmental phthalates, but it can be used in water purification.

There has been considerable interest, as discussed in the last section, in treatment methods for water contaminated with organic chemicals. Ozonolysis is one procedure which can be used. Two groups have reported recently on the oxidative behavior of phthalates under different conditions. Baillod, Faith, and Masi [17] studied the fate of specific pollutants during wet oxidation and ozonation. They found that the major product of the wet oxidation of DMP was acetic acid, whereas a significant product of its ozonization was oxalic acid. Il'in and coworkers [113] studied the kinetics of the reaction of ozone with PAEs. In carbon tetrachloride the ozonization reactions for DBP and DEHP were first order in PAE and in ozone and had activation energies of 9.8 and 9.9 kcal/mol, respectively. Neither of these reactions would be of importance in "natural" environments, however.

Metabolism and Biodegradation

In most ecosystems the principal route for phthalate disappearance will be biodegradation. Virtually all organisms have some capability for accelerating the hydrolysis of PAEs through relatively nonspecific esterases. Furthermore, some organisms can further convert phthalate monoesters and phthalic acid into still other products. There are a wide variety of pathways by which this metabolism occurs, probably indicating the relative *unimportance* of PAEs in natural environments before the recent past. There is still a great deal of controversy surrounding this subject, but much progress has been made in recent years.

Microbial Metabolism. As in the case of chemical degradation, the principal interest in biodegradation has derived from the need to treat wastewater and to determine the biodegradability of synthetic organic chemicals. Saeger and Tucker [203] reported a classic biodegradability study of this type. They investigated the biodegradation of PAEs (BGBP, BBP, DEHP, DUDP and a mixed phthalate plasticizer with an average side-chain mass corresponding to DiNP) in river water and a semicontinuous activated sludge reactor. They found that the half-life for degradation of PAEs in river water was quite rapid for some (BBP $\simeq$ BGBP $<$ 2 da) but much slower for others (DUDP = 31 da $<$ DEHP $<$ DiNP; experiment terminated after 35 days). In the single injection method of reacting with activated sludge the half-lives for all compounds dropped greatly; they were less than four hours for both BBP and BGBP. The more resistant PAEs were tested by a semicontinuous method in which the PAE is added at regular intervals (1–250 mg/da) to a 1.5 L vessel and the percent biodegradation is measured after steady state was reached. Both MBP and BGBP were 100% degraded at all concentrations used and BBP was also rapidly degraded. In contrast, DEHP was about 75% degraded at a 5 mg/da feed rate, "DiNP" was 50% degraded. With DEHP no water soluble aromatic compounds were observed, suggesting that they were more rapidly degraded than the parent ester.

Studies of this type have also been reported by later workers (e.g., [93]), and they are important for determining the expected level of phthalate contamination in the human environment. Tabak and his coworkers [227] reported their results on the biodegradability of 96 organic priority pollutants using a static-culture flask-screening technique. The pollutants were tested at 5 and 10 mg/L for seven days and then the culture was subcultured and challenged again with the pollutant for three separate 7 day periods. This allowed direct determination of the ability of the waste water inoculum to adapt to the pollutant. The results with the six phthalates divided them into two groups. DEP, DMP, DBP, and BBP were 100% degraded in the original culture and all three subcultures. DEHP and DOP were only about 50% degraded in the original culture but were more than 90% degraded by the third subculture. Thus, all of the tested PAEs can be removed from water by standard biological treatment, but DOP and DEHP are more persistent.

Still another type of system for carrying out the biodegradation of organic substances has recently been patented by Kurane, Suzuki, and Takahara [137]. They disclose a microbiological method for the biochemical treatment of various substances. The microorganisms are cultivated on a hydrophilic support in a biological reactor. A reaction substrate is fed to the vessel and brought into contact with the microorganisms, whereby the microorganisms are allowed to propogate within the support and act upon the reaction substrate to induce decomposition or synthesis. The method has obvious implications for wastewater treatment.

Other biodegradability studies have emphasized the search for particular organisms which are capable of metabolizing phthalates so that the biochemical pathways could be better understood. A classic study of this type was conducted by Engelhardt, Wallnöfer, and Hutzinger [62]. They divided the PAE-metabolizing microorganisms which they found into four classes on the basis of the method by which they were found and the extent to which they were able to utilize PAEs and phthalic acid (PA). Class a consisted of one fungus (*Penicillium lilacinum*) and three bacteria that were isolated from soil by using DBP as a carbon-enrichment source for culturing them. Class b was three bacterial strains (*Corynebacterium petrophilum, Arthrobacter hydrocarboglutamicus, Mycobacterium phlei*) selected by screening 40 defined stock cultures for their ability to degrade DBP. Both classes c and d were also selected from these stock cultures, but on their ability to grow on media with PA as the enrichment. Organisms in class c could also convert MBP to PA.

A more recent study was that of Sakagami and coworkers [204] who described the isolation of PAE-degrading bacteria from sea water and sediment. The biodegrading ability of the isolated microorganisms was tested by TLC. One strain of *Pseudomonas* and three strains of *Bacillus* were shown to decompose PAEs. Diheptyl phthalate (DHpP) was used as the substrate in studies of the effects of solvents and concentration on biodegradability.

Mathur and Rouatt [154] investigated a bacterium, tentatively identified as the virtually ubiquitous saprophyte *Serratia marcescens* Bizio, which was isolated for its ability to utilize the plasticizer DEHP as sole carbon and energy source. The organism was able to grow on PA but not on 2-ethylhexanol. They found the esterase is constitutive, but the enzyme(s) that mediated PA utilization appeared to be adaptive. It is unclear how seriously this contention should be taken, however, in light of the fact that phthalates are so widespread in both human and natural environments that a facultative PAE esterase might appear constitutive simply because the organism is always activated for this reaction. Kurane, Suzuki, and Takahara [137] reported that the esterase and the phthalate and protocatechuate (Scheme 2, compound *10*) oxidizing enzyme of *Nocardia erythropolois* were all inducible by their immediate substrates; the esterase was also inducible by olive oil. They found that a number of commercial esterases and lipases were able to hydrolyze DEHP.

A third general type of biodegradability study involves testing the parameters which control biodegrading ability. Kollig and Hall [131] described a set of experiments in which they used the rate of DEP biodegradation as one measure of

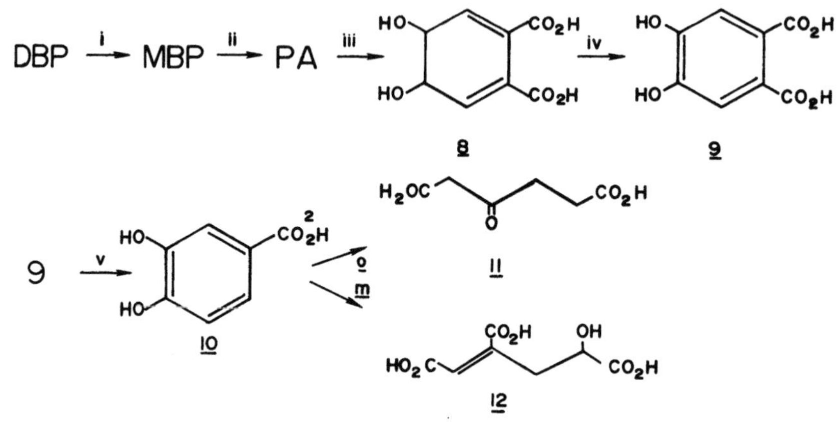

Scheme 2. Microbial metabolism of di-*n*-butyl phthalate

stress in a controlled aqueous ecosystem. They found that acidifying a continuous flow microcosm stimulated a decrease in alkalinity and a change in the diurnal variation in hydrogen ion activity that indicated that the system was being severely stressed. Removal of excess acid input was accompanied by a return to the normal diurnal pH cycle. However, even in this severely stressed system the rate of DEP degradation was not significantly altered, and a number of other parameters used for monitoring environmental stress (including nutrient concentrations, oxygen production, total biomass and algal community structure) were not good indicators of the level of system stress.

In addition to the studies of "biodegradability" and of "biodegrading ability", there is a completely distinct class of studies which attempt to characterize in some detail the biochemical pathways followed in PAE metabolism. In the study mentioned above, Engelhardt, Wallnöfer, and Hutzinger [62] utilized the strains which they isolated to investigate the metabolism of DBP and related PAEs. They used several methods for analyzing the metabolites (TLC isolation and subsequent MS, UV, and TLC R_f identification) and they provide tables of spectral and chromatographic data for the metabolites found. All microorganisms studied could hydrolyse DBP to MBP almost quantitatively, but there were differences in the subsequent metabolic pathways.

Recently Eaton and Ribbons [56–58] have described the pathways for PAE metabolism in *Micrococcus* sp. They report both ring-opened and hydroxylated products from these compounds. Scheme 2 illustrates the pathway. Steps i and ii, the hydrolysis of DBP and MBP, respectively, are found in many organisms from many phyla. However, the ring oxidation reaction appears to be more unusual. Steps iii, iv, and v together constitute an oxidative decarboxylation which yields a product (*10*) identified previously by Engelhardt and his collaborators in their study. Protocatechuic acid (*10*) is cleaved by *Micrococcus* in two different places leading to two different ring-opened products (*11, 12*) which can be further metabolized. Eaton and Ribbons isolated mutant strains which were unable to utilize phthalate and they identified several different defects related to steps in the pathway.

Table 16. Biodegradative index for PAEs[a]

Organism	Compound	Concentration (ppb)	Exposure (days)	Biodegradative index	Ref.
Invertebrates					
Oster	DBP	100	1	0.6	[275]
(Crassostrea virginica)		500	1	0.5	[275]
	DBP	100	1	0.6	[275]
		500	1	0.5	[275]
	DEHP	100	1	0.3	[275]
		500	1	0.3	[275]
Shrimp	DMP	100	1	1.0	[275]
(Penaeus aztecus)		500	1	0.4	[275]
	DBP	100	1	32.0	[275]
		500	1	17.2	[275]
	DEHP	100	1	1.0	[275]
		500	1	0.7	[275]
Fish					
Channel catfish *(Ictalurus punctatus)*	DEHP	1	1	7.1	[222]
Fathead minnow	DEHP	2–62	28	2.0	[275]
(Pimephales promelas)		2–62	56	1.7	[275]
Sheepshead minnow	DMP	100	1	5.8	[275]
(Cyprinodon variegatus)	DBP	100	1	5.9	[275]
	DEHP	100	1	3.5	[275]
		500	1	23.9	[275]

[a] Biodegradative index = (Σ metabolites)/unmetabolized PAE

Benckiser and Ottow [21] studied the metabolism of DBP by *Pseudomonas pseudocaligenes* under anaerobic conditions with nitrate as the only electron acceptor. They analyzed culture extracts (20 da at 30 °C) by TLC and MS and showed that MBP and PA are the only products, suggesting that the ester is hydrolyzed and the butanol moiety served as the carbon source for growth and denitrification. Dinitrogen and nitrous oxide were detected by GC. Thus, in contrast to aerobic metabolism, phthalate was not degraded further if nitrate was the only electron acceptor.

Invertebrate Metabolism. Only a few studies have been made of the metabolism of PAEs by invertebrates. One of these is the study by Wofford and his collaborators [275] which was discussed in the preceding major section. They investigated the metabolism of phthalate esters by oyster (*Crassostrea virginica*), brown shrimp (*Penaeus aztecus*), and in the vertebrate, sheepshead minnow (*Cyprinodon variegatus*). They defined a biodegradative index based on the ratio of metabolized to unmetabolized PAE isolated after the administration of radiolabelled PAE for one day. Their results are tabulated in Table 16. They found that the indices did not vary significantly with chain length or concentration but did vary significantly with species. Oysters were less efficient than fish or shrimp. The principal metabolites were the phthalate monoester and a group of unidentified polar metabolites (not present in oysters) which were probably conjugates. Concentrations were checked by GC/ECD and liquid scintillation counting. The authors note that studies of this type are complicated by the possibility of bacterial degradation and, in the case of DMP, by volatilization. They carried out microfiltration of their solutions to reduce the initial bacterial populations.

In a more biochemical vein, Hudson and coworkers [111] isolated and purified thirtyfold a PAE-hydrolyzing-enzyme from the brine shrimp, *Artemia* sp. The activity was assayed by hydrolysis of radiolabelled DBP; the product is MBP. The enzyme activity increases during larval development parallel to the increasing acute toxic action of DBP which implies that MBP rather than DBP is the actual toxin. The enzyme is not stimulated by 10 mM sodium cholate, is stable days to weeks at 4 °C (indefinitely at −70°) and is distinct from previously reported trypsin- and chymotrypsin-like enzymes. It would be interesting to determine whether or not the PAEs are the natural substrate for this enzyme.

Vertebrate Metabolism: Fish. The presence of high concentrations of PAEs in some aqueous systems has prompted a number of studies of the toxicity and metabolism of PAEs into fish. These studies are relatively easy to carry out, since aquatic organisms inhabit the pollutant solution rather than requiring injection or force-feeding. The results of Stalling, Hogan, and Johnson [222] and of Wofford and his collaborators [275] are summarized in Tables 16 and 17. Stalling, Hogan, and Johnson analyzed the metabolites found by TLC. They found much un-

Table 17. DEHP metabolites in fish [a]

Organism	Exposure (days)	Concentration (ppb)	Metabolite	Relative %
Channel catfish	1	1	DEHP	14
			MEHP	66
			PA	4
			MEHP-con.	14
			PA-con.	0.3
			unident.	2
Fathead minnows	28	2–62	DEHP	50
			MEHP	37
			PA	5
			MEHP-con.	1
			PA-con.	3
			unident.	4
	56		DEHP	60
			MEHP	29
			PA	5
			MEHP-con.	1
			PA-con.	4
			unident.	1

[a] Source: Ref. 222. The suffix "-con" indicates a conjugated derivative; "unident." indicates other, unidentified metabolites

metabolized DEHP in both catfish and fathead minnows. In addition to MEHP and PA, there were conjugates (presumably glucuronides) of MEHP and PA plus a small amount of polar material which remained at the origin during the TLC separation.

Melancon and Lech [160] described an in vitro system for performing metabolic investigations. They report that trout liver homogenates metabolized DEHP to MEHP without added NADPH and to MEHP and more polar metabolites with NADPH. The DEHP-hydrolyzing activity was associated with the microsomes and the 100,000 *g* supernatant fractions, whereas DEHP oxidation was associated only with microsomes. They also analyzed the metabolites by autoradiographic TLC; they did not identify the polar metabolites which they observed.

Sanborn and coworkers [205] investigated the metabolism of DOP in a variety of aquatic organisms. They calculated a half-life of 5 days for DOP disappearance from freshwater. In trying to establish how the process is controlled, they utilized inhibitors of various biochemical processes. They found that piperonyl butoxide (PB) and tri-o-cresyl phosphate (TOCP) both reduced the formation of MOP in mosquitoes and caterpillar larvae about 2-fold. PB is a mixed function oxygenase (MFO) inhibitor, while TOCP is an esterase inhibitor requiring MFO activation. Prior inhibition of microsomes from the mosquito fish (*Gambusia affinis*) with paraoxon had a marked effect on the rate of the hydrolysis of DOP to MOP (about a 23-fold reduction) which provides evidence of a nonoxidative pathway. In the insect studies PB reduced overall metabolic levels to a small extent, while the primary effect of TOCP appears to be an altered product distribution. These results are highly suggestive but do not present a clear picture of the process as yet.

Vertebrate Metabolism: Mammals. The biodegradation of PAEs in mammals is relevant to their toxicological action, and the common occurrence of PAEs in the human environment has meant that this topic has not been neglected. Furthermore, as we will see in the next major section, there is a special pertinence to such studies since medical practices such as kidney dialysis and blood transfusion (and less common practices such as artificial organ implantations) expose patients to much higher than normal concentrations of PAEs. Kluwe [128] has reviewed the pharacokinetics of PAEs and Thomas and coworkers [233, 234] have discussed the metabolism of DEHP and MEHP. Metabolic studies have also taken on a particular heatedness recently because of disputes over the relevance of rodent data to the prediction of human toxicity.

The great majority of these studies have been carried out in rodents. For example, Williams and Blanchfield [271] studied the retention, excretion and metabolism of radiolabelled DEHP in male rats. Virtually all of the radioactivity is excreted in the urine or feces within 48 h and no radiolabelled CO_2 is produced. The relative amounts of various metabolites and the route by which they are eliminated from the body are both dose dependent. When low levels of DEHP are given, 90%–97% of the activity is excreted in the urine with 2%–5% unchanged radiolabelled DEHP found in the feces. Metabolic products are excreted in the feces with a diet containing 2,000 ppm DEHP, but with a diet of 10 ppm DEHP all metabolic products are excreted in the urine. With the highest level of DEHP 30%–50% of the activity was in the feces, mostly as unchanged DEHP, but 5%–10% was an uniden-

Table 18. Mammalian metabolites of MEHP[a]

Designation	Alcohol moiety	Designation	Alcohol moiety
MEHP	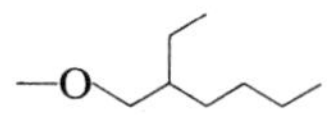	VI	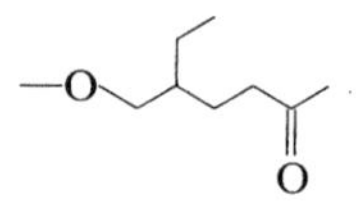
I[b]	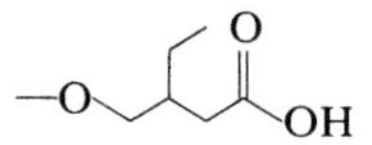	VII	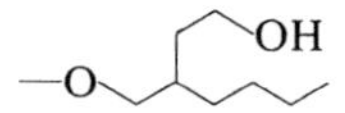
II[b]	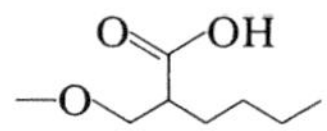	VIII	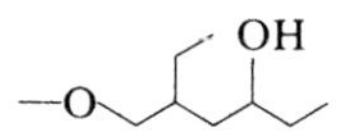
III[b]	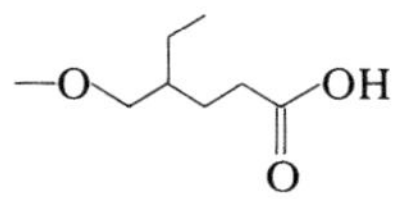	IX	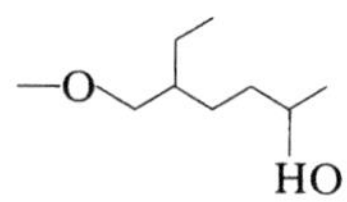
IV[b]	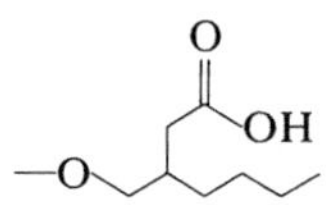	X	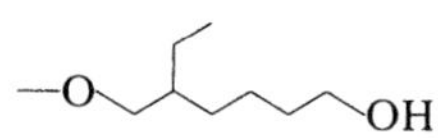
V[b]	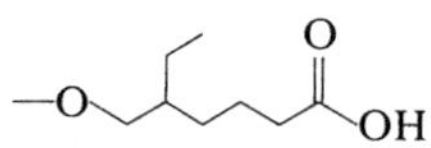		

[a] Based on information in Ref. 4. These metabolites are all based on MEHP (structure I with $R^1 = H$ and $O—R^2$ = the alcohol moiety shown with MEHP in the Table). In these compounds R^2 has been metabolically altered to the structures listed here. In vivo the free acid group is often conjugated (glucuronidated) to alter its solubility properties; see Table 20 for the extent of glucuronidation in various organisms

[b] Prior to structure determination the metabolites are esterified by treatment with diazomethane. Albro and his coworkers [4] report these carboxylic acid groups as methyl esters. There is no evidence that they are so esterified in vivo, and they have been left unsubstituted here

tified metabolite. There was no evidence for hydroxylation of the aromatic ring in rats.

Gaunt and Butterworth [74] performed an autoradiographic study of orally administered DEHP in male mice. A single dose (6.72 mg/animal) of ^{14}C-DEHP was administered and its excretion and organ distribution were monitored. The radioactivity was excreted in urine and feces mainly in the first 24 h and completely within 3–5 days. Lindgren and coworkers [147] described a similar study in pregnant and nonpregnant female mice. Pretreatment with DEHP, sodium phenobarbital or 3-methylchoanthrene caused a relative increase in the activity in brown fat.

Albro and his coworkers have carried out a vigorous research program on the metabolism of PAEs, especially DEHP, in mammals [2–4, 105, 191, 192]. They recently reviewed their results on DEHP metabolism in a number of mammalian species [2]. Their reported excretion rates are similar to those above and to those reported in Autian's 1973 review [15], although their study conditions were different from earlier studies. Eighty percent of the radioactive DEHP had been eliminated in the urine and feces of male rats after 12 days. At low doses the DEHP is almost

Table 19. Proportion of DEHP metabolites found in various mammals[a]

Metabolite[b]	Rat %	Mouse %	Man %
Unmetabolized DEHP	–	0.5	–
MEHP	tr	18.6	18.3
I	17.2	16.8	–
II	2.0	1.0	1.8
III	1.2	0.4	–
IV	3.3	0.8	5.3
V	51.3	1.1	12.1
VI	2.6	14.9	11.9
VII	2.6	7.2	8.1
VIII	–	–	36.2
IX	13.3	12.3	0.1
X	0.6	2.2	4.9
PA	1.8	12.4	0.1

[a] Based on data in Ref. 4.
[b] Structures for the various metabolites listed are shown in Table 18

Table 20. Free versus conjugated MEHP metabolites in mammals[a]

Organism	% Unconjugated metabolites
Rat	100
Hamster	85
Mouse	36
Guinea pig	35
Green monkey	20
Human	20

[a] Based on data in Ref. 4

entirely metabolized, but at doses above about 500 mg/kg rat livers began to show unmetabolized DEHP. Table 18 lists DEHP metabolites identified in these and earlier studies [2–4, 191, 192].

The studies of Albro and his coworkers have revealed a number of differences in the way in which different mammals metabolize DEHP. Table 19 gives a comparison of metabolites found in humans, rats, and mice in their investigations. In addition to major differences in the product spectrum produced in metabolism, there is another distinction between rats and the other mammals studied: rats excrete virtually all their metabolites in the free, as opposed to conjugated (mostly glucuronides), chemical from (Table 20). They hypothesize that the product spectrum in rats is quite different because of the lack of conjugation; the mildly oxidized products which dominate the primate metabolites would not be sufficiently water-soluble to be excreted easily. On the basis of these results Peck and Albro

[191] argue that a great deal of care must be taken when extrapolating results from rodent studies to either lower concentrations (at which levels unmetabolized DEHP would not be found) or to primate species (for which the metabolite spectrum is quite different).

Concluding Remarks

Biodegradation is a more important mechanism for the removal of PAEs from both natural and human environments than chemical degradation. Chemical degradation and biodegradation proceed initially by simple hydrolysis to the monoester. In mammals, however, a broad spectrum of products are produced, and some of these are found in other vertebrates. The liver is the primary organ involved in this production and fatty tissue is the only region which regularly shows any accumulation of PAEs. Microbial degradation of phthalates can be quite efficient, leading to ring hydroxylation and finally to cleavage of the aromatic system. The rates of biodegradation can be quite high, but it is notably smaller for large phthalates (DEHP, DiNP, DUP) than for smaller phthalates such as DMP, DEP, and DBP. Thus, as noted in the section on transport and distribution, sediment sinks tend to be dominated by DEHP which is both abundant and resistant to hydrolysis. Our knowledge of the normal products of environmental biodegradation of PAEs is still quite slight, although in particular systems this information is beginning to accumulate.

Toxicology and Biological Effects

An enormous literature has accumulated on the biological effects of phthalates, primarily in mammalian species. The following will attempt to summarize the principal areas of current research rather than to detail this information.

In considering the biological effects of phthalates, we should keep in mind what has *not* occurred with phthalates. They have been in use for half a century, and yet there have been no reports of major disasters due to human or environmental exposure to PAEs. In all the literature on the biological effects of phthalates, there is nothing which corresponds to the Minamato disease (mercury poisoning), the Seveso disaster (dioxin contamination), or long-term problems with exposure such as those attributed to Agent Orange or asbestos. While this record should provide some comfort, it should not be allowed to provide a false sense of security. Better planned laboratory or field experiments and better executed epidemiological studies might reveal effects which we have not yet seen. Indeed, the recent report [251, 252] by the United States National Toxicology Program that high concentrations of DEHP are carcinogenic in rodents may have significant future effects on the use and disposal of these materials.

In the following sections we will use several toxicological abbreviations and quantities which may not be familiar to all readers; definitions for these are listed in Table 21.

Table 21. Toxicological abbreviations

Measures of lethal effects:	
EC_{50}	Median concentration showing an effect; the observed effect is stated
LC_{50}	Median lethal concentration
LC_{Lo}	Lowest lethal concentration
LD_{50}	Median lethal dose
LD_{Lo}	Lowest lethal dose
TC_{Lo}	Lowest concentration with toxic effect; the observed (nonlethal) toxic effect is stated
TD_{Lo}	Lowest dose with toxic effect; the observed (nonlethal) toxic effect is stated
Organisms employed:	
hmn	Human
mus	Mouse
gpg	Guinea pig
rbt	Rabbit
Routes of compound administration:	
ihl	Inhalation
ipr	Intraperitoneal
ivn	Intravenous
orl	Oral
scu	Subcutaneous
skn	Skin

Nonmammalian Studies

Plants. Very little is known concerning the effects of PAEs on the normal growth and development of plants. We have already mentioned that aquatic plants accumulate large amounts of phthalates during microcosm studies [221], but there is no report of their toxicity under these conditions. Gledhill and his coworkers [93] reported the growth inhibiting concentration (EC_{50}) of BBP for several algal species (Table 22); most of the compounds had values in the high ppb range, although there is one striking exception in which the plants tolerated BBP at the parts per thousand level. Wilson and coworkers [272] examined the effect of PAEs on the red-tide organism, a primitive alga of the dinoflagellate group, *Gymnodinium breve*. They found that there was a steady falloff in lethality with side-chain length, so that DMP was the most toxic and DEHP was essentially nonlethal to the communities. In addition, they found that the EC_{50} for growth inhibition was about a third the LC_{50} for most of these compounds. DEHP exhibits a growth-inhibiting effect, but only at concentrations ($EC_{50} = 3.1\%$) that far exceed its water solubility.

Even fewer studies have considered the effects of these materials on land plants. Lokke [148] found that 1.5 $\mu g/cm^2$ caused chlorosis on *Sinapis alba* leaves. Overcash, Shea, Weber, and Miles [184–186, 215] reported studies of the effect of DBP on corn and soybean germination and growth. In corn they found that germination was not significantly affected when it was grown in soil containing concentrations of up to 20,000 ppm, but soybean was affected at 200 ppm. In the corn studies they observed plant height and shoot weight reductions at 2,000 ppm DBP, but not at 200. Their investigations were related to the potential use of sludge from waste treatment plants as a soil enrichment. As we have seen, such sludges usually con-

Table 22. Acute lethality of PAEs toward aquatic organisms

Organism	Compound	Exposure (h)	Toxic level (ppm)	Ref.
Algae				
Dunaliella tertiolecta[a]	BBP	96	1.0	[93]
Microcystis aeruginosa[a]	BBP	96	1,000	[93]
Navicula pelliculosa[a]	BBP	96	0.6	[93]
Red-tide dinoflagellate	DMP	96	155	[272]
(*Gymnodinium breve*)[b]	DEP	96	33	[272]
	DPP	96	3.9	[272]
	DBP	96	0.3	[272]
	DEHP	96	> 10% w/v	[272]
Red-tide dinoflagellate	DMP	96	75	[272]
(*Gymnodinium breve*)[a]	DEP	96	4.6	[272]
	DPP	96	1.8	[272]
	DBP	96	0.1	[272]
	DEHP	96	3.1%	[272]
Selenastrum capricornatum[a]	DMP	96	39.8	[244]
	DEP	96	85.6	[244]
	BBP	96	0.1	[244]
	BBP	96	0.4	[93]
Skeletonema costatum[a]	DMP	96	65.5	[244]
	DEP	96	26.1	[244]
	BBP	96	0.2	[244]
	BBP	96	0.6	[93]
Invertebrates				
Brine shrimp	DMP	96	>100	[141]
(*Palaemonetes pugio*)[b]	DBP	96	30	[141]
	DEHP	28 da	> 1.0	[141]
Crayfish (*Orconectes nais*)[b]	DBP	96	> 10.0	[156]
Mysid shrimp	DMP	96	73.7	[244]
(*Mysidopsis bahia*)[b]	DEP	96	7.6	[244]
	BBP	96	9.6	[244]
	BBP	96	0.9	[93]
Scud (*Gammarus psedolimnaeus*)[b]	DBP	96	2.1	[156]
	DEHP	96	> 32.0	[206]
Water flea (*Daphnia magna*)[b]	DMP	48	33	[144]
	DMP	48	33.0	[244]
	DEP	48	52	[144]
	DEP	48	52.1	[244]
	BBP	48	92	[144]
	BBP	48	92.3	[244]
	BBP	48	3.7	[93]
	DEHP	48	11	[144]
	DEHP	48	9.4	[244]
	DEHP	chronic[c]	< 0.003	[156]
Vertebrates				
Bluegill	DMP	96	49.5	[244]
(*Lepomis macrochirus*)[b]	DEP	96	98.2	[244]
	DBP	96	1.2	[244]
	DBP	96	0.7	[157]

Table 22 (continued)

Organism	Compound	Exposure (h)	Toxic level (ppm)	Ref.
	BBP	96	1.7	[93]
	BBP	96	43.3	[244]
	DEHP	96	>770	[244]
Channel catfish *(Ictalurus punctatus)* [b]	DBP	96	2.9	[157]
Fathead minnow *(Pimephales promelas)* [b]	BBP	96	5.3 [c]	[93]
	BBP	96	2.1 [d]	[93]
	DBP	96	1.3	[157]
Rainbow trout *(Salmo gairdnerii)* [b]	BBP	96	3.3	[93]
	DBP	96	6.5	[157]
	DEHP	chronic [e]	0.004	[159]
Sheepshead minnow *(Cyprinodon variegatus)* [b]	DMP	96	58.0	[244]
	DMP	96	21	[107]
	DEP	96	29.6	[244]
	DEP	96	22	[107]
	BBP	96	445	[244]
	BBP	96	3.0	[93]
	BBP	96	360	[107]
	DEHP	96	>550	[107]

[a] EC_{50} for growth inhibition
[b] LC_{50}
[c] 160 mg/L hardness as calcium
[d] 40 mg/L hardness as calcium carbonate
[e] LC_{Lo}

tain much lower concentrations of DBP, although DEHP and other large phthalates might be present under some conditions.

Invertebrates. Invertebrate studies have been essentially confined to aquatic species, since the principal motivation has been the influence of environmental pollutants. However, the plant studies just presented suggest that there might be significant effects on populations of soil microorganisms in areas in which sludge feeding is used.

Table 22 lists lethal concentrations for BBP, DBP, and DEHP in aquatic invertebrates. Again, the values are in the high ppb levels and there are significant variations among the different species. It is important to note that the concentrations reported here are limited by the water solubility of the various compounds. These solubility limits (for pure water) are listed in Table 2; for seawater they will be lower. Thus, for example, an LC_{50} for BBP of greater than 3 ppm, for DBP of greater than 10 ppm or of even 0.04 ppm for DEHP is not environmentally meaningful. Such concentrations mean that undissolved or micellar PAE is present in the test chamber during the experiment. Indeed, in their study of the effects of PAEs on marine dinoflagellates, Wilson and his coworkers [272] note that they could observe the undissolved phthalate, and a similar observation was made by Laughlin and collaborators [141].

These studies can be complex to interpret. A study by Medlin [158] illustrates the difficulties. She examined the joint effects of DBP (0%–100% saturated solution) and salinity (14‰–36‰) on the growth of the diatom *Skeletonema costatum* and found that the two variables interacted strongly. DBP was inhibitory to the organisms only near saturation and at low salinities; above 22‰ it had no effect at any level up to saturation.

In addition to the lethal and growth-inhibiting effects reported in Table 22, there are other important toxic effects of phthalates. Bringmann and Kühn [29, 30] have reported environmental assays for a variety of organic compounds (including PAEs) using their action on the mobility of the waterflea (*Daphnia magna*) and their effect on flagellates, ciliates and on bacteriovorous and saprozoic protozoans. Sanders, Mayer, and Walsh [156, 204] found that continuous exposure of the waterflea to 3 ppb DEHP reduced reproductive ability by 60%. This reduction was dose-related and they estimated an EC_{50} of 2.5 ppb for this effect. In a more recent study, though, Brown and Thompson [31] reported that they found no reproductive effect of DEHP and DiDcP on *Daphnia magna* at exposures in the range of 80–980 ppb and 67–697 ppb, respectively. In the latter study the number of offspring per adult was more than ten times higher *at all concentrations* than the controls used in the former study, indicating that there may have been problems with population viability in that earlier study.

Sugawara [223] reported that DMP, DEP, DBP, DHP, and DEHP all affected the hatchability of brine shrimp (*Artemia salina*) eggs in concentrations in the low ppm range. A plot of hatchability versus PAE molecular weight is paraboloid with minimum hatchability at DBP. However, this study was run at a nominal concentration of 50 ppm, which is far above the aqueous solubility of DHP and DEHP; their actual water concentrations were not measured.

Laughlin and his coworkers [141] found that there was a sizeable effect on the actual PAE concentration produced by introducing organisms into the experimental tank. In their study on the effect of DMP, DBP, and DEHP on the survival and development rate of grass shrimp larvae (*Palaemonetes pugio*) they found that the measured phthalate concentration was 9%–91% different from the nominal concentration of added phthalate. They corrected their data accordingly. DEHP had no effect on these parameters up to its solubility limit, but DMP and DBP were lethal in concentrations at the ppm level and markedly slowed development at lower concentrations. In addition, lethality and development interacted in such a way that lethality was greatest at the approximate period of the first two molts. This finding contrasted with their earlier study of the mud crab *Rhithropanopeus harrissii* [140]; they found no developmental effects of DMP and DBP even up to concentrations of 1,000 ppb.

Additional information on the nonlethal effects of phthalates, especially at the ppb level, is needed in order to determine the seriousness of these effects under actual environmental conditions. However, such studies need to be made with a great deal more concern for the effects of the solubility limits of these compounds than has been the case in the past. The effect of oily phthalate droplets on the physiology of small aquatic organisms has not been studied directly, but it is quite possible that some of the effects reported in the literature are indirect studies of exactly this point.

Fish. A number of workers have investigated the lethality of PAEs toward fish and these results are summarized in Table 22. The values of LC_{50} are generally in the low ppm range for those phthalates which have been tested and which are soluble to that level. Again, there are some studies which report values of LC_{50} which exceed the solubilities of the particular compounds of interest. Concentrations found are generally below those encountered in water samples, and the toxicity of the phthalates is generally below that of many other industrial pollutants. For example, Heitmuller and Hollister [137] reported on tests for the acute toxicity of 54 industrial chemicals to sheepshead minnows (*Cyprinodon variegatus*). They found that these compounds varied widely with 1,2,4,5-tetrachlorobenzene being most toxic (96 h $LC_{50} = 0.8$ ppm). Most compounds, including DMP, DEP, and DBP, had an LC_{50} in the 10–500 ppm range and were considered to be slightly toxic to practically nontoxic; ten chemicals (including DEHP) had no apparent effect even at the highest tested concentration. The significance of their results for BBP and DEHP is unclear, since the toxic levels are significantly higher than the solubility of these compounds in distilled water, and thus well above their solubility in seawater. The authors note that "many of the chemicals were insoluble in seawater and either floated upon the water surface or formed globules on the bottoms of the test containers." Furthermore, there is a large discrepancy between their value for BBP (>360 ppm) and the value of 3.0 ppm reported by Gledhill and coworkers [93] for the same compound; their value is in closer accord with the value reported by the EPA [243, 244] of 445 ppm.

There have been very few studies of the metabolic effects of PAEs in fish. One exception is a study on the effects of DEHP on steroid metabolism in the Atlantic cod, *Gadus morhua,* by Freeman and collaborators [66]. They administered dietary doses of 0, 10, 100, and 1,000 ppm for 121 days to both male and female cod. They found no significant difference in steroid metabolism in the testes and head kidneys of male fish maintained on the 0 and 1,000 ppm diets, but female fish showed a significant alteration of steroid (i.e. cortisol and 11-deoxycortisol) biosynthesis in both the ovaries and head kidneys. Thus, synthesis of 11-deoxycortisol in the two groups of females fed high doses of DEHP was more than twice that in the two groups fed low doses of DEHP, but there was no difference in cortisol synthesis among the four dietary DEHP groups. These data suggest that further studies in this area would be useful to elucidate the metabolism of xenobiotics in fish.

The difference in male and female metabolism is quite important, since it may point to alterations in the reproductive potential of these organisms. Mayer and Sanders [156] reported that there was indeed a direct reproductive effect of dietary DEHP on reproduction in guppies and zebra fish. Their results are presented in Table 23. The largest effect with zebra fish is the sensitivity of the fry to DEHP leading to reduced survival. In the case of guppies the principal effect is an increase in spontaneous abortions.

Mammalian Studies

Lethality. In contrast to the situation with nonmammalian species, there has been a steady stream of studies on the toxic effects of PAEs to mammals. Such studies are to be expected for such a large and industrially important group of compounds,

Table 23. Reproductive effects of DEHP in fish[a]

Species	Reproductive variable	DEHP concentration (ppm)		
		0	50	100
Zebra fish *(Brachydamio rerio)*	Spawn count	6	8	14
	Eggs/spawn	20.3	15.2	10.1
	% Fry survival	51.1	31.7	11.5
Guppies *(Poecilia reticulatus)*	Fry/female	33		29
	% Abortions	0		8

[a] Source: Ref. 156. DEHP was administered in the food for 90 days

Table 24. Mammalian lethality of phthalates and related compounds[a]

Compound	Route of administration	Animal	Toxic dose[b]
Phthalate esters			
benzyl butyl	ipr	mus	3,160
bis(2-butoxyethyl)	orl	gpg	6,000[c]
bis(2-ethylhexyl)	orl	rat	31,000
	ipr	rat	30,700
	ivn	rat	250
	orl	mus	30,000
	ipr	mus	14,000
	orl	rab	34,000
	skn	rab	25,000
	skn	gpg	10,000
bis(1-methylheptyl)	ipr	mus	14,000
butyl decyl	orl	rat	21,000
	skn	rbt	16,000
decyl hexyl	orl	rat	49,000
decyl octyl	orl	rat	45,000
diallyl	orl	rat	770
	ipr	mus	700
	orl	rbt	700[c]
	skn	rbt	2,800[c]
dibutyl	orl	rat	12,000
	ipr	rat	3,050
	ipr	mus	3,570
didecyl	skn	rbt	17,000
diethyl	ipr	rat	5,058[c]
	ipr	mus	2,749
	orl	rbt	1,000[c]
	scu	gpg	3,000[c]
dihexyl	orl	rat	30,000
	skn	rbt	20,000
diisobutyl	orl	rat	20,000
	ipr	rat	3,749
	orl	mus	13,000

Table 24 (continued)

Compound	Route of administration	Animal	Toxic dose[b]
	ipr	mus	3,990
	skn	gpg	10,000
di(methoxyethyl)	ihl	rat	1,595[e]
	orl	rat	2,750[c]
	orl	pgp	1,600
dimethyl	orl	rat	6,900
	ipr	rat	3,375
	orl	mus	7,200
	orl	rbt	4,400
	orl	gpg	2,400
	ihl	cat	10,000[d]
dinonyl	orl	rat	2,000
dioctyl	ipr	mus	65,000
methoxyethyl	ipr	mus	300
3-methyl-1-pentyn-3-yl	orl	mus	400
mono(2-ethylhexyl)	orl	rat	1,340
Related Compounds			
Phthalic acid	orl	rat	7,900
	ipr	mus	1,670
Phthalic anhydride	orl	rat	4,020
	orl	mus	2,000
	orl	gpg	100

[a] Source: Ref. 250. See Table 21 for abbreviations
[b] Value given is the LD_{50} in mg/kg unless otherwise noted
[c] LD_{Lo} in mg/kg
[d] LC_{Lo} in ppm; exposure was for 16 h
[e] TC_{Lo} in mg/m^3

and the results on lethality have been very conveniently summarized in the US National Institute for Occupational Safety and Health (NIOSH) *Registry of the Toxic Effects of Chemical Substances* (RTECS) [250]. Tables 24 and 25 summarize the RTECS data on lethal and nonlethal effects, respectively. These data have not been critically evaluated by NIOSH, and they are not fully consistent. However, they do provide a very good overview of these compounds. The lethal dose is generally measured in thousands of mg/kg, which means that a fairly large does must be imbibed (or injected) to be fatal in most animals studied.

A typical investigation of lethality is represented by the work of Timofievskaya [237]. He investigated an industrial phthalate plasticizer whose side chains were in the C6–C8 range. The PAE was administered by gavage to mice and rats; the LD_{50} values were 42,000 and 58,000 respectively. Clinical manifestations include disrupted coordination, slow movement, adynamia and death; clinical symptoms disappeared after a few hours. Inhalation of the plasticizer for 4 h was nontoxic in mice. The phthalate produced some nervous system effects during skin resorption tests, but it was a skin irritant only after repeated application. On the basis of these investigations a safe exposure level of 1 mg/m^3 was recommended.

Short-term feeding and injection studies may be misleading, though. Autian [15, 16] and his collaborators [142, 143] reported that repeated dosing has a marked effect on the observed lethality of PAEs. For example [15] an initial acute LD_{50} of 38.35 mL/kg (DEHP) declined to 1.37 mL/kg after ten weeks, suggesting that the ester was producing a cumulative response. They found in a later study [142, 143] that the magnitude of this effect was strongly dependent on the PAE used. They studied twelve PAEs and found that for all but two the acute toxicity was about 2–4 times the chronic value, but for DEHP and DBP the values were 28 and 22, respectively. They do not speculate on the cause for this decline, but it does suggest that the high values for acute lethality may not present the whole picture.

Circulatory and Hepatic Effects. Many of the published studies on sublethal effects of PAEs have focused on DEHP, the most widely used phthalate ester. Thomas and his coworkers reviewed the biological effects of DEHP in 1978 and 1982 [233, 234], and Autian reviewed general PAE toxicity in 1973 [15]. Table 25

Table 25. Mammalian toxicity (nonlethal) of phthalic acid esters[a]

Compound	Route of administration	Animal	TD_{Lo} (mg/kg)	
Teratogenicity[b]				
DAP	scu	rbt	1,000	
DBP	ipr	rat	874	
DEP	ipr	rat	1,232	
DEHP	ipr	rat	30,000	
	orl	mus	7,500	
DiBP	ipr	rat	1,125	
DMEP	ipr	rat	1,122	
DMP	ipr	rat	1,014	
DOP	ipr	rat	15,000	
Other toxic effects				
DBP	orl	hmn	140	eye effects
DEP	ihl	hmn	1,000	irritation[c]
	eye	rbt	112[d]	irritation
DEHP	orl	man	143	gastrointestinal
	skn	rbt	500[d]	mild irritation
	eye	rbt	500[d]	irritation
DHP	eye	rbt	500[d]	irritation
DMP	eye	rbt	119[d]	irritation
DOP	skn	rbt	500[d]	mild irritation
	eye	rbt	5[d]	severe irritation
DTDP	skn	rbt	10[d]	mild irritation
DUP	eye	rbt	100[d]	mild irritation

[a] Source: Ref. 250. See Table 21 for abbreviations
[b] The compound was administered on days 5–15 of gestation in pregnant rats or on day 8 in pregnant mice
[c] TC_{Lo} in mg/m^3
[d] Total dose administered in mg

summarizes studies in the NOISH Registry. The conclusions of the 1978 Thomas review [233] are worth quoting:

> "DEHP is very poorly soluble in water, disappears rapidly from the blood, undergoes some degree of biotransformation, and is excreted by the kidney quite rapidly. There is little evidence of any substantial accumulation of DEHP in tissues."
>
> "While low levels of DEHP have been detected in the ecosystem, there are no reports of it adversely affecting wildlife. Animal experiments using high doses of DEHP reveal some deleterious effects upon the reproductive system, but there is little or no evidence that suggests that they are carcinogenic. Some transient growth inhibitory actions caused by DEHP have been noted in certain cell systems in vitro, but the biological significance of such findings remains to be elucidated. Only further investigations will lead to finally establishing the complete toxicological profile of DEHP."

These conclusions are those which might have been reported by any careful reviewer in the late 1970s, and much of what Thomas and his colleagues said would still hold firm. For example, their conclusions about the metabolism and distribution of phthalates are still generally accepted. In 1982 Kluwe reviewed the pharmacokinetics of PAEs in mammalian species [128]. He concluded that PAEs are generally well-absorbed from the gastrointestinal tract following oral administration. Hydrolysis to the corresponding monoester, with release of an alcoholic substituent, largely occurs prior to intestinal adsorption of longer-chain alkyl derivatives such as DEHP. PAEs are widely distributed in the body, with the liver being the major initial repository organ. Clearance from the body is rapid and there is only a slight cumulative potential. PAEs with short side chains, such as DMP, can be excreted in an unchanged form or following complete hydrolysis to PA, but long sidechain compounds such as DEHP are primarily converted to polar derivatives of the monoesters by oxidative metabolism prior to excretion.

A principal motivation for early studies was concern for patients receiving blood donations. Early papers published in the field include those by Calley, Autian, and Guess [37], by Nematollahi, Guess, and Autian [179], by Guess, Jacob, and Autian [100], by Jaeger and Rubin [117, 118], by Marcel [150], and by Marcel and Noel [151, 152]. These and subsequent studies definitely established that PAE plasticizers were leached from PVC blood bags by stored blood or plasma and that this material was transmitted to the patient. However, they also found that DEHP was rapidly cleared from the patient's plasma and that it was excreted without significant longterm storage or apparent toxic effect.

In 1982 Kevy and Jacobson [125] reexamined the hepatic effects of DEHP leached from PVC blood bags, using primates (rhesus monkeys). The average cumulative amount infused in one year was comparable to that received by human patients undergoing transfusion. They found histological abnormalities and abnormalities in the hepatic scan and in the clearance kinetics of injected dyes. These effects persisted for up to 26 months after transfusion. They note that patients undergoing maintenance hemodialysis receive a yearly dose of DEHP which is 10–20 times that which produced hepatotoxicity in the transfused rhesus. Thus, there is still cause for serious concern that safer materials be found to replace these plasticizers for medical use, although none has yet been found.

In the course of these investigations, a reasonably consistent picture of the site of action of most PAEs in the body has emerged. Susceptible sites are lipids and lipid metabolism, the liver, the testes, and to a lesser extent the kidneys.

A number of investigators have found that PAEs have a marked but selective action on the lipid content of rats. There is a clear consensus from these studies that DEHP fed to rats leads to a reduction in serum triglycerides without an increase in serum cholesterol [8, 167]. 2-Ethylhexanol, which would be released when DEHP is hydrolyzed, does produce a decrease in the amount of serum cholesterol [167]. The monoesters act differently; Oishi and Hiraga [182] found that male rats administered dietary MBP, MiBP or MEHP for one week showed an increase in nonesterified fatty acid and a decrease in both triglycerides and total cholesterol. These metabolite studies would seem to indicate that both the monoester and the alcohol moiety are rapidly metabolized following hydrolysis of the dialkyl ester. Both Moody and Reddy [167] and Oishi and Hiraga [182] report that the change in serum lipid concentration is also accompanied by an apparent change in hepatic peroxisome activity and in lipoperoxide concentrations.

Bell and his coworkers reported a number of investigations on the effects of PAEs on lipid metabolism. In one of his most recent papers [20] he reported on the effects of dietary DEHP (1,000 to 10,000 ppm) on lipid metabolism in various tissues, cells and organelles in rats, rabbits, and pigs. He noted a variety of effects including the inhibition of cholesterologenesis in liver, testes and adrenal glands; the inhibition of cholesterol production in the brain and liver of fetal rats from DEHP-fed dams; decreased plasma cholesterol; decreased hepatic phospholipid and triglyceride synthesis, increased fatty acid oxidation in isolated liver mitochondria and a transient decrease in fatty acid oxidation in isolated heart mitochondria. In vitro studies revealed that DEHP inhibited cholesterologenesis in rat liver and inhibited adenine nucleotide translocase in rat heart mitochondria.

Many of the PAE effects described above [20, 125] are related to liver function, and this is a commonly observed principal effect of these compounds. However, the actual hepatic effect depends on the phthalate administered. For example, Lake and coworkers [139] reported on the effects of DCHP at doses of 500–2,500 mg/kg/da for 7 days to young male rats. The animals showed liver enlargement and induction of some parameters of hepatic xenobiotic metabolism. However, the observed induction resembled sodium phenobarbitone rather than that observed with polynuclear aromatic hydrocarbons. DCHP metabolites (MCHP and cyclohexanol) were also inducers. They concluded that DCHP is a weak drug-type inducer of hepatic xenobiotic metabolism in rats and it has hepatic effects different from those observed with DEHP.

A variety of enzymes and metabolic pathways are affected by PAEs. Recently Melnick and Schiller [161] reported on investigations of the effects of MBP, DBP, MEHP, and DEHP on energy-dependent potassium ion uptake, respiration rates and succinate-cytochrome c reductase activities of isolated rat liver mitochondria. The energy-coupling processes, active potassium ion transport, and oxidative phosphorylation were most affected by DBP and MEHP, while MBP had a moderate effect on energy coupling and DEHP had no apparent effect. The potency of succinate-cytochrome c reductase activity inhibition was MEHP > DBP > MBP $\simeq$ DEHP. They concluded that phthalate esters affect mitochondrial activities by altering the permeability properties of the inner membrane and by inhibiting succinate dehydrogenase activity.

Shibata and coworkers [216] reported other pathways which were affected at high doses of PAEs. They found that weanling male Wistar rats given a starch-sugar-amino acid diet which was free of nicotinic acid and low in tryptophan had weight gains of 60 g after 33 days, but the addition of 1% DBP to the diet increased these gains to 78 g. The liver weights were 5.5 and 7.2 g, respectively. The DBP diet also increased liver ascorbic acid, *N*-methylnicotinamide, nicotinic acid, quinolinic acid, NAD, and quinolate phosphoribosyltransferase, but nicotinamide methyltransferase was decreased.

Many parameters of liver function have come under close scrutiny. For example, Agarwal, Agarwal, and Seth [1] studied the interaction of DEHP with the pharmacological response and metabolic aspects of ethanol in mice. They administered oral (1.5–7.5 g/kg) or intraperitoneal (3.7–19.9 g/kg) doses of DEHP once or twice daily for seven days. They report that a single administration resulted in a significant increase in the ethanol-induced sleeping time associated with an inhibition of alcohol dehydrogenase activity in the liver. Repeated oral doses of DEHP produced significant reductions in the ethanol-induced sleeping time and increases in the activities of alcohol and aldehyde dehydrogenases, whereas repeated intraperitoneal doses of DEHP significantly increased the sleeping time and decreased alcohol dehydrogenase, without any perceptible effect on the aldehyde dehydrogenases. In all cases in vitro enzyme inhibition by MEHP was very pronounced.

Seth [214] also found that DEHP produces changes in liver weight and in morphological and biochemical parameters, including xenobiotic response. It alters the activity of microsomal drug-metabolizing enzymes and ethanol metabolism. Oral and intraperitoneal administrations produce a time- and route-dependent effect on the hepatic cytochrome c content and the activity of aminopyrine *N*-demethylase, aniline hydroxylase, alcohol dehydrogenase and high and low K_m aldehyde dehydrogenase. Under in vitro conditions, DEHP has no effect on aminopyrine *N*-demethylase or aniline hydroxylase, but MEHP and 2-ethylhexanol significantly inhibits their activity at concentrations ranging from 2.5 to 15.0 m*M*. Oral administration of DMP and DBP also inhibits these last two enzymes in vivo.

In another study of xenobiotic metabolism, Walseth, Toftgaard, and Nilsen [260] considered the effects of DMP, DBP, and DEHP on cytochrome P-450 mediated metabolism in rat liver and lung, serum enzymic activities and serum protein levels. Intraperitoneal doses (3.8 m*M*/kg) were administered to rats for five days. Only minor effects were seen after treatment with DEHP or DMP, but DBP produced significant changes. It increased liver cytochrome P-450 but decreased lung concentrations by about 40%. It also decreased the lung concentration of cytochrome b_5 and NADPH-dependent cytochrome c reductase activity by about 30%. They found that the direction of benzo[*a*]pyrene metabolism was changed and the formation of 2- and 3-hexanol metabolites from hexane were increased in liver microsomes after DBP treatment. All PAEs decreased the lung metabolism of benzo[*a*]pyrene. The cytochrome P-450 enzyme system in the lung was 10 times more effective than that in the liver for the metabolism of hexane. DBP produced only minor effects in serum enzyme activities but a significant decrease in serum albumin level.

Warren, Lalwani, and Reddy [262] have discussed the activity of DEHP as a rodent hepatic peroxisome proliferator. This property is not shared by all mammalian species and, as we shall see, they consider this crucial for understanding the carcinogenicity of DEHP. As supporting evidence for the peroxisome proliferator hypothesis they cite data from their earlier work [167, 199] including increased fatty acid β-oxidation, elevated H_2O_2 levels, accumulation of peroxidized lipofuscin, a disproportionately small increase in catalase, and elevated peroxisomal uricase activity which accompany peroxisome proliferation in hepatocytes. They also state that peroxisome proliferators induce hepatomegaly and hypotriglyceridemia in rodents.

Furuta, Miyazama, and Hashimoto [70] have also investigated the liver peroxisome effects of DEHP. They stimulated the biosynthesis of enzymes of peroxisomal β-oxidation by feeding male Wistar rats a diet containing DEHP for two weeks. They found a 31–33 fold increase in individual enzymes of the hepatic peroxisomal β-oxidation system. In vitro experiments showed that the rate of synthesis was enhanced by a factor of 16–20 while the rate of degradation decreased 1.7- to 1.9-fold. The increase in enzyme is primarily due to an increased rate of synthesis produced by an increased amount of mRNA coding for these enzymes.

Thus, there is a complex suite of hepatic effects of DEHP and other PAEs in rats, but the underlying biochemistry and physiology are poorly understood, and the effects are ordinarily observed only under very high dosages. Using high dosages is itself a problem, since these compounds are extraordinarily insoluble in aqueous solution. Thomas [233] discussed the problem introduced by the carriers which are used for injection of these compounds, and Olsen and coinvestigators [183] recently reported a practical example of this problem. They examined the kidney toxicity of plasticizers using 48 h hypothermal perfusion of dog kidneys as an in vitro model. They found that there was considerable change in lactate dehydrogenase activity *with or without* added PAE. Polysorbate 80 was used to solubilize the plasticizers for perfusion, and they concluded that the plasticizers had only a slight neurotoxicity additional to that of Polysorbate 80.

Reproductive Effects. In addition to the hepatic effects, much effort has gone into determining the reproductive effects of these compounds. Three types of effects have been studied in some depth: testicular alterations, spontaneous abortions and other antifertility effects, and teratogenicity.

A number of workers have found that there is distinct testicular atrophy induced in rodents by administering high doses of PAEs. In addition, this atrophy is usually accompanied by testicular zinc depletion. Foster and coworkers [64] reported on the structure-activity requirements for the induction of testicular atrophy and zinc content depletion using isomeric monobutyl phthalates. Phthalate monoesters were administered to immature rats for 6 days; the *tert*-butyl ester was inactive but the other isomers produced a significant decrease in testicular mass and a marked atrophy of the majority of the seminiferous tubules with a diminution in spermatocytes and spermatogonia. No abnormalities were detected in sections of prostate or seminal vesicles. Diesters of iso- and terephthalates did not affect testicular mass. The active compounds altered the interaction of zinc with the testes and kidneys. The authors concluded that there is an important relationship between the structure of the butyl side chain of the phthalate monoesters, the position of the ester side chain on the ring, and the production of testicular atrophy.

The effect is not due to the monoesters only, though. Thomas, Curto, and Thomas [235] found that both DEHP and MEHP adversely affect rodent testes at high doses. Rat gonadal zinc levels can be decreased by DEHP but not MEHP, and the prostate gland is especially sensitive to PAE-induced zinc depletion. PAE-induced gonadal changes were more evident in rats than in mice. They found that MEHP alters the uptake of ^{65}Zn in rat gonads and accessory sex organs.

Other PAEs are also effective. In addition to the butyl ester studies mentioned, Lake and coworkers [139] found that while DCHP did not affect renal and testicular mass, histological evidence of testicular damage was obtained at the highest dose. However, MCHP produced marked testicular atrophy. Gangolli [73] reported that the testicular effects observed for DEHP, namely, decrease in organ weight and histological changes in the seminiferous tubules, can also be produced by DBP, DPeP, and DHP. The corresponding monoesters (formed in vivo as a result of nonspecific esterases in the intestinal mucosa and other tissues) were equally effective. The phthalate-induced testicular injury was accompanied by a decrease in the zinc content of the gonads and by increased urinary excretion of zinc. He reported that in vitro exposure of rat seminiferous tubule cells to monophthalates resulted in a dose-related detachment of germinal cells from Sertoli cells similar to the in vivo results. Cater and collaborators [42] found that the testicular effect of DBP was apparently due to the MBP formed in vivo, and they also found that co-administration of zinc reduced the testicular damage. The damage was found only at relatively high doses (2,000 mg/kg to young male rats for four days).

The complexity of this subject is very well brought out by the results of Gray and collaborators [98] on the species differences observed in the testicular toxicity of phthalate esters. Oral administration of DBP produced severe seminiferous tubular atrophy in rats and guinea pigs, but only focal atrophy in mice. Hamsters showed no testicular changes with DBP and only minor changes with DEHP and DPeP. The rate of intestinal monohydrolysis of DEHP was significantly lower in hamsters than in rats. MEHP did produce focal seminiferous tubular atrophy in hamsters, but MBP did not. The decrease in testicular zinc concentration and the increase in urinary zinc excretion found in rats with DEHP and DBP was not observed in hamsters.

The changes produced are not simply histological. Bainova, Antov, and Ilieva [18] reported on the effect on rat testes of skin doses of BBP and a second industrial plasticizer at 0.5–4 mL/kg. These levels did not cause any damage to the testes of albino Wistar rats, but the highest dose caused deviations in sperm mobility and increased the activity of succinyl dehydrogenase, glucose 6-phosphate dehydrogenase, lactate dehydrogenase, ATPase, and alkaline phosphatase. Thus, the testicular effects of PAE mono- and diesters extend even to the level of molecular biology.

The quotation from Thomas' 1978 review [233] is indicative of the concern for the effects of PAEs on the reproductive system. This concern had earlier been raised in Autian's 1973 review [15] of the toxicology of phthalates when he reported that, on the basis of early embryotoxicity studies [51, 220], all six PAEs tested have a low but measurable toxicity as determined by reduction in fetal weight and by gross and skeletal abnormalities, and DMEP is the most embryotoxic ester studied

[220]. More recently he [16] has reported further on the antifertility effects and dominant lethal assays for mutagenic effects of cumulatively administered DEHP. The preliminary results indicated that antifertility effects occurred with as little as three subcutaneous doses of 1 mL/kg each. Thus, the sensitization arising from repeat injections was found here as well as in the lethality studies.

Parkhie, Webb, and Norcross [187] have recently reinforced Autian's earlier comments about the embryopathy and teratogenicity of DMEP. They administered a single intraperitoneal injection (0.6 mL/kg) of DMEP to pregnant Wistar rats on day 10, 11, 12, 13 or 14 of gestation. Phthalate-treated rats showed a high incidence (12%–79%) of fetal deaths and resorptions and a significant reduction in fetal weight. Other effects observed were hydrocephalus interna, a congenital malformation of the brain, and congenital skeletal malformations. These effects were not observed in the saline-injected control population. They found that DMEP is rapidly transferred across the placenta without prior degradation and that there is a significant decrease in the zinc content of the fetus following its administration.

The teratogenicity and fetotoxicity of DEHP were investigated by Tomita and collaborators [239]. They administered DEHP orally to pregnant mice in dosages of 0.05–30.0 mL/kg on day 6, 7, 8, 9 or 10 of gestation. A single administration of over 100 µL/kg on day 7 of gestation decreased the numbers and the body weight of living fetus, whereas no significant changes in the numbers of living fetus with no gross and skeletal abnormalities were observed compared with those of the control group when 50 µL/kg of DEHP was administered. The observed fetotoxicity was dose-dependent. The fetal LD_{50} and maximum nonteratogenic dosage of DEHP in its single, oral administration were both determined to be about 600 mg/kg. The maximum nonfetolethal dosage was 64 mg/kg.

Shiota and Nishimura [217] also reported on their recent investigations of the teratogenicity of DEHP and DBP. DEHP and DBP were mixed with the diet at 0.05, 0.1, 0.2, 0,4, and 1.0 wt% and fed to pregnant ICR mice throughout gestation. Effects observed were suppression of maternal weight gain and fetal resorptions; both of these parameters increased at 0.2, 0.4, and 1.0% DEHP and 1.0% DBP. All implanted ova died early in rats fed 0.4 and 1.0% DEHP. External malformations were increased significantly by 0.2% DEHP, and 1.0% DBP showed borderline significance. The major malformations were neural tube defects, suggesting that the PAEs affect neural tube closure in developing embryos. These results indicate that a high dose of DEHP and DBP might be embryotoxic and teratogenic in mice; the maximum nonembryotoxic doses of PAEs in mice were more than 2,000 times the estimated level of human intake through the food chain (which Tomita and coworkers [238] estimated at approximately 0.03 mg/kg/da DEHP consumed from commercial foodstuffs). The authors suggest that the current "normal" exposure level of PAEs does not pose an imminent threat to human fetal development. They do not, however, consider the question of sensitization arising from chronic exposure.

Few studies on the maternal effects other than fetotoxicity have been published. One exception is the recent study on the reproductive effects of some environmentally persistent organic substances by Bjellin, Maartensen, and Widmark [22]. They investigated the effects of these substances on blood flow in pregnant guinea pigs.

DBP and DEHP were among the compounds studied and neither produced a significant change in microflow; however some embryos were more affected than others.

At present there is no firm evidence – apart from patients receiving chronic blood or plasma transfusions and possibly some industrial workers – that the toxic levels found in these investigations of the hepatic and reproductive effects are reached by normal exposures. However, the apparent sensitization of both lethal and reproductive effects [15, 16] is of some concern, and investigations are continuing.

Mutagenicity and Carcinogenicity. The one point in the picture of the biological effects of DEHP which has changed radically since the 1978 Thomas review quoted above is that it can no longer be said that "there is little or no evidence that suggests that they are carcinogenic." The status of PAE carcinogenicity testing prior to 1982 was briefly reviewed by Wilbourn and Montesano [269], and they found that the mammalian studies reported to that time were inconclusive because of a variety of experimental weaknesses which obscured the significance of the results. However, in 1981 that situation was radically altered when the US National Toxicology Program [251, 252] reported their results on the carcinogenicity of phthalates and related compounds. The study involved feeding rats and mice DEHP at the maximally tolerated doses and half maximally tolerated doses over a period of 103 weeks. All animals that died during the study and all survivors at the end of 2 yr were examined grossly and microscopically for the presence of tumors [129]. They concluded that under the conditions of this bioassay DEHP was carcinogenic for F344 rats and B6C3F1 mice, causing an increased incidence of hepatocellular carcinomas in female rats and male and female mice, and an increased incidence of hepatocellular carcinomas and neoplastic nodules in male rats.

Later that year, in discussions held in conjunction with a conference on the carcinogenicity of phthalates and related compounds, the NTP discussants reported [252] that phthalamide has a low acute toxicity and up to a 3% dose (w/w) in rats and mice gave no evidence of carcinogenicity. Phthalic anhydride also has a low acute toxicity, and testing with doses up to 5% in rats showed little evidence of carcinogenicity. On the other hand, there is strong evidence for the carcinogenicity of DEHP based on the dose-related excess of liver tumors in both sexes of rats and mice. For BBP there is possible leukemia induction in female rats and definite testicular and thymic atrophy. For bis(2-ethylhexyl) adipate (another plasticizer and possible substitute), the discussants concluded that further evaluation is required because there were a large number of early deaths in males administered low doses. On the basis of the portion of the adipate study completed there is an apparent excess of liver tumors.

Thus, the apparent low toxicity of PAEs has now been called into question on the basis of the apparent carcinogenicity of the larger phthalates. There has been, then, a spate of activity in this area by governmental, industrial, and academic scientists to determine the full extent of this problem.

Currently a large amount of "carcinogen" screening is actually carried out as "mutagen" screening using the Ames *Salmonella typhimurium* or related assays. For example, Blevins and Taylor [23] reported on the mutagenicity screening of

Table 26. Phthalate esters tested for mutagenicity by the Ames/*Salmonella* assay[a]

Compound	Solvent	Dose range (μg/plate)	
		min	max
BBP	DMSO	100	11,550
DAP	DMSO	1	10,000
DBP	DMSO	100	10,000
DEHP	DMSO	100	10,000
DEP	DMSO	10	10,000
DiBP	95% EtOH	100	10,000
DiDcP	95% EtOH	100	10,000
DMP	DMSO	33	6,666
DOP	DMSO	100	10,000
DTDcP	Acetone	100	10,000
2-Ethylhexanol	DMSO	3	333
Phthalamide	DMSO	33	1,500
Phthalic anhydride	DMSO	1	666

[a] Source: Ref. 281. All assays were negative. DMSO is dimethylsulfoxide; EtOH is ethanol

twenty-five cosmetic ingredients (including phthalates) with the *Salmonella*/microsome test. Of the compounds tested, only DEA Laureth Sulfate gave positive mutagenic results in the plate incorporation test. Choudhury [44] investigated the mutagenicity of diesel emission particulates by the Ames/*Salmonella* assay and found phthalates in an active fraction which, however, also contained several polynuclear aromatic hydrocarbons. A similar screen was reported by Parry and collaborators [188] who monitored substances in marine waters for genetic activity. They found that tissue extracts of the mussel *M. edulis* from 6 of 8 sites sampled in Great Britain showed the presence of a mutagenically active chemical in screenings with *S. typhimurium* and a variety of *Escherichia coli* strains. They provisionally identified the mutagen as DEHP on the basis of GC, IR, and NMR data.

Most reported mutagenicity tests involve pure compounds, and there have been conflicting results. Zeiger and his collaborators [281] summarized the large series of tests run in the NTP's Environmental Mutagenesis Test Development Program. The compounds tested are listed in Table 26. All of these chemicals were judged to be nonmutagenic in the *his*$^+$ reversion assay. They note that DEP has been found to be mutagenic by other authors, but that DEHP is agreed to be nonmutagenic.

However, Tomita and coworkers [240] reported that DEHP and MEHP show mutagenic activity in *Salmonella* TA-100 assay with and without S-9 rat liver microsomal activation, respectively, and they also reported extensively on other mutagenicity assays. Thus, MEHP, but not DEHP, exerted a dose-dependent DNA damaging effect on *Bacillus subtilis* in Rec-assay. MEHP not only produced mutations in *E. coli* WP2 B/r try$^-$ but also caused sister chromatid exchange in Chinese hamster V79 cells. It also induced 8AG/6TG-resistent gene mutations and chromosomal aberrations in V79 cells. Transplacental administration of DEHP or

MEHP to Syrian golden hamster embryos was carried out by dosing gravid animals with DEHP or MEHP on day 11 of gestation, followed by the cultivation of embryonic cells for 15–20 days. Both DEHP and MEHP induced 8AG/6TG resistant mutation, chromosomal aberration and morphological transformation in the embryonic cells of the Syrian golden hamster.

Most mutagen assays have not found activity from DEHP, although the same cannot be said for MEHP. Phillips, James, and Gangolli [196] investigated the clastogenic activity of DEHP, MEHP, PA, and 2-ethylhexanol in cultured Chinese hamster ovary (CHO) cells and found that only MEHP caused chromosome damage. However, MEHP was without effect in the sister chromatid exchange and hypoxanthine-guanine phosphoribosyl transferase mutation test in CHO cells. These results conflict somewhat with those of Tomita and his coworkers [240].

Kozumbo, Kroll, and Rubin [134] assessed the mutagenicity of phthalate esters using the Ames test. DMP and DEP produced a positive dose-related mutagenic response with *Salmonella* TA100, but only in the *absence* of activation by S-9 liver enzymes. DBP, DEHP, MEHP, and BBP, as well as the dimethyl isophthalates and terephthalates and trimethyl mellitate, were not mutagenic with TA100 or TA98 in the presence or absence of S-9 enzymes. Extracts of 24-h urine of rats injected intraperitoneally with DMP were not mutagenic to TA100 at levels up to 8 equivalent-mL of urine per plate. In vitro studies revealed that the S-9 associated esterase hydrolyzed DMP to MMP and eliminated its mutagenicity. They made the interesting observation that skin cancer, rather than other forms, could be a hazard in rats because of the absence of esterase activity.

Seed [212] reported on *Salmonella* TA100 tests for the mutagenic activities of eight PAEs and related compounds in the 8-azaguanine resistance assay. DMP, DEP, DBP, and 2-ethylhexanol were found to be mutagenic. Their mutagenicities were dose-dependent but weak; no dose-response curve exceeded 3.5X background. DMP was also tested by histidine reversion and showed a similar level of mutagenicity. DEHP, DOP, DAP, DiBP, and DiDcP were not mutagenic.

Thus, the results reported recently have disclosed a very confused picture of the basis for the observed carcinogenicity. Indeed, the unusual metabolism of DEHP in rodents (see above) and their liver activity has led to controversy over the applicability of these rodent studies to the assessment of DEHP carcinogenicity. Warren, Lalwani, and Reddy [262] cite the fact that DEHP is both a peroxisome proliferator and a hepatic carcinogen. They argue that peroxisome proliferators *as a class* are hepatocarcinogenic in rodents, but they are not mutagenic or DNA-damaging in the *Salmonella*/microsomal or lymphocyte ^{3}H-thymidine assays. They propose that peroxisome proliferation itself initiates neoplastic transformation of hepatic parenchymal cells by increasing intracellular rates of DNA-damaging reactive oxygen production. Although this argument has not been subjected to extensive experimental verification, it has considerable force and illustrates the scientific proplems involved.

The full import of the NTP findings has yet to be determined. The doses involved are quite high and the arguments raised about their applicability to humans are serious. Both the National Toxicology Program and the Chemical Manufacturer's Association (CMA) are heavily involved in following up the NTP results. At the end of the conference on phthalates (published as volume 45 of *Environmen-*

tal Health Perspectives) [252], the discussants summarized the status of their current research as follows:

"The testing program of the CMA is designed to indicate whether additional phthalate esters need to undergo carcinogenicity testing in animals, while that of the NTP is directed more toward determining the mechanism of DEHP carcinogenicity and to evaluating the other toxic effects of phthalate esters... To date there is no concensus [sic] as to whether or not DEHP is active (positive) in short term tests... Second, chemicals can cause cancer via nongenetic mechanisms, a likely one for DEHP (based on biological effects) is promotion... There is no single standard test for liver tumor promotors nor is there any definitive indication of the mechanism of tumor promotion in the experimental systems currently available... it is clear from the data presented at this conference and the resulting discussion that the evaluation of chemical safety is a very complex issue... The emphasis at this conference has clearly been on the carcinogenic potential of phthalate esters. Other toxic effects, such as teratogenicity and infertility, however, should not be ignored in assessing the risk of human exposures to plasticizers."

American Regulatory Aspects

It is fair to say that the NTP report on the carcinogenicity of DEHP and BBP has completely altered the environment within which decisions about PAEs will be made in the future. However, it is not the case that these compounds were not regulated in the past.

The Environmental Protection Agency, as part of its responsibilities under the Clean Water Act, has listed 129 priority pollutants, including six phthalates [123, 245, 247]. These six are DMP, DEP, DBP, DEHP, DOP, and BBP. Furthermore, they have established ambient water quality criteria for four of these six [244] as listed in Table 27.

The EPA also has responsibility for enforcement of the Toxic Substances Act (TOSCA). In 1980 six phthalate esters of undisclosed structure became the first compounds barred from manufacture by the EPA under TOSCA [7]. The EPA informed the manufacturer that the premanufacture notice did not contain enough information about the toxicity and environmental fate of these compounds, and the company decided to withdraw them rather than to carry out further testing. The article stated that according to the EPA assistant administrator there was "no

Table 27. Regulation of phthalates and related compounds[a]

Compound	Regulatory agency	Medium standard
Air standards[a]		
phthalic anhydride	OSHA	2 ppm
dimethyl phthalate	OSHA	5 mg/m^3
dibutyl phthalate	OSHA	5 mg/m^3
bis(2-ethylhexyl) phthalate	OSHA	5 mg/m^3
Water standards[b]		
dimethyl phthalate	EPA	160 mg/L
diethyl phthalate	EPA	60 mg/L
dibutyl phthalate	EPA	5 mg/L
bis(2-ethylhexyl) phthalate	EPA	10 mg/L

[a] Source: Ref. 250. Time-weighted average concentrations
[b] Source: Ref. 244

information that the new compounds are any worse than the phthalate ester plasticizers already in use... 'Congress clearly didn't tie our hands just because similar chemicals are already being marketed.'" The administrator also stated that "EPA is also concerned with the possible hazards of the phthalate esters in use and eventually may call for further testing of these compounds as well."

The US Occupational Safety and Health Administration (OSHA) is charged with setting safe working limits for a number of environmental contaminants, and it has promulgated air standards for the working day for four phthalates (DMP, DEP, DBP, and DEHP) and phthalic acid. NIOSH has published standard analytical methods for two of these compounds (DBP and DEHP) [230]. These air quality standards were primarily established because of short-term, nonlethal effects of these compounds, including irritation of the nose, upper respiratory tract, and stomach and possible sensitization [198]. Recently reported OSHA health hazard evaluations employing these standards include a Ford Motor company plant in California which was checked for DBP following worker respiratory complaints (<0.1 mg/m^3 was found in all areas checked) [198] and a supermarket chain meat-wrapping plant which was checked for DCHP following similar complaints. In this latter case it was found that the heating of the PVC meat wrap did release DCHP and concentrations of up to 0.42 mg/m^3 DCHP were found, all associated with atmospheric particulates [65]. Respirators for use in phthalate-contaminated atmospheres have been described [130].

There is no doubt that future certification of phthalates will deal directly with the carcinogenicity issue. The US Consumer Products Safety Commission (CPSC) is currently carrying out a study of the safety of phthalate plasticizers in consumer products, especially children's items such as toys, pacifiers, and crib mattresses [242]. The exposure and hazard represented by these compounds is being studied as a basis for possible regulatory action; only a preliminary report has been filed to date [242]. A news report of this action [102] prompted a response by a CMA spokesman about the details of the report [163], indicating the complexity of the scientific, regulatory and economic issues involved in this area.

Recently, the NIH and EPA have moved to speed up information transfer in this complex area by establishing a joint clearinghouse:
International Clearinghouse on Phthalate Esters
National Toxicology Program
Landow Building
National Institutes of Health
Bethesda, MD 20205

Summary

Phthalic acid diesters are an immensely important class of industrial organic compounds. They have chemical and physical properties – low volatility, high lipophilicity, chemical stability and low color – which make them useful in providing flexible materials for which there is a steady demand. Although there are some substitutes, none have proven to be so widely usable or so innocuous. Barring major shifts in plastics technology or new information concerning their toxicological or environmental hazard, phthalates will continue to be produced in large quantities for decades to come.

The analytical chemistry of environmental phthalates has been pushed to the parts per trillion (ng/L) level in water, sediment and biota and to the ng/m^3 level in air. Gas chromatography, high performance liquid chromatography and combined gas chromatography/mass spectrometry are all in common use in these studies. The most important continuing concern in the area is the ubiquity of phthalates in the laboratory and environment. Contamination of samples is a common problem, so that the consistent production of reliable analyses requires constant attention to the materials and methods used in the sampling, isolation, fractionation, and characterization of the compounds. A second problem which arises in some samples is the problem of adequately fractionating the sample to separate phthalates from other high molecular weight, polar organic compounds such as the polychlorinated biphenyls. Solutions to these problems have been found, and an increasing number of laboratories are able to carry out this difficult analysis at trace and ultratrace levels.

Phthalates are everywhere. A few analyses show that particular samples have a phthalate level below some quite low detection limit, but the great majority of samples – air, water, sediment and biota – show significant concentrations of one or more of these compounds. The principal source of these compounds is human activity, although there is evidence for minor amounts of natural phthalate production. Individual plasticized materials may contain up to 67% plasticizers and, despite their low solubility and volatility, these materials are slowly released to the environment. Transport of these materials is global, and the heavier ones such as bis(2-ethylhexyl) phthalate (DEHP), are persistent on a time scale of months in water and even longer in sediments. Thus, sediments in both Europe and the United States show a growth in phthalate concentrations which is parallel to their production growth over the past half century.

The principal mechanism for phthalate disappearance is biodegradation, especially in microorganisms, but also in higher organisms. Chemical degradation rates (photodegradation, abiotic hydrolysis, oxidation) are slow in natural systems, although they can be accelerated by ozonation or UV photolysis in treatment plants. Virtually all organisms studied are capable of accelerating the hydrolysis of phthalate diesters to monoesters and even to phthalic acid. In addition, some microorganisms are able to metabolize the aromatic ring portion of the molecule, and mammalian species (and probably other organisms) can oxidize the alcohol moiety of the monoester metabolite. The toxicological significance of these metabolites is of current concern.

The acute toxicity of phthalates is quite low; only a diet which largely consists of phthalates produces any significant hazard. There has long been some concern about the reproductive effects of relatively large doses of phthalates, but the significance of this in actual environments has been hard to assess. Of continuing concern is the presence of phthalates in medical devices such as blood bags and intravenous solution bags. Patients undergoing chronic transfusion or long-term intravenous administration of nutrients or medicines are exposed to much larger doses of phthalate esters than would be obtained from the diet, and there is evidence that the potency of this route of administration is higher. There is no satisfactory substitute at present for these plasticizers.

Recently the US National Toxicology Program reported that DEHP shows carcinogenic activity when administered in high doses to both mice and rats. However, DEHP does not appear to be a potent carcinogen when tested by in vitro sytems, and some workers have contended that the observed carcinogenicity is a peculiarity of rodents and should not be used to guide decisions concerning human exposure. Three important differences in DEHP metabolism between humans and rodents are cited in arguing that, for this compound at least, rodents are not good model systems for human carcinogenesis:

1) at high doses, rodents are exposed to unmetabolized DEHP, so it is impossible to extrapolate the results in this range to lower concentrations where this is not the case;

2) there are striking differences between the metabolites observed in rodents and those observed in humans, and it is this difference which is important, and

3) rodent carcinogenesis is always observed when compounds are administered which, like DEHP, stimulate the liver peroxisome system, but this system is not stimulated by DEHP in primates.

The prima facie merits of these arguments cannot be denied, and they are being actively pursued. At present the question of human carcinogenesis must be considered still open.

Thus, the phthalate esters are an extremely important class of environmental contaminants which will continued to be produced and monitored for decades. Although there are still questions concerning their long-term safety, there have been no major incidents which have marred their overall record as useful and innocuous products of the age of organic chemicals.

Acknowledgements

The authors would like to acknowledge the generous support of the US National Science Foundation of our research on environmental phthalates over the past few years, the technical and editorial comments of Dr. Lee Ray, and the invaluable contribution of Marilyn B. Hollar whose word processing abilities improved both the production speed and quality of the final manuscript. His coauthors wish to acknowledge the enormous contribution – both in searching the literature and in preparing the manuscript – of Dr. Jack E. Leonard. We thank those who have provided us with manuscripts and reprints on this topic and we apologize to our colleagues whose work may have been unintentionally omitted; the magnitude of the task prevented a comprehensive citation of all those who have contributed to work in this area over the past two decades.

References

1. Agarwal, D.K., Agarwal, S., Seth, P.K.: Biochem. Pharmacol. *31*, 3419 (1982)
2. Albro, P.W., Corbett, J.T., Schroeder, J.L., Jordan, S., Matthews, H.B.: EHP, Environ. Health Perspect. *45*, 19 (1982)
3. Albro, P.W., Hass, J.R., Peck, C.C., Odom, D.G., Corbett, J.R., Bailey, F.J., Blatt, N.E., Barrett, B.B.: Drug Metab. Dispos. *9*, 223 (1981)
4. Albro, P.W., Jordan, S.T., Schroeder, J.L., Corbett, J.T.: J. Chromatogr. *244*, 65 (1982)
5. Aldrich Chemical Company: 1982–1983 Catalog Handbook of Fine Chemicals, Milwaukee, WI, 1982

6. Alfredson, T., Tallman, L.: Analysis of Phthalates by High Performance Liquid Chromatography, Palo Alto, CA, Varian Instrument Group, LC116, 1981
7. Anonymous: Chem. Eng. News (5 May), 5 (1980)
8. Aoyama, Y., Yasui, K., Otaki, R., Yoshida, A., Ashida, K.: Nutr. Rep. Int. *26*, 395 (1982)
9. Atlas, E.: Synthetic organics in the Gulf of Mexico – a Review, in: Proc. Symp. Environ. Res. Needs in the Gulf of Mexico (GOMEX) (ed.) Atwood, D.K., p. 131, Miami, FL, NOAA/ERL 1981
10. Atlas, E., Foster, R., Giam, C.S.: Environ. Sci. Technol. *16*, 183 (1982)
11. Atlas, E., Giam, C.S.: Science *211*, 163 (1981)
12. Atlas, E.L., Giam, C.S.: Phthalates and related plasticizers, in: Mass Spectrometry in the Environmental Sciences (Theory and Applications) (eds.) Karasek, F.W., Hutzinger, O., Safe, S., New York, Plenum Press 1983
13. Atlas and Giam, unpublished
14. Atlas, E., Martinez, G., Giam, C.S.: Chemical characterization of ocean-dumped waste materials, in: Ocean Dumping of Industrial Wastes (eds.) Bostwick, H.K., Kester, D.R., Park, P.K., New York, Plenum 1981
15. Autian, J.: EHP, Environ. Health Perspect. *4*, 3 (1973)
16. Autian, J.: EHP, Environ. Health Perspect. *45*, 115 (1982)
17. Baillod, C.R., Faith, B.M., Masi, O.: Environ. Prog. *1*, 217 (1982)
18. Bainova, A., Antov, G., Ilieva, P.: Dokl. Bolg. Akad. Nauk *35*, 121 (1982)
19. Balabanova-Radonova, E., Stefanova, M., Kumanova, B.: Dokl. Bolg. Akad. Nauk *35*, 773 (1982)
20. Bell, F.P.: EHP, Environ. Health Perspect. *45*, 41 (1982)
21. Benckiser, G., Ottow, J.C.G.: Appl. Environ. Microbiol. *44*, 576 (1982)
22. Bjellin, L., Maartensen, L., Widmark, E.: Rapp.-Naturvardverket (Swed.), SNV PM 1552, Proj. Foerekomst Eff. Persistenta Org. Aemnen Miejoen, 119, 1982; Chem. Abstr. 97:209799q (1982)
23. Blevins, R.D., Taylor, D.E.: J. Environ. Sci. Health, Part A. *A17*, 297 (1982)
24. Bove, J.L., Dalven, P., Kukreja, V.P.: Int. J. Environ. Anal. Chem. *5*, 187 (1978)
25. Bove, J., Dalven, P.: Int. J. Environ. Anal. Chem. *10*, 189 (1981)
26. Branson, D.R.: ASTM Spec. Tech. Publ. *657*, 55 (1978)
27. Breger, I.A.: J. Am. Oil Chem. Soc. *43*, 197 (1966)
28. Breger, I.A.: Chem. Eng. News (Jan. 17), 19 (1972)
29. Bringmann, G., Kühn, R.: GWF, Gas-Wasserfach: Wasser/Abwasser *122*, 308 (1981)
30. Bringmann, G., Kühn, R.: Z. Wasser-Abwasser-Forsch. *15*, 1 (1982)
31. Brown, D., Thompson, R.S.: Chemosphere *11*, 417 (1982)
32. Brown, D., Thompson, R.S.: Chemosphere *11*, 427 (1982)
33. Brownlee, B., Strachan, W.M.J.: in: Identification and Analysis of Organic Pollutants in Water, (ed.) Keith, L.H., p. 661, Ann Arbor, Michigan, Ann Arbor Sci. 1976
34. Buescher, M.D., Rutledge, L.C., Wirtz, R.A.: Mosq. News *42*, 428 (1982)
35. Burns, G., Musial, C.J., Uthe, J.F.: J. Assoc. Off. Anal. Chem. *64*, 282 (1981)
36. Callahan, M., Slimak, M., Gabel, N., May, I., Fowler, C., Freed, R., Jennings, P., Durfee, R., Whitmore, F., Maestri, B., Mabey, W., Holt, B., Good, C.: Water Related Fate of 129 Priority Pollutants, U.S. EPA 440/4-79-029B, NTIS PB80-204381 (1980)
37. Calley, D., Autian, J., Guess, W.L.: J. Pharm. Sci. *55*, 158 (1966)
38. Canada, Minister of Health and Welfare: Phthalic Acid Esters (Environmental and Health Impact), Can. Environ. Health Dir., Ottawa, Ontario, 1980
39. Carlberg, G.E., Martinsen, K.: Comm. Eur. Communities, [Rep.] *EUR 7623*, Anal. Org. Micropollut. Water, 42, 1982
40. Carmignani, G.M., Bennett, J.P.: Aquaculture *7*, 89 (1976)
41. Carmignani, G.M., Bennett, J.: Aquaculture *8*, 291 (1976)
42. Cater, B.R., Cook, M.W., Gangolli, S.D., Grasso, P.: Toxicol. Appl. Pharmacol. *41*, 609 (1977)
43. Cautreels, W.L.M., Van Cauwenberghe, K.A., Guzman, L.A.: Sci. Total Environ. *8*, 79 (1978)
44. Choudhury, D.: Environ. Sci. Technol. *16*, 102 (1982)
45. Cifrulak, C.F.: Soil Sci. *107*, 63 (1979)
46. Clement, R.E., Karasek, F.W.: J. Chromatogr. *234*, 395 (1982)
47. Corcoran, E.F.: EHP, Environ. Health Perspect. *3*, 13 (1973)
48. Corcoran, E.F., Curry, R.W.: Rev. Biol. Trop. *26 (Supp. 1)*, 125 (1978)
49. Daniel, J.W., Bratt, H.: Toxicology *2*, 51 (1974)

50. Derenbach, J.B., Ehrhardt, M., Osterrhot, C., Petrick, G.: Marine Chem. *6*, 351 (1978)
51. Dillingham, E.O., Autian, J.: EHP, Environ. Health Perspect. *3*, 81 (1973)
52. Dmitriev, M.T., Zarubin, G.P., Mishchikin, V.A.: Gig. Sanit. 55 (1982); Chem. Abstr. 98:77378k (1982)
53. Dobbs, A.J., Cull, M.R.: Environ. Pollut. Ser. B *3*, 289 (1982)
54. Dobbs, A.J., Williams, N.: Chemosphere *12*, 97 (1983)
55. Dymerski, P., Palmer, A., Dahlgran, J., Insalco, S.: Case history and use of a mobile GC/MS laboratory, in: Hazardous Material Spills Conference Proceedings (ed.) Ludwigson, J., p. 15, Rockville, MD, Governmental Institution, Inc. 1982
56. Eaton, R.W., Ribbons, D.W.: J. Bacteriol. *151*, 48 (1982)
57. Eaton, R.W., Ribbons, D.W.: J. Bacteriol. *151*, 465 (1982)
58. Eaton, R.W., Ribbons, D.W.: Arch. Microbiol. *132*, 185 (1982)
59. Eglinton, G., Simoneit, B.R.T., Zoro, J.A.: Proc. Roy. Soc. B *189*, 415 (1975)
60. Ehrhardt, M., Derenbach, J.: Marine Chem. *8*, 339 (1980)
61. Ellis, D.D., Jone, C.M., Larson, R.A., Schaeffer, D.J.: Arch. Environ. Contam. Toxicol. *11*, 373 (1982)
62. Engelhardt, G., Wallnöfer, P.R., Hutzinger, O.: Bull. Environ. Contam. Toxicol. *13*, 342 (1975)
63. Esteve, N., Hogg, J., Julia, E.: Comun. Jorn. Com. Esp. Deter., 13th, p. 109 (1982)
64. Foster, P.M.D., Lake, B.G., Cook, M.W., Thomas, L.V., Gangolli, S.D.: Adv. Exp. Med. Biol. *136A* (Biol. React. Intermed.-2, Chem. Mech. Biol. Eff. Pt. A), 445 (1982)
65. Frederick, L., Wilcox, T., Mosely, C.: Health Hazard Evaluation 79-011-1011, King Souper's Meat Plant, Denver, CO., NTIS PB82-182551 (1982)
66. Freeman, J.C., Sangalang, G.B., Burns, B.G., McMenemy, M.: Can. Tech. Rep. Fish. Aquat. Sci. *990*, 198 (1981)
67. Freitag, D., Geyer, H., Kraus, A., Viswanathan, R., Kotzias, D., Attar, A., Klein, W., Korte, F.: Ecotoxicol. Environ. Saf. *6*, 60 (1982)
68. Frische, R., Esser, G., Schaoenborn, W., Klöpffer, W.: Comm. Eur. Communities, [Rep.], *EUR 7549*, Environ. Qual. Life, 50 (1982)
69. Fuller, M.: J. Soc. Cosmet. Chem. *33*, 51 (1982)
70. Furuta, S., Miyazama, S., Hashimoto, T.: J. Biochem. (Tokyo) *92*, 319 (1982)
71. Galyashin, V.N., Rozynov, B.V.: Izv. Akad. Nauk SSSR, Ser. Khim., 280 (1982); Chem. Abstr. *96*:180547x (1982)
72. Galyashin, V.N., Rozynov, B.V.: Izv. Adad. Nauk SSSR, Ser. Khim., 581 (1982); Chem. Abstr. *96*:217062p (1982)
73. Gangolli, S.D.: EHP, Environ. Health Perspect. *45*, 77 (1982)
74. Gaunt, I.F., Butterworth, K.R.: Food Chem. Toxicol. *20*, 215 (1982)
75. Geyer, H., Viswanathan, R., Freitag, D., Korte, F.: Chemosphere *10*, 1307 (1981)
76. Giam, C.S.: Trace analysis of phthalates (and chlorinated hydrocarbons) in marine samples, in: Strategies for Marine Pollution Monitoring (ed.) Goldberg, E.D., p. 61, New York, John Wiley & Sons 1976
77. Giam, unpublished data
78. Giam, C.S., Atlas, E.: Naturwissenschaften *67*, 508 (1980)
79. Giam, C.S., Atlas, E.L., Chan, H.S.:Proc. 4th Jt. Conf. Sensing Environ. Pollut., p. 707, 1977
80. Giam, C.S., Atlas, E.L., Chan, H.S., Neff, G.S.: Rev. Int. Oceangr. Med. *47*, 79 (1977)
81. Giam, C.S., Atlas, E., Chan, H.S., Neff, G.S.: Atmos. Environ. *14*, 65 (1980)
82. Giam, C.S., Chan, H.S.: Control of Blanks in the Analysis of Phthalates in Air and Ocean Biota Samples, NBS Spec. Publ. (U.S.) *422*, 701 (1976)
83. Giam, C.S., Chan, H.S., Hammergren, T., Neff, G.S., Stalling, D.L.: Anal. Chem. *48*, 78 (1976)
84. Giam, C.S., Chan, H.S., Neff, G.S.: Anal. Chem. *46*, 2319 (1974)
85. Giam, C.S., Chan, H.S., Neff, G.S.: Proc. Int. Conf. Environ. Sensing Assess. *2*, 21-1-1 (1975)
86. Giam, C.S., Chan, H.S., Neff, G.S.: Anal. Chem. *47*, 2225 (1975)
87. Giam, C.S., Chan, H.S., Neff, G.S.: Concentrations and fluxes of phthalates, DDTs, and PCBs to the Gulf of Mexico, in: Marine Pollutant Transfer (eds.) Windom, H.L., Duce, R.A., p. 375, Lexington, MA, D.C. Heath, 1976
88. Giam, C.S., Chan, H.S., Neff, G.S.: Phthalate ester plasticizers: A new class of marine pollutant, in: Pollutant Effects on Marine Organisms (ed.) Giam, C.S., p. 32, Lexington, MA, D.C. Heath and Co. 1977
89. Giam, C.S., Chan, H.S., Neff, G.S.: Marine Pollut. Bull. *9*, 249 (1978)

90. Giam, C.S., Chan, H.S., Neff, G.S., Atlas, E.L.: Science *199*, 419 (1978)
91. Giam, C.S., Goodwin, T.E., Yano, T.K.: J. Chem. Soc. Perkins Trans. II, 831 (1978)
92. Giam, C.S., Wong, M.K.: J. Chromatogr. *72*, 283 (1972)
93. Gledhill, W.E., Kaley, R.G., Adams, W.J., Hicks, O., Michael, P.R., Saeger, V.W., LeBlanc, G.A.: Environ. Sci. Technol. *14*, 301 (1980)
94. Godly, E.W., Mortlock, A.E.: Analyst (London) *98*, 493 (1973)
95. Gonzalez-Vila, F.J., Saiz-Jimenez, C., Marin, F.: J. Environ. Qual. *11*, 251 (1982)
96. Gordon, A.W., Gordon, M.: Trans. Ky. Acad. Sci. *42*, 149 (1981)
97. Graham, P.R.: EHP, Environ. Health Perspect. *3*, 3 (1973)
98. Gray, T.J.B., Rowland, I.R., Foster, P.M.D., Gangolli, S.D.: Toxicol. Lett. *11*, 141 (1982)
99. Gross, F.C., Colony, J.A.: EHP, Environ. Health Perspect. *3*, 37 (1973)
100. Guess, W.L., Jacob, J., Autian, J.: Drug Intell. *1*, 120 (1967)
101. Gurka, D.F., Laska, P.R., Titus, R.: J. Chromatogr. Sci. *20*, 145 (1982)
102. Hanson, David: Chem. Eng. News (Feb. 28), 36 (1983)
103. Harris, J.C., Cohen, M.J., Grosser, Z.A., Hayes, M.J.: Evaluation of solid sorbents for water sampling, EPA-600/2-80-1983, NTIS PB81-106585 (1981)
104. Harris, J.C., Cohen, M.J., Hayes, M.J.: Collection and recovery of organics from water using XAD-2 and XE-347 resins, EPA-600/8-81-018, NTIS PB81-179855 (1981)
105. Harvan, D.J., Hass, J.R., Albro, P.W., Friesen, M.D.: Biomed. Mass Spectrom. *7*, 242 (1980)
106. Hattori, U., Kuge, Y., Nakagawa, S.: Mizu Shori Gijutsu *16*, 951 (1975)
107. Heitmuller, P.T., Hollister, T.A.: Bull. Environ. Contam. Toxicol. *27*, 596 (1981)
108. Hites, R.A.: EHP, Environ. Health Perspect. *3*, 17 (1973)
109. Hites, R.A.: J. Chromatogr. Sci. *11*, 570 (1973)
110. Hoffman W.A., Jr., Lindberg, S.E., Turner, R.R.: Environ. Poll., Ser. B *4*, 219 (1982)
111. Hudson, R.A., Giancarlo, T., II, Austerberry, C.F., Bagshaw, J.C.: Toxicol. Lett. *10*, 389 (1982)
112. Hwang, S.T.: AIChE Symp. Ser. *77*, 304 (1981)
113. Il'in, V.A., Buinov, V.N., Samilova, O.V., Shereshovets, V.V.: Zh. Prikl. Khim. (Leningrad) *55*, 1658 (1982)
114. Isaaq, H.J., Klose, J.R., Reitz, G.A.: J. Liq. Chromatogr. *5*, 1069 (1982)
115. Ishida, M., Suyama, K., Adachi, S.: J. Chromatogr. *189*, 421 (1980)
116. Ishiguro, S., Inoue, Y., Hosogane, T.: J. Chromatogr. *239*, 651 (1982)
117. Jaeger, R.J., Rubin, R.J.: Science *170*, 460 (1970)
118. Jaeger, R.J., Rubin, R.J.: New England J. Med. *287*, 1114 (1972)
119. Jordan, E.C., Inc.: Fate of priority pollutants in publicly owned treatment works – 30 day study, EPA-440/1-82/302, NTIS (1982)
120. Kameyama Candle Co.: Jpn. Kokai Tokkyo Koho JP 81,125,499. Issued 1 Oct. 1981
121. Kartseva, V.M.: Zashch. Rast. (Moscow) 26 (1982); Chem. Abstr. 97:67762w (1982)
122. Katou, T., Hanai, Y., Sato, H.: Yokohama Kokuritsu Daigaku Kankyo Kagaku Kenkyo Senta Kiyo *8*, 63 (1982); Chem. Abstr. 97:60411x (1982)
123. Keith, L.H.: AICHE Symp. Ser. *77*, 249 (1981)
124. Kennicutt, M.C., II., Jeffrey, L.M.: Mar. Chem. *10*, 367 (1981)
125. Kevy, S.V., Jacobson, M.S.: EHP, Environ. Health Perspect. *45*, 57 (1982)
126. Klindukhov, V.P., Sufgova, N.Z., Glebovskaya, E.A., Mel'tsanskaya, T.N., Kulikova, E.M., Solov'eva, I.L., Belyaeva, L.S., Bikkenina, D.A., Sarbeeva, L.I.: Tr. Vses. Neft. Nauchno-Issled. Geologorazved. Inst. *353*, 58 (1974); Chem. Abstr. 83:150132x (1975)
127. Klöpffer, W., Rippen, G., Frische, R.: Ecotoxicol. Environ. Saf. *6*, 294 (1982)
128. Kluwe, W.M.: EHP, Environ. Health Perspect. *45*, 3 (1982)
129. Kluwe, W.M., McConnell, E.E., Huff, J.E., Haseman, J.K., Douglas, J.F., Hartwell, W.V.: EHP, Environ. Health Perspect. *45*, 129 (1982)
130. Kolesar, E.S., Jr.: Automated calculation of protection factors for the di-2-ethylexyl phthalate respirator quantitative fit test instrument, NTIS AD-A115327 (1982)
131. Kollig, H.P., Hall, T.L.: Water Air Soil Pollut. *17*, 225 (1982)
132. Kotzias, D., Parlar, H., Korte, F.: Naturwissenschaften *69*, 444 (1982)
133. Kovar, P.: Acta Hydrochim. Hydrobiol. *10*, 449 (1982)
134. Kozumbo, W.J., Kroll, R., Rubin, R.J.: EHP, Environ. Health Perspect. *45*, 103 (1982)
135. Kraus, G., Schierhorn, M., Van, T.T.H., Tucek, E.: Fresenius' Z. Anal. Chem. *312*, 242 (1982)
136. Krishen, A.: Anal. Chem. *43*, 1130 (1971)
137. Kurane, R., Suzuki, T., Takahara, Y.: Agric. Biol. Chem. *44*, 529 (1980)

138. Kurane, R., Suzuki, T., Takahara, Y.: U.S. Patent No. 4,332,904, 1 June 1982
139. Lake, B.G., Foster, J.R., Collins, M.A., Stuggerfield, C.R., Gangolli, S.D., Srivastava, S.P.: Acta Pharmacol. Toxicol. *51*, 217 (1982)
140. Laughlin, R.B., Jr., Neff, J.M., Giam, C.S.: Effects of polychlorinated biphenyls, polychlorinated naphthalenes, and phthalate esters on larval development of the mud crab *Rhithropanopeus harrisii,* in: Pollutant Effects Marine Organisms (ed.) Giam, C.S., p. 95, Lexington, MA, D.C. Heath and Co. 1977
141. Laughlin, R.B., Jr., Neff, J.M., Hrung, Y.C., Goodwin, T.C., Giam, C.S.: Water Air Soil Pollut. *9*, 323 (1978)
142. Lawrence, W.H., Malik, M., Turner, J.E., Singh, A.R., Autian, J.: Toxicol. Appl. Pharmacol. *29*, 87 (1974)
143. Lawrence, W.H., Malik, M., Turner, J.E., Singh, A.R., Autian, J.: Environ. Res. *9*, 1 (1975)
144. LeBlanc, G.A.: Bull. Environ. Contam. Toxicol. *24*, 684 (1980)
145. Leyder, F., Boulanger, P.: Bull. Environ. Contam. Toxicol. *30*, 152 (1983)
146. Liao, W., Christma, R.F., Donald, J.J., Millington, D.S., Hass, J.R.: Environ. Sci. Technol. *16*, 403 (1982)
147. Lindgren, A., Lindquist, N.G., Lyden, A., Olsson, T., Ullberg, S.: Toxicology *23*, 249 (1982)
148. Lokke, H., Bro-Rasmussen, F.: Chemosphere *10*, 1223 (1981)
149. Lue-Hing, C., Lordi, D.T., Kelada, N.P.: AIChE Symp. Ser. *77*, 144 (1983)
150. Marcel, Y.L.: EHP, Environ. Health Perspect. *3*, 119 (1973)
151. Marcel, Y.L., Noel, S.P.: Lancet (7636), 35 (1970)
152. Marcel, Y.L., Noel, S.P.: Chem. Phys. Lipids *4*, 417 (1970)
153. Mathur, S.P.: J. Environ. Qual. *3*, 189 (1974)
154. Mathur, S.P., Rouatt, J.W.: J. Environ. Qual. *4*, 273 (1975)
155. Matsuda, K., Schnitzer, M.: Bull. Environ. Contam. Toxicol. *6*, 200 (1971)
156. Mayer, F.L., Jr., Sanders, H.O.: EHP, Environ. Health Perspect. *3*, 153 (1973)
157. Mayer, F.L., Jr., Stalling, D.L., Johnson, J.L.: Nature *238*, 411 (1972)
158. Medlin, L.K.: Bull. Environ. Contam. Toxicol. *25*, 75 (1980)
159. Mehrle, P.M., Mayer, F.L.: Trace Subst. Environ. Health *10*, 519 (1976)
160. Melancon, M.J., Jr., Lech, J.J.: Drug Metabol. Dispos. *5*, 29 (1977)
161. Melnick, R.L., Schiller, C.M.: EHP, Environ. Health Perspect. *45*, 51 (1982)
162. Men'shikova, T.A.: Gig. Sanit. *36*, 23 (1971), cited in ref. 244
163. Mieure, J.P.: Chem. Eng. News, May 16, 1983
164. Mieure, J.P., Dietrich, M.W.: J. Chromatogr. Sci. *11*, 559 (1973)
165. Milkov, L.E., Aldyreva, M.V., Popova, T.B., Lopukhova, K.A., Markarenko, Yu.L., Malyar, L.M., Shakhova, T.K.: EHP, Environ. Health Perspect. *3*, 175 (1973)
166. Miller, H.C., James, R.H., Dickson, W.R., Neptune, M.D., Carter, M.H.: ASTM Spec. Tech. Publ. *760*, 240 (1981)
167. Moody, D.E., Reddy, J.K.: Toxicol. Lett. *10*, 379 (1982)
168. Mori, S.: J. Chromatogr. *129*, 53 (1976)
169. Morita, M.: Kokuritsu Kogai Kenkyusho Kenky Hokoku *36*, 151–155 (1982); Chem. Abstr. 97:194056u (1982)
170. Morita, M., Nakamura, H., Mimura, S.: Water Res. *8*, 781 (1974)
171. Müller, G.: Heavy metals and other pollutants in the environment: a chronology based on the analysis of dated sediments, in 3rd Int. Conf. Heavy Met. Environ., p. 12, Edinburgh, UK, CEP Consultants, Ltd., 1982
172. Müller, G., Dominik, J., Reuther, R., Malisch, R., Schulte, E., Acker, L., Irion, G.: Naturwissenschaften *67*, 595 (1980)
173. Müller, R.K.: Pharmazie *37*, 416 (1982)
174. Murray, H.E., Ray, L.E., Giam, C.S.: Bull. Environ. Contam. Toxicol. *26*, 769 (1981)
175. Murray, H.E., Ray, L.E., Giam, C.S.: Chemosphere *10*, 1327 (1981)
176. Musial, C.J., Uthe, J.F., Sirota, G.R., Burns, B.G., Gilgan, M.W., Zitko, V., Matheson, R.A.: Can. J. Fish. Aquat. Sci. *38*, 856 (1981)
177. Nazir, D.J., Alcaraz, A.P., Bierl, B.A., Beroza, M., Nair, P.P.: Biochemistry *10*, 4228 (1971)
178. Neely, W.B.: Environ. Toxicol. Chem. *1*, 259 (1982)
179. Nematollahi, J., Guess, W.L., Autian, J.: J. Pharm. Sci. *56*, 1446 (1967)
180. Neumayr, V.: in: Bewertung Chem. Stoffe Wasserkreislauf (ed.) by Aurand, K., p. 37, Berlin, Schmidt 1981

181. Ogner, G., Schnitzer, M.: Science *170*, 317 (1970)
182. Oishi, S., Hiraga, K.: Toxicol. Lett. *14*, 79 (1982)
183. Olsen, P.R., Larsen, N.A., Faurby, V., Krarup, T., Larsen, S.: Scand. J. Urol. Nephrol. *16*, 187 (1982)
184. Overcash, M.R., Weber, J.B., Miles, M.L.: Behavior of organic priority pollutants in the terrestrial system: di(*n*-butyl)phthalate ester, toluene, a 2,4-dinitrophenol, NTIS PB-82-224544 (1982)
185. Overcash, M.R., Weber, J.B., Miles, M.L.: Report of the Water Resources Research Institute of the University of North Carolina, *171*, 103pp. (1982)
186. Overcash, M.R., Weber, J.B., Miles, M.L.: Behavior of Organic Priority Pollutants in the Terrestrial System: Di-n-butyl phthalate ester, toluene, and 2,4-dinitrophenol. Rayleigh, NC, Water Recources Research Institute of the University of North Carolina, NTIS PB82-224544 (1982)
187. Parkhie, M.R., Webb, M., Norcross, M.A.: EHP, Environ. Health Perspect. *45*, 89 (1982)
188. Parry, J.M., Al-Mossawi, M.A.J., Danford, N., Ballantine, J.: The monitoring of substances in marine waters for genetic activity, in: Symposium on Carcinogenicity of Polynuclear Aromatic Hydrocarbons in the Marine Environment (eds.) Richards, N.L., Jackson, B.L., EPA-600/9-82-013, NTIS PB82-263864 (1982)
189. Payne, W.R., Jr., Benner, J.E.: J. Assoc. Off. Anal. Chem. *64*, 1403 (1981)
190. Peakall, D.B.: Residue Rev. *54*, 1 (1975)
191. Peck, C.C., Albro, P.W.: EHP, Environ. Health Perspect. *45*, 11 (1982)
192. Peck, C.C., Albro, P.W., Hass, J.R., Odom, D.G., Barrett, B.B., Bailey, F.J.: Clin. Res. 25:101A (1978)
193. Peters, T.L.: Anal. Chem. *54*, 1913 (1982)
194. Peterson, J.C.: Ph.D. Dissertation, University of Maryland, College Park, Maryland, 1980
195. Peterson, J.C., Freeman, D.H.: Environ. Sci. Technol. *16*, 464 (1982)
196. Phillips, B.J., James, T.E.B., Gangolli, S.D.: Mutat. Res. *102*, 297 (1982)
197. Phillips, H.F., Breger, I.A.: Geochim. Cosmochim. Acta *15*, 51 (1958)
198. Pryor, P., Whorton, D.: Health Hazard Evaluation 80-094-840 of Ford Motor Company, San Jose, CA, NTIS PB82-172214 (1982)
199. Reddy, J.K., Azarnoff, D.L., Hignite, C.E.: Nature *283*, 397 (1980)
200. Reinhard, M., McCarty, P.L.: Detection of organic materials at Water Factory 21, in: The Chemistry of Water Reuse (ed.) Cooper, W.J., vol. 2, p. 55, Ann Arbor, MI, Ann Arbor Sci. 1982
201. Rhoades, J.W., Thomas, R.E., Johnson, D.E., Tillery, J.B.: U.S. EPA-600/4-81-063; NTIS PB81-232167, Washington, DC, 1981
202. Roll, D.B., Douglas, J.D., Petersen, R.V.: J. Pharm. Sci. *63*, 1628 (1974)
203. Saeger, V.W., Tucker, E.S.: Appl. Environ. Microbiol. *31*, 29 (1976)
204. Sakagami, Y., Tanaka, K., Watanabe, K., Odagiri, M., Mori, K., Murooka, H.: Nippon Suisan Gakkaishi *48*, 633 (1982); Chem. Abstr. 97:20451d (1982)
205. Sanborn, J.R., Metcalf, R.L., Yu, C.C., Lu, P.Y.: Arch. Environ. Contam. Toxicol. *3*, 244 (1975)
206. Sanders, H.O., Mayer, F.L., Jr., Walsh, D.F.: Environ. Res. *6*, 84 (1973)
207. Schauenberg, H., Schlitt, H., Knoppel, H.: Comm. Eur. Communities [Rep.], *EUR 7623*, 193 (1982)
208. Schouten, M.J., Peereboom, J.W.C., Brinkman, U.A.Th.: Int. J. Environ. Anal. Chem. *7*, 13 (1979)
209. Schwartz, H.E., Anzion, C.J.M., Van Vliet, H.P.M., Peereboom, J.W.C., Brinkman, U.A.Th.: Int. J. Environ. Anal. Chem. *6*, 133 (1979)
210. Schwarz, F.P., Miller, J.: Anal. Chem. *52*, 2162 (1980)
211. Sdika, A., Cabridenc, R., Hennequin, C.: Comm. Eur. Communities [Rep.] *EUR 7623*, 24 (1982)
212. Seed, J.L.: EHP, Environ. Health Perspect. *45*, 111 (1982)
213. Selover, S.J., Gunderson, E.C., Beeman, C.W., Boland, P.A., Coulson, D.M.: Sampling and Analytical Methods for Airborne Dialkyl Phthalates, presented at The 184th National Meeting of the American Chemical Society, 1982
214. Seth, P.K.: EHP, Environ. Health Perspect. *45*, 27 (1982)
215. Shea, P.J., Weber, J.B., Overcash, M.R.: Bull. Environ. Contam. Toxicol. *29*, 153 (1982)
216. Shibata, K., Motooka, K., Murata, K., Iwai, K.: J. Nutr. Sci. Vitaminol. *28*, 173 (1982)
217. Shiota, K., Nishimura, H.: EHP, Environ. Health Perspect. *45*, 65 (1982)
218. Shuckrow, A.J., Pajak, A.P.: U.S. EPA-600/9-81-0026, Washington, DC, 1981
219. Shuckrow, A.J., Pajak, A.P.: J. Hazard. Mater. *7*, 37 (1982)
220. Singh, A.R., Lawrence, W.H., Autian, J.: J. Pharm. Sci. *61*, 51 (1972)

221. Södergren, A.: Environ. Pollut., Ser. A *27*, 263 (1982)
222. Stalling, D.L., Hogan, J.W., Johnson, J.L.: EHP, Environ. Health Perspect. *3*, 149 (1973)
223. Sugawara, N.: Toxicol. Appl. Pharmacol. *30*, 87 (1974)
224. Sugiyama, H., Tanaka, K.: Kenkyu Hokoku – Kanagawa-ken Kogai Senta *4*, 33 (1982); Chem. Abstr. 97:187915v (1982)
225. Sullivan, K.F., Atlas, E.L., Giam, C.S.: Environ. Sci. Technol. *16*, 428 (1980)
226. Sullivan, K.F., Atlas, E.L., Giam, C.S.: Anal. Chem. *53*, 1717 (1981)
227. Tabak, H.H., Quave, S.A., Mashni, C.I., Barth, E.F.: J. Water Pollut. Control Fed. *53*, 1503 (1981)
228. Taborsky, R.G.: J. Agric. Food Chem. *15*, 1073 (1967)
229. Takeuchi, T., Ishii, D.: J. Chromatogr. *218*, 199 (1981)
230. Taylor, D.G. (ed.): NIOSH Manual of Analytical Methods, vol. 2, Washington D.C., 1977[2]
231. Tepper, L.B.: EHP, Environ. Health Perspect. *3*, 179 (1973)
232. Thomas, G.H.: EHP, Environ. Health Perspect. *3*, 13 (1973)
233. Thomas, J.A., Darby, T.D., Wallin, R.F., Garvin, P.J., Martis, L.: Toxicol. Appl. Pharmacol. *45*, 1 (1978)
234. Thomas, J.A., Northrup, S.J.: J. Toxicol. Environ. Health *9*, 141 (1982)
235. Thomas, J.A., Curto, K.A., Thomas, M.J.: EHP, Environ. Health Perspect. *45*, 85 (1982)
236. Thrun, K.E., Harris, J.E., Rechsteiner, C.E., Sorlin, D.J.: Methods for Level 2 Analysis by Organic Compound Category, EPA 600/7-81-029, NTIS PB81-179855 (1981)
237. Timofievskaya, L.A.: Gig. Sanit. 91 (1982); Chem. Abstr. 97:121632u (1982)
238. Tomita, I., Nakamura, Y., Yagi, Y.: Ecotoxicol. Environ. Saf. *1*, 275 (1977)
239. Tomita, I., Nakamura, Y., Yagi, Y., Tutikawa, K.: EHP, Environ. Health Perspect. *45*, 71 (1982)
240. Tomita, I., Nakamura, Y., Aoki, N., Inui, N.: EHP, Environ. Health Perspect. *45*, 119 (1982)
241. Tomson, M.B., Dauchy, J., Hutchins, S., Curran, C., Cook, C.J., Ward, C.H.: Water Res. *15*, 1109 (1981)
242. U.S. Consumer Products Safety Commission: Status Report on Phthalates in Consumer Products, submitted to the Commission, 19 Jan. 1983
243. U.S. Environmental Protection Agency: In-depth Studies on Health and Environmental Impacts of Selected Water Pollutants. Contract 68-01-4646 (1978), cited in Ref. 240
244. U.S. Environmental Protection Agency, Phthalate Esters: Ambient Water Quality Criteria, NTIS PB-296804 (1978)
245. U.S. Environ. Protection Agency: Fed. Regist. *44*, 69464, 3 Dec. 1979
246. U.S. Environ. Protection Agency: Fed. Regist. *44*, 75029, 18 Dec. 1979
247. U.S. Environ. Protection Agency: Fed. Regist. *47*, 38780, 2 Sept. 1982
248. U.S. Internat. Trade Commission: Synthetic Organic Chemicals, Production and Sales, 1979, Washington, DC, 1980
249. U.S. Internat. Trade Commission: Imports of Benzenoid Chemicals and Products, 1978, Washington, DC, 1979
250. U.S. Nat. Inst. Occupational Safety and Health: Registry of Toxic Effects of Chemical Substances, 1980, Washington, DC, 1982
251. U.S. Nat. Toxicology Progr.: Report, NTIS: PB-82-184011 (1982)
252. U.S. Nat. Toxicology Progr. EHP, Environ. Health Perspect. *45*, 149 (1982)
253. Van Vliet, H.P.M., Boorsman, Th.C., Frei, R.W., Brinkman, U.A.Th.: J. Chromatogr. *185*, 483 (1979)
254. Veith, G.D., Austin, N.M., Morris, R.T.: Water Res. *8*, 43 (1979)
255. Veith, G.D., DeFoe, D.L., Bergsted, B.V.: J. Fish. Res. Board Can. *36*, 1040 (1979)
256. Veith, G.D., Kuehl, D.W.: Anal. Chem. *53*, 1132 (1981)
257. Volk, F., Wunsch, G., Miller, D.: Propellants, Explos., Pyrotech. *7*, 53 (1982)
258. Waddell, W.J., Marlowe, C., Miripol, J.E., Garvin, P.S.: Toxicol. Appl. Pharmacol. *39*, 339 (1977)
259. Waggot, A.: Trace organic substances in the River Lee, in: The Chemistry of Water Reuse (ed.) Cooper, W.J., vol. 2, p. 55, Ann Arbor, MI, Ann Arbor Sci. 1982
260. Walseth, F., Toftgaard, R., Nilsen, O.G.: Arch. Toxicol. *50*, 1 (1982)
261. Warner, J.S., Hidy, B.J., Jungclaus, G.A., McKown, M.M., Miller, M.P., Riggin, R.M.: ASTM Spec. Tech. Publ. *760*, 40 (1981)
262. Warren, J.R., Lalwani, N.D., Reddy, J.K.: EHP, Environ. Health Perspect. *45*, 35 (1982)
263. Weast, R.C. (ed.): CRC Handbook of Chemistry and Physics, CRC Press, Boca Raton, FL, 1980[61]

264. Webster, R.D.J., Nickless, G.: Proc. Anal. Div. Chem. Soc. *13*, 333 (1976)
265. Wechsler, C.J.: Environ. Sci. Technol. *14*, 428 (1980)
266. Wechsler, C.J.: Atmos. Environ. *15*, 1365 (1981)
267. Weissermel, K., Arpe, H.-J.: Industrial Organic Chemistry, p. 335, Weinheim, Germany, Verlag Chemie, 1978
268. Werner, A.C.: Ind. Eng. Chem. *44*, 2736 (1952)
269. Wilbourn, J., Montesano, R.: EHP, Environ. Health Perspect. *45*, 127 (1982)
270. Williams, D.T.: J. Agric. Food Chem. *21*, 1128 (1973)
271. Williams, D.T., Blanchfield, B.J.: Bull. Environ. Contam. Toxicol. *11*, 371 (1974)
272. Wilson, W.B., Giam, C.S., Goodwin, T.E., Aldrich, A., Carpenter, V., Hrung, Y.C.: Bull. Environ. Contam. Toxicol. *20*, 149 (1978)
273. Wiseman, P.: An Introduction to Industrial Organic Chemistry, London, p. 130, Applied Science Publishers, Ltd., 1979[2]
274. Wittcoff, H.A., Reuben, B.G.: Industrial Organic Chemicals in Perspective, Part I: Raw Materials and Manufacturing, p. 107, New York, John Wiley & Sons, 1980
275. Wofford, H.W., Wilsey, C.D., Neff, G.S., Giam, C.S., Neff, J.M.: Ecotoxicol. Environ. Saf. *5*, 202 (1981)
276. Wolfe, N.L., Burns, L.A., Steen, W.C.: Chemosphere *9*, 393 (1980)
277. Wolfe, N.L., Steen, W.C., Burns, L.A.: Chemosphere *9*, 403 (1980)
278. Wolfe, N.L., Steen, W.C., Burns, L.A.: Phthalate Ester Hydrolysis: Linear Free Energy Relationships, EPA 600/J-80016, NTIS PB81-129579 (1981)
279. Yamasaki, H., Kuwata, K.: Bunseki Kagaku *26*, 1 (1973)
280. Yamasaki, H., Kuwata, K.: Taiki Osen Gakkaishi *16*, 417 (1981); Chem. Abstr. 96:204582h (1982)
281. Zeiger, E., Haworth, S., Speck, W., Mortelmans, K.: EHP, Environ. Health Perspect. *45*, 99 (1982)
282. Zitko, V.: Int. J. Environ. Anal. Chem. *2*, 241 (1973)
283. Zoeteman, B.C., Harmsen, J.K., Linders, J.B.H.J., Morra, C.F.H., Slooff, W.: Chemosphere *9*, 231 (1980)

Thallium

J. Schoer

Institut für Sedimentforschung, Universität Heidelberg
D-6900 Heidelberg, Federal Republic of Germany

Introduction

Thallium was discovered in 1861 by the English chemist William Crookes, who, while examining residues from a sulfuric acid plant, noticed a green line in the emission spectrum never seen before. The studied material contained an unknown element. The green line resembled the colour of spring vegetation and so he named the new element "thallium" from the Greek "thallos" for a young twig. Metallic thallium was first prepared by the French scientist C. Lamy in 1862 [395].

In the following years the toxic characteristics of the metal were recognized but badly underestimated. Thallium compounds could even be found in medicine cabinets. Thallous acetate was used, for example, as a depilatory and in cases of tuberculosis as an antihydroticum. Applied doses were near the range of acute toxicity and cases of poisoning did occur. Munch [268] reported on the frequency of thallium poisoning after medical treatment and noted that from 692 investigated cases, 31 were mortal.

The direct application to exposed body parts also led to serious health problems. During the late 1920's, "Koremlu-Cream", a depilatory cream containing nearly 5% thallium, entered the US market. Within months a thallium poisoning occurred. Altogether there were 51 cases registered, and those affected suffered from lasting effects such as problems of vision including blindness even after disuse [301]. Although as early as 1937 marketing of such products for the removal of hair was discontinued further cases were still reported.

Since the 1920's, thallium became widely used in industry. During mining, ore processing and product application, numerous cases of industrial thallium poisoning occurred. In the 1950's, Swedish workers in 4 different production facilities were affected by rodenticides containing thallium. In West Germany, an employee of a pigment factory was reported to show indications of thallium poisoning, and in the United States 15 labourers suffered from consequences incurred in handling gravity solutions prepared from thallium salts [133, 316].

It is assumed that the number of persons afflicted is much higher than indicated because the symptoms of a chronic thallium poisoning are not easily determined and an accurate diagnosis is rarely possible [166].

Others fell victim to mishaps. In California, 32 people showed the effect of acute poisoning after partaking of food-stuffs prepared from thallium-spiked flour in 1932; 6 of them died [174, 270]. Between 1953 and 1959, more than 130 children in Texas alone were reported to have shown effects of thallium from contact with rodenticides [312].

Thallium pollution is less widespread than that by other heavy metals but it has nevertheless caused local problems. The contamination of the environment with thallium results chiefly from nonferrous mines and metal works, from coal combustion, from steel and manganese production and from cement plants [302, 364]. Other authors described thallium emission from sulfuric acid plants, iron works, brick works and mining effluents [52, 416]. The hazard of thallium pollution should therefore be judged on the basis of the amounts released into the environment rather than on the basis of production.

Smith and Carson [364] have estimated that in the USA alone nearly 1,000 tons of thallium are released into the environment each year from which 350 tons are traces in vapours and dusts, 60 tons are bound in nonferrous metals and more than 500 tons are contained in fluid and solid wastes; the latter are, to a certain extent, deposited in special storage sites.

The uncontrolled outflow of thallium from settling basins of base-metal mining has led to the pollution of a number of water bodies. In two rivers draining such mining areas in New Brunswick, Canada, water concentrations of 1–80 μg Tl/l were reported. Algae and moss samples from these rivers contained thallium in concentrations ranging from 9.5 to 162 μg/g dry weight [418].

In 1979, a serious situation developed in Lengerich, Federal Republic of Germany, where a heavy thallium contamination of the environment was caused by a cement plant. Leaves fell from trees in June, sheep died without any outward signs of disorder or disease, rabbits and horses were reported to loose fur and hair. In samples of green cabbage and in grains thallium values of 45 μg/g and 9.5 μg/g, respectively, were found [302]. The thallium concentration in beef tissue was in the range from 0.4 to 1.7 μg/g [151] with the result that no permission was given for the marketing of certain feed supplies for animals and foodstuffs for humans. Epidemiological investigations, however, indicated that only as slight chronic poisoning of hypersensitive persons might have occurred [46].

In addition to liquid waste discharges and stack emissions, thallium is also brought onto agricultural soils as trace component of sewage sludge from water purification plants [157] and in potassium fertilizers [364].

Although there has been no report of persons seriously threatened by exposure to thallium released by man, it must nevertheless be included in the group of potentially dangerous heavy metals due to its strong toxic effects and the high quantity reaching the environment annually.

Production and Use

Although several minerals containing thallium as an essential element are known, thallium deposits with a sufficiently high ore concentration to warrant mining operations have not been found.

On the other hand, thallium occurs in relatively high concentrations in some sulfidic deposits. During the processing of such ores it is enriched on dust particles due to the volatility of most thallium compounds. Almost all thallium used in the world today is produced from such dusts.

Several methods have been developed to extract thallium from ores, flue dusts and other materials. In one process applied to cadmium- and thallium-bearing flue dust resulting from the smelting of lead-zinc sulfide ores, the dust is leached with sulfuric acid, and cadmium and thallium sulfates are electrolyzed for the deposition of cadmium. When the ratio of thallium to cadmium in the electrolyte reaches 1:10, new cathodes are introduced. The electrolysis is continued and a cadmium-thallium alloy containing 5%–20% thallium is deposited. This alloy is converted to hydroxides by treatment with hot water and steam. Then, sodium carbonate is added to precipitate the cadmium. The cadmium-free filtrate containing thallium carbonate is treated with hydrogen sulfide to precipitate thallium sulfide. The washed thallium sulfide is dissolved in sulfuric acid and boiled free of hydrogen sulfide. The resulting solution of pure thallous sulfate is electrolyzed to recover pure metallic thallium sponge. The sponge is then pressed, melted and cast.

Thallium metal can also be prepared from thallium sulfate solution by precipitation on zinc plates arranged vertically around the sides of a vessel. After heating for several hours, thallium sponge is deposited on the zinc. The spongy mass is separated from the plates, thoroughly washed with water and compressed or fused into sticks.

Another method uses zinc dust cementation of thallium, cadmium, and indium from a leached solution of pyrite cinders. The cementite is redissolved and converted into a zinc-free mercury amalgam. Subsequent fractional distillation yields an indium-containing thallium fraction which is further refined to thallium sulfate.

High-purity thallium is produced by a two- or multistage electrolysis, amalgam electrolysis, or fractional dissolution and precipitation. Purification can also be performed by the zone-refining process [9].

World production of thallium is only 10–12 t/a [201]. Information on its use is difficult to obtain and sometimes contradictory. Until the 1960's thallium compounds were used primarily as rodenticides, and the consumption in the USA is clearly characterized by this application.

$$\text{Raw Material} \xrightarrow{H_2SO_4} CdSO_4 + Tl_2SO_4 \xrightarrow{\text{Electrolysis}}$$

(From Smelting Pb- and Zn-Sulfide Ores)

$$\text{5–20\% Tl-Cd Alloy} \xrightarrow{\text{Hot } H_2O} \begin{matrix} TlOH \\ Cd(OH)_2 \end{matrix} \xrightarrow{Na_2CO_3}$$

$$CdCO_3 + TlOH \xrightarrow[\text{2) } H_2S]{\text{1) Filtration, } -CdCO_3} Tl_2S\downarrow \xrightarrow{H_2SO_4}$$

$$Tl_2SO_4 \text{ Solution} \xrightarrow{\text{Electrolysis}} \text{Pure Tl} \xrightarrow[\text{from } H_2SO_4]{\text{Crystallization}}$$

$$Tl_2SO_4$$

Fig. 1. Production of thallium from lead/zinc roast dusts [287]

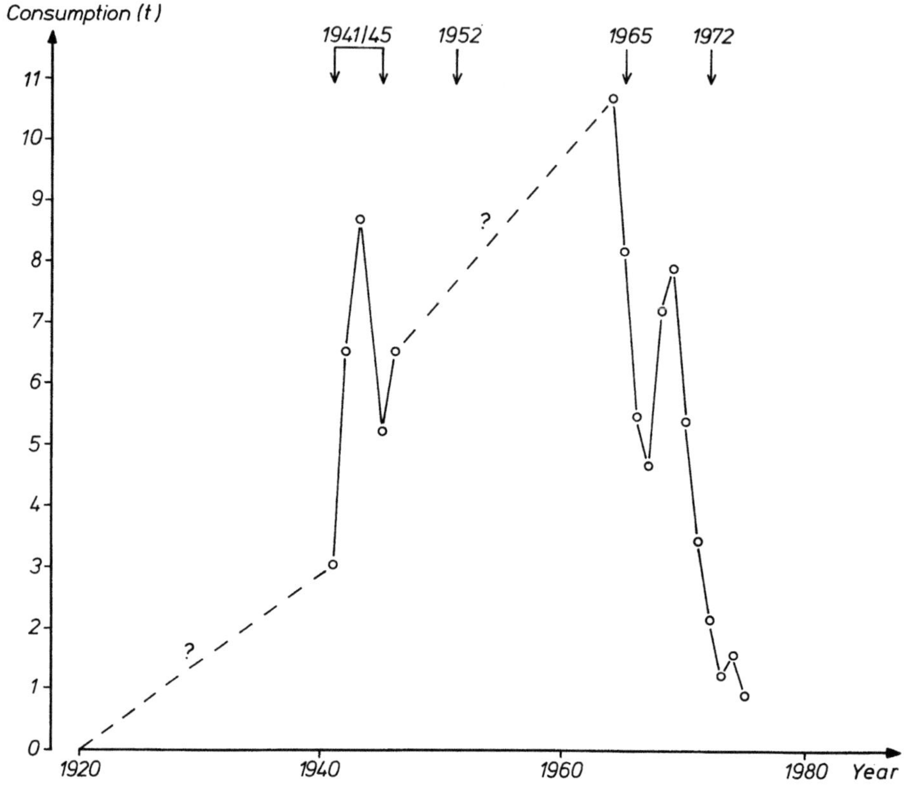

Fig. 2. US consumption of thallium between 1920 and 1980 [24, 84]

Figure 2 shows the consumption of thallium between 1920 and 1980. The temporary increase in consumption during the Second World War was due to a greater need for the protection of stored and transported foodstuffs during the war years. Inspite of increased amounts of foodstuffs to be protected, the production rate of thallium remained presumedly constant during the 1950's as a result of the marketing of thallium-containing rodenticides being held back since 1952. The thallium concentration in these substances was reduced in 1960 and they were outlawed for private users in 1965 which led to a decline in consumption. In 1972, the application of thallium-containing preparations for these purposes was prohibited entirely.

Greenspoon [139] reported conflicting thallium consumption data of 13 t/a for the mid-fifties. Robinson [321] presumes that this deviation probably results from the increased research using radiation equipment and from sophistication of military defence systems, particularly infrared detection. Exact data for the years 1945 to 1965 cannot be given since consumption rate was classified information.

Thallium is further used in alloys. Lead, silver or gold-based bearing alloys containing between 1% and 50% thallium have high endurance limits, very low frictional coefficients, and good resistance to acids. In addition, insoluble anodes used in the electrolytic extraction of copper contain 10% thallium.

Furthermore, thallium also finds application in the electric and electronic industry, mercury vapour lamps and deep-temperature thermometers. The electrical conductivity of "thallofide" (incompletely oxidized Tl_2S) is very sensitive to changes in both the temperature and intensity of illumination. Because of their sensitivity to infrared rays, thallofide photocells have replaced selenium ones [388].

Thallium is used as a catalyst in the synthesis of hydrocarbons and olefins as well as in polymerisations and epoxidation reactions. Although no exact data on these processes are available, it is expected that this use of thallium will increase in the future [364].

Pigment and anti-knock material production as well as electroplating procedure require only minimum amounts of thallium.

Thallous compounds have interesting optical characteristics that can be used for a number of purposes. Added to glass, thallium halides increase the refractive index. This seems to offer prospects for the future since thallium is not only used in decorative glasses and spectrometry but also imparts technically desired characteristics to fiber glass. Since fiber glass is on the threshold of widespread use in communication systems, an increase in thallium consumption must be expected. To satisfy this new demand, a Japanese electrolysis plant should have started thallium production with a capacity of 12 t/a in October 1982 [18].

Chemistry

Elemental Thallium

Thallium with the atomic number 81 and the atomic weight 204.39 occurs next to lead in Group III B of the Periodic Table. There are two naturally occurring stable isotops, Tl-203 and Tl-205, their relative abundancies being 29.46 and 70.54 percent, respectively.

The density of the bluish-white to silvery-white metal is 11.85 g/cm^3 at 20 °C, the melting point 303 °C and the boiling point 1,457 °C.

Thallium occurs in three crystal modifications: below 230 °C it is hexagonal close-packed, above 230 °C it is cubic body-centered and at high pressures it is cubic face-centred, the triple point being at 110 °C and 30 kbars.

The electron configuration of thallium is (Xe) $4f^{14}\,5d^{10}\,6s^2\,6p$. In contrast to the other Group III B elements – boron, aluminum, gallium, and indium – its s-electrons display a low propensity to be released or bound covalently. Therefore, this metal chiefly occurs in a monovalent form. The oxidation potential of the reaction $Tl(I) \rightarrow Tl(III)$ is +1.21 V [313].

Ionic radii of thallium in various coordination polyhedra are summarized in Table 1 [351].

Besides the ionic radii, the electronegativity constant of the monovalent form ($EN_{Tl\text{-}I}$: 2.04; $EN_{Tl\text{-}III}$: 1.44) is also very similar to those of higher alkali metals. These properties and its position in the Periodic Table of Elements impart a lithophilic and chalcophilic character to thallium, and under certain conditions it is even siderophilic [387].

Table 1. Ionic radii of thallium in various coordination polyhydra

	Coordination No.	Ionic radius
Tl–(I)	VI	1.50 A
	VIII	1.60 A
	XII	1.76 A
Tl–(III)	VI	0.88 A
	VIII	1.00 A

Table 2. Physical properties of some inorganic thallium compounds [364, 392]

Compound	Formula	Solubility		Temperature (°C)	
		g/l (20°C)	Product	Melting	Boiling
Thallium bromide	TlBr	0.48	2.6×10^{-6}	480	815
Thallium carbonate	Tl_2CO_3	52.3		273	d
Thallium (I) chloride	TlCl	3.4	1.5×10^{-4}	430	720
Thallium (III) chloride	$TlCl_3 \cdot 4H_2O$	826		37	d
Thallium ferrocyanide	$Tl_4(Fe(CN)_6) \cdot 2H_2O$	3.7	3.8×10^{-14}	–	–
Thallium fluoride	TlF	786	–	327	655
Thallium hydroxide	TlOH	350	–	139 d	–
Thallium jodide	TlJ	0.06	3.6×10^{-8}	440	823
Thallium nitrate	$TlNO_3$	86.7	–	206	430
Thallium oxalate	$Tl_2(C_2O_4)$	15.3	–	–	–
Thallium (I) oxide	Tl_2O	–	–	300	1,080
Thallium (III) oxide	Tl_2O_3	–	–	717	875 O_2
Thallium phosphate	Tl_3PO_4	5.0	6.6×10^{-9}	–	–
Thallium sulfate	Tl_2SO_4	46.4	–	632	d
Thallium sulfide	Tl_2S	$3 \cdot 10^{-6}$; –0.695	10^{-24}–7×10^{-20}	448.5	d

In moist air at normal temperature, thallium is rapidly coated with a gray film of thallium(I) oxide which changes to brown Tl_2O_3 at 100 °C whereas it is unaffected by absolutely dry oxygen or nitrogen and by water containing no air or oxygen. Water with dissolved air or oxygen rapidly corrodes the metal to give TlOH.

Thallium is readily dissolved in acid solutions until its concentration reaches the solubility limit of the salts formed. Thallium does not dissolve in basic solutions such as alkalis.

Inorganic Thallium Compounds

Various inorganic thallium compounds do exist; some physical properties of ecologically important compounds are listed in Table 2.

Thallium(I) halides are chiefly ionic, stable in air and incompletely dissociated in water; with the exception of the fluoride they are poorly soluble. During heating the chloride starts to sublime near 360 °C; the vapour pressure as a function of tem-

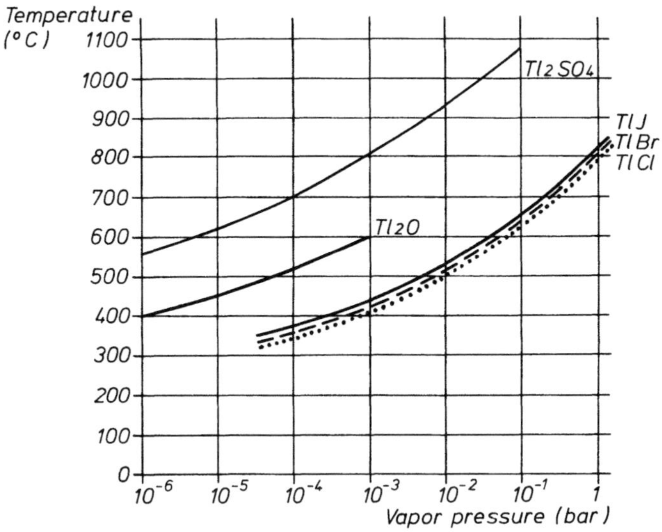

Fig. 3. Vapour-pressure curves of some thallium(I) compounds

perature can be calculated by the formula:

$$\log p = -5492.85/T + 5.0093 \,.$$

Tl(III) halides are unstable and lose halogen on heating. For example, at 40 °C the monochloride is formed from the trichloride. They are easily soluble in water and hydrolyze immediately.

Thallium-oxygen compounds exist as mono- and trioxides as well as poorly defined mixed oxides and peroxide, Tl_2O_4. Black thallium monooxide, Tl_2O, is obtained by oxidation of elementary thallium in air, by heating the hydroxide at 100 °C or the trioxide at 700 °C. It is strongly hygroscopic and immediately reacts with water to form the monohydroxide [397].

Thallium trioxide can be prepared by oxidation of an alkaline Tl(I) solution with hydrogen peroxide or by treatment of an aqueous Tl(I) nitrate solution with chlorine or bromine and subsequent alkaline oxidation with oxygen.

From the carbonates only the monovalent thallium compound is known. Thallium(I) carbonate hydrolyzes immediately and acts as a base. The water-soluble thallium(I) nitrate is formed when elementary thallium is dissolved in diluted nitric acid; thallium(I) sulfate can be prepared by treatment of the nitrate with diluted sulfuric acid.

Generally, inorganic Tl(I) compounds are more stable than the Tl(III) analogues. Some of them can be evaporated without dissociation as shown in Fig. 3.

Organic Thallium Compounds and Thallium Complexes

In contrast to inorganic thallium compounds, covalent organothallium compounds are stable only in the trivalent form. They are produced via Grignard reactions of from aryldiazonium- or diphenylthallium salts [251].

Table 3. Equilibrium constants of thallium complexes

Central atom	Ligand		Log equilibrium constant
Tl^+	OH^-	hydroxide	$K_1=0.42–0.85$
Tl^{3+}	OH^-	hydroxide	$K_1=-0.2$
Tl^+	$Fe(CN)_6^{4-}$	cyanoferrate (II)	$K_1=3.0–3.22$
Tl^+	$Fe(CN)_6^{3-}$	cyanoferrate (III)	$K_1=1.83^c$
Tl^+	NO_3^-	nitrate	$K_1=0.31–0.44$
Tl^{3+}	NO_3^-	nitrate	$K_1=0.18–0.41$
Tl^+	S^{2-}	sulfide	$K_1=-20$
Tl^+	SO_4^2	sulfate	$K_1=1–2$
Tl^+	Cl^-	chloride	$K_1=0.47–0.78$
Tl^{3+}	Cl^-	chloride	$K_1–K_4=8.1;\ 5.5;\ 2.2;\ 2.2$
Tl^+	PO_4^{3-}	phosphate	$K_1=2.41$
Tl^+	$C_2H_2O_4$	oxalic acid	$K_1=2.03$
Tl^+	$C_2H_4O_2$	acetic acid	$K_1=-0.11$
Tl^{3+}	$C_2H_4O_2$	acetic acid	$K_1=15.4$
Tl^+	$C_2H_8N_2$	ethylenediamine	$K_1=0.3–0.4$
Tl^+	$C_6H_8O_7$	citric acid	$K_1=0.65–1.04$
Tl^+	$C_6H_9O_6N$	nitrilotriacetic acid	$K_1=0.44–0.74$
Tl^{3+}	$C_{10}H_{16}O_8N_2$	EDTA	$K_1=22.5$
Tl^+	$C_4H_6O_6$	tartaric acid	$K_1=1.39$

Thallium shows only a low propensity to form complex compounds. Table 3 compiles some complexing constants [364].

The logarithms of most constants of complex formation between organic substances and thallium are in the range 0.5–2.0. Tl(I) is not complexed by humic acids whereas Tl(III) forms stable complexes of the TlX_4^-- or TlX_6^{3-}-type.

Moreover, Tl(III) undergoes methylation reactions to form stable dimethylthallium salts [416].

The most complete review articles of thallium chemistry have been published by Lee [215, 216].

Analytical Methods

Improvement of classical analytical techniques, introduction of new methods and a combination of both with enrichment or separation processes provide suitable methods for solving all analytical problems encountered in the quantitative determination of thallium.

In this respect preconcentration procedures are very important since thallium in environmental samples can only rarely be determined with sufficient accuracy directly from the sample or digestion solution.

Atomic Absorption Spectroscopy (AAS)

Atomic absorption spectroscopy is the most widely applied method of quantitative thallium determination in various media. Depending on the concentration and on the sample matrix, various enrichment and partitioning steps must be carried out before actual measurements are performed.

Heinrichs [156] suggests the volatilization technique, which was developed by Geilmann and Neeb [126], for the determination of thallium traces. In this technique thallium is volatilized from the sample material at temperatures above 1,000 °C, condensed on a cooling rod, digested in concentrated nitric acid, diluted with water and pipetted directly into the tube of the flameless AAS apparatus. The precision of this method is reported to range from 2% to 30%, depending on the thallium concentration. The detection limit is 1 ng/g and the sensitivity 1×10^{-11} g/1% adsorption on the 277 mn line.

In addition to volatilization techniques, other procedures can be applied for preconcentration, for example, the extraction of organo thallium complexes.

Thallium in higher concentrations can be determined directly from digestion solutions [390]. At low concentrations of the matrix ions there is no interference; at matrix ion concentrations greater than 1% interferences, which generally lead to a systematic increase in the measured values, may occur. In such cases, the application of the addition method or compensation of the interference factors, e.g. with alkaline buffer solutions, is recommended.

In direct measurements the detection limits of thallium in the digested solution are 0.02 µg/ml (flame) or 0.02 ng (tube). The 1%-sensitivity is 0.5 µg/ml or approx. 0.1 ng [408].

Photometry

Photometry is a classical analytical technique for the determination of thallium [110]. In recent years, this method has been developed to a thallium-specific routine procedure for large numbers of samples [309].

Trivalent thallium forms coloured complexes with many organic reagents which may be utilized in photometry. The complex of thallium(III) with brilliant

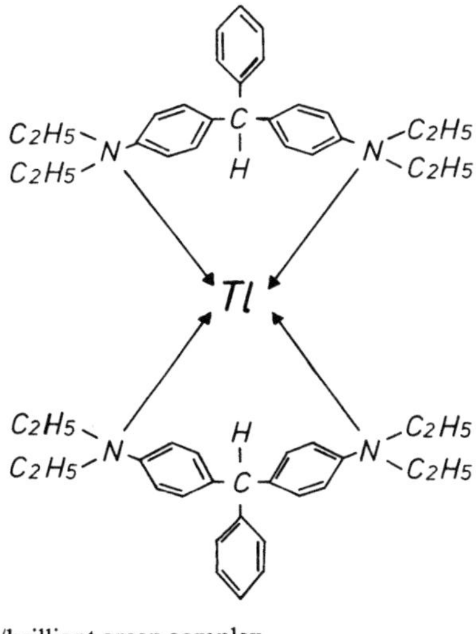

Fig. 4. Configuration of the Tl/brilliant green complex

green is used for the determination of thallium because of its selectivity and high molar extinction coefficient of 100,000 l/mol × cm [309].

Oxidation of monovalent to trivalent thallium is mostly performed with bromium. After the addition of brilliant green, the complex illustrated in Fig. 4 may be extracted with organic solvents.

The extinction curve shows maxima at 435 and 627 nm, the latter being much more sensitive. In this context, there is some difference of opinion as to the validity range of the Lambert Beer's Law [182, 239]. The detection limit is approximately 5 μg/ml · cm. Other useful complex reagents have been reviewed by Snell [366].

Voltametry

Electrochemical procedures have rapidly been developed in recent years. One of the most sensitive methods for the quantitative detection of heavy metals is differential pulse inverse voltametry (DPIV) which is the preferred method for the determination of thallium in water samples [198].

In addition to its sensitivity, DPIV also offers the advantage of determining simultaneously several metals from one solution. Difficulties which arise when the potentials of various metals coincide can generally be eliminated with special analytical procedures. For example, if Cu, Cd, Pb, and Tl occur together, the concentrations of Cd and Cu are first determined in acidic solution, and subsequently the Pb and Tl concentrations in alkaline solution. In the latter determination Cd is precipitated and the coincidence of the potentials of Pb and Tl is thus eliminated since the Pb peaks have been transferred to more negative potential values [198]. Alternative coincidence eliminations involve the use of complexing agents (see Neeb [276] and Christmann et al. [69]).

In potable water, thallium concentrations as low as 1 ng/ml can be directly detected by DPIV. Data on the principal detection limit or reproducibility have not been reported so far.

Neutron Activation Analysis

Neutron activation analysis has been applied to the determination of thallium in rocks, sediments, biological tissues, and water samples [243, 245, 274]. NAA is especially useful when, in addition to thallium, a large number of other elements is to be analysed.

Detection limits are determined by irridation time, neutron flux, preparation of samples, matrix and measurement time. They usually lie in the ng-range. Several important data for activation analysis are compiled in Table 4.

From Table 4 it is obvious that Tl-204 is chiefly used for the determination of trace amounts of thallium. Since this nuclide is chiefly a β-radiator, thallium must be isolated before or after irradiation. Depending on the matrix ions, various procedures have been employed, for instance precipitaton or extraction [274] as well as ion exchangers [48]. In spite of the enrichment involved, a neutron flux of 1–5×10^{13} n/cm^2 s lasting several days is required if the error in the ng-range is to be held under 5%.

The γ-emitter Hg-203, which arises from Tl-203 by electron capture, can be used for the determination of thallium without previous separation if rather high

Table 4. Physical data on the activation analysis of thallium [33]

Parameter	Tl-203		Tl-205
Isotope abundance	29.5		70.5%
Activation cross section			
thermal neutrons	8		0.1
resonance neutrons	129		0.5
Product	Tl-204 (97.6%) /	Hg-203 (2.6%)	Tl-206
half-life	3.78 a	46.6 d	4.2 m
decay, energy, abundance	γ = 0.071 MeV; 0.074%	0.279 MeV; 81.5%	1.163 MeV; 0.08%
	β^-_{max} = 0.766 MeV; 100%	0.214 MeV 0.214%	1.52 MeV 99.9%

amounts of thallium are to be analysed and if the sample is free of mercury since Hg-203 can be created from Hg-202 by (n,γ) reaction.

X-Ray Fluorescence Analysis

If relatively large sample amounts are available (approx. 10 g), thallium can be quantitatively determined in the lower µg/g range by X-ray fluorescence without chemical separation and enrichment techniques [314]. Calibration is performed by use of the addition method, the thallium L-α line being applied for measurements.

With a statistical certainity of 99.7%, detection limits of 0.34–0.64 µg/g are indicated for plant samples after a measurement time of 10 minutes.

Field-Desorption Mass-Spectroscopy (FDMS)

Mass-Spectroscopy has been used for many years especially for the determination of thallium in biologic tissue [405]. Today, quantitative analysis of sample amounts in the milligram scale is possible in the ultratrace range [3].

Since in the analysis of biological samples no other pretreatment besides homogenisation and centrifugation is necessary, FDMS is very useful, primarily in pharmacokinetic studies.

FDMS analysis is mostly carried out by isotopic dilution analysis. The sample is mixed with a known amount of thallium with an isotopic composition different to that of the natural metal. The observed change in isotope ratio depends on the thallium concentration in the sample. The amount of natural thallium in the sample can be calculated from the altered peaks according to the formula:

$$\mathrm{Tl} = \frac{\alpha(\mathrm{Tl}_{enr.})}{W} \, \frac{x_1 - x_2}{0.705R - 0.295},$$

where α is the weight and Tl_{enr} the thallium concentration of the added solution, W the sample weight, R the peak relation, and X_1 and X_2 are the percentages of Tl-203, and Tl-205 in the artificial isotopic solution.

An analysis time of 20 min is necessary to reach a precision of 10% in the pg range.

Gas Chromatography

Monovalent thallium is precipitated with cyclopentadiene as thallium cyclopentadienide (TlC_5H_5) from aqueous alkaline solutions. After extraction with dichloromethane TlC_5H_5 can be detected gas chromatographically.

With an input volume of 2 µl a detection limit of 1 µg is possible [38] but due to improved columns and detectors the detection limit may be significantly lowered in the near future.

Occurrence and Geochemistry

Because of its lithophilic and chalcophilic character thallium occurs ubiquitously. Although its crustal abundance is only 0.3 to 0.6 g/t, it exceeds those of mercury, silver and gold. In contrast to these elements, however, thallium does not occur in a native form and is rarely enriched in specific minerals. Several characteristic thallium sulfide, selenide and oxide minerals occur in nature but they are extremely rare; in general thallium occurs as a trace element in other minerals.

Table 5 lists some of the important thallium containing minerals.

As trace element in other minerals, thallium is mainly associated with potassium, rubidium and cesium in magmatic, metasomatic and pegmatic low-temperature formations, with iron and various nonferrous metals in meso- and epithermal deposits as well as in supergenetic manganese precipitations [321].

From the data of 2,000 magmatic, metamorphic, and sedimentary rock samples and rock-forming minerals of the earth's crust and mantle (Table 6), Heinrichs et al. [160] developed a geochemical transport model for thallium.

The earth's mantle contains very low concentrations of thallium compared to the crust. This is explained by the release of thallium from energetically unfavourable lattice sites during partial melting of the mantle material. Thallium is trans-

Table 5. Chemical composition, paragenetic associations and locations of thallium minerals occurring in nature [346]

Mineral	Chemical Composition	Paragenesis	Comments
Lorandite	$TlAsS_2$	Realgar, orpiment, colloform pyrite	Found in Macedonia
Picotpaulite	$TlFe_2S_3$		
Chalkothallite	Cu_3TlS_2		
Vrbaite	$Hg_3Tl_4As_8Sb_2S_{20}$	Realgar, orpiment, colloform pyrite	Macedonia, 29%–32% Tl
Hutchinsonite	$(Pb, Tl)_2(Cu, Ag)As_5S_{10}$	Sulfides of As, Zn, Pb, Fe	18%–33% Tl
Bukovite	$Cu_{3+a}Tl_2FeSe_{4-a}$		
Wallisite	$PbTlCuAs_2S_5$		
Hatchite	$PbTlAgAs_2S_5$		
Crookesite	$(Cu, Tl, Ag)_2Se$	Selenides	Found in Skrikerum, Sweden
Avicennite	Tl_2O_3		

Table 6. Average abundance of potassium, rubidium and thallium in rock samples [160]

	Potassium (%)	Rubidium (mg/g)	Thallium (μg/g)
Continental crust			
– upper	2.52	142	749
sedimentary rocks	1.77	100	413
magmatic rocks	3.07	197	980
metamorphic rocks	2.13	97	592
– lower	1.85	54	232
– average	2.19	98	491
Upper mantle	0.04	1.2	11

Table 7. Thallium concentrations in uncontaminated geochemical samples

Sample	Thallium concentration (μg/l)	Ref.
Air (Nebraska, US average)	0.2[a]	[376]
Seawater	0.01	[237]
	0.02	[79]
Fresh water	0.002–0.04	[154]
Sediments:		
German rivers	0.01–0.38	[244]
Deep sea: Clays,	0.8	
Carbonates	0.16	[79]
Soils	0.1–0.6	[338]
from Posidonian slates	6.5	[304]
Shale	1.2	[42]
Limestone	0.14	[42]
Sandstone	0.36	[42]
Coal	0.1–2.3	[79]
Manganese nodules (average)	100	[354]
Galena	1,000	[16]
Sphalerite	5,000	[16]
Pyrite	100	[16]

[a] Concentration in ng/m^3

ported into the crust via magma as reflected by the high thallium concentrations of magmatic rocks.

The isomorphous substitution of thallium for potassium occurs readily, especially in the later-stage pegmatitic alkali feldspars and micas.

In addition, increased thallium concentrations occur in sulfidic deposits which, however, can vary widely from 1 to 10,000 μg/g. High concentrations are chiefly encountered in recent deposits which are precipitated from colloidal solutions in shallow depths and at low temperatures. The occurrence of thallium has even been suggested as a possible indicator of conditions favourable for the formation of such deposits [321, 364].

During rock weathering processes, thallium is largely retained by adsorption, e.g. on clay minerals. On the other hand, a loss of thallium relative to Rb in the formation of argillaceous rocks from igneous rocks is indicated. Thallium is transported in the univalent state and removed from solutions by precipitation or adsorption [79, 354].

Upon oxidation of sulfidic deposits, most thallium dissolves. It either coprecipitates later with oxides (manganese oxides contain up to 2,300 μg/g thallium [364]) or is adsorbed on clay minerals such as illite and biotite or deposited under calm and anaerobic conditions in sediments [309]. The thallium concentrations in different geochemical samples are listed in Table 7.

Emission Sources

Emissions of thallium and its compounds do not occur chiefly where the metal is recovered, treated or consumed. Rather, the major source of environmental thallium are processes in which the metal is involved only as a trace element of raw materials handled in large quantities such as coal, sulfidic ores or cement.

The thallium mass flow in these processes has been established to a certain degree. However, available data have been derived from relatively few single investigations and are not the result of systematic study. In light of this fact, the data of Carlson and Smith [364] should be considered as a rough estimate of possible thallium emissions.

Detailed results on thallium emissions are reported from the cement industry. After a quantitative estimation of other emitters, the behaviour of thallium in the cement production processes, the role of dust recycling, and potential measures for reducing the thallium content in emission will be discussed.

Combustion of Fossil Fuels

The thallium content of coal varies over several orders of magnitude depending on the type of coal, location of coal deposits and the portion of sulfidic components. Several examples are compiled in Table 8.

Table 8. Thallium concentrations in several coal samples

Type of coal	Origin	Average Tl conc. (μg/g)	Ref.
Lignite	FR Germany	0.011	[155]
Coal	–	0.55	[231]
Coal	–	0.59	NBS-Standard
Coal	Illinois, USA	0.66	[364]
Soft coal (bituminous)	east coast, USA	0.73	[364]
Anthracite coal	west coast, USA	0.05	[364]
Anthracite coal (with pyrite coatings)	–	4.0	[364]
Anthracite coal (with high sulfide portions)	–	135.0	[364]

For estimations of thallium emissions from coal burning an average value of 0.6 μg Tl/g coal can realistically be assumed.

The high thallium content of coal with a significant sulfidic portion and the low quantity in anthracite coal indicates that in coal thallium is not organically bound but rather incorporated in sulfides [395]. Separation of the sulfidic components often decreases the thallium content of coal significantly.

Thallium volatilizes during the combustion and coking of coal as chloride, sulfate or oxide, depending on the accompanying ions, temperature and other parameters. It recondenses in cooler parts of the system onto the surface of ash particulates where its concentration increases with decreasing particle size. An amount of 76 μg/g thallium has been detected in "air-borne" fly ash with a diameter of 1–5 μm [77]. Modern facilities remove between 90% and 99.9% of the fly ash produced. Upon recycling of these ashes, high concentrations of volatile heavy metals occur and Tl emissions increase until input and emissions are balanced.

Research groups in the USA determined the thallium content of various input, transformation and emission substances such as dusts, bottom ash, precipitator ash, scrubber solids and liquids and flue gas etc. to establish flow diagrams of coal-burning plants (Fig. 5).

In spite of the limited data available they nevertheless allow a rough estimate of thallium emission from coal-burning power plants projected on a universal

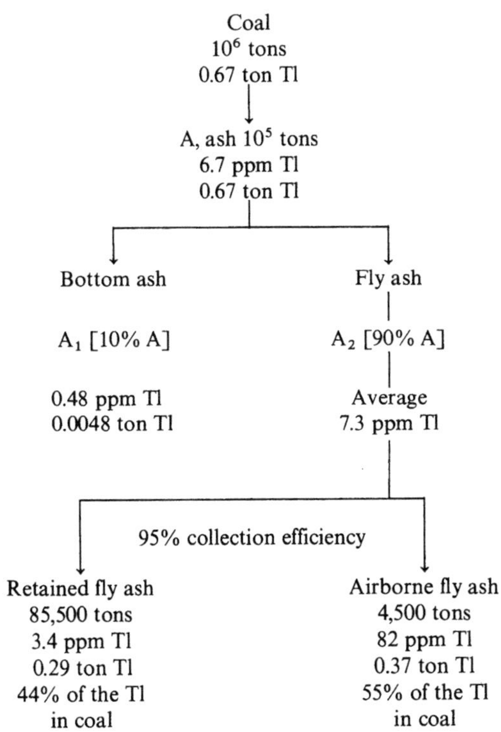

Fig. 5. Thallium material balance in a coal-burning process [364]

scale. The annual values calculated are for the USA 140 t, for Western Europe 80 t and for the whole world 600 t.

The thallium content in raw oil is significantly lower than in coal; the averadge concentration is given as 5 ng/g [283]. During raw-oil cracking and fractionation, Tl is enriched in the tar and asphalt fractions. There are no accurate data avaible on the thallium concentration in the various phases, which makes the computation of emissions during the burning of light heating oil or in motor vehicles impossible.

The content of thallium in oil slates is similar to that found in coal. The fate of thallium during processing is uncertain. Because of the large quantities involved, research on the release of thallium during mining and processing is imperative.

Iron and Steel Production

Thallium contents in oxidic iron deposits are much lower than those of sulfidic iron compounds. Pure oxidic, iron ore only contains a few μg Tl/g. However, iron oxides from polymetallic deposits can reach thallium concentrations of up to 100 μg/g.

It is estimated that during smelting 72% of the total thallium input is retained in raw iron. The remainder is found in furnace gases and dusts, of which approximately 1% is emitted. This corresponds to an emission of nearly 1 t in the United States where about 250 mill. t raw iron were produced in 1970. Another source of thallium emissions by raw iron production is the coal used in this process which amounts up to one-third of the world consumption. Using the same data for steelmaking processes, Smith and Carson [364] have calculated an additional emission rate of 5 t thallium per year.

Depending on their genesis, manganese oxides show varying contents of thallium. In super- and hypergenetic depositories the thallium content is up to 0.7%. In reductive electrosmelting, dusts containing up to 3,000 μg/g thallium have been collected [26]. It has been estimated that 120,000 t of such dusts are produced each year in the USA. Assuming that 60% is not released, 140 t of thallium are emitted into the environment from US plants alone, an amount which is of the same order of magnitude estimated for thallium emissions originating from coal combustion [364].

Thallium contents of 100 μg/g have also been found in manganese nodules [131]. However, only little additional information on this thallium source is available and emissions occurring in the processing of these nodules have not been sufficiently studied.

Nonferrous-Metal Smelting

Estimates of thallium emissions during smelting of nonferrous metals are based on US industry data where, during the 1940's and 50's, a program was initiated to determine the supplies of minor but strategic elements. In this program a large number of samples taken from nonferrous metal production sites were analyzed, in most cases also for the determination of thallium.

At that time, problems associated with sampling were still unknown or unrecognized as was the influence of certain operating conditions on the thallium content in samples taken at various process stages. Furthermore, analytical methods were much less developed than today. In spite of these deficiencies, the presented data allow a rough estimation of the various emission pathways and their magnitude.

Most nonferrous metals are generally recovered from their sulfides. The thallium content in lead and zinc sulfides has been reported to range from 2 to 2.2 μg/g [321]. It is 5 times higher in the concentrates of these sulfides which are used as raw materials for the production of pyro- and hydrometallurgic zinc [183].

In 1960, Soviet scientists investigated the fate of thallium in zinc/cadmium recovery processes. Approximately 40% of the thallium of roasted zinc concentrates remain in the end product, while 50% are contained in solid wastes and less than 2% as an inpurity in the by-product cadmium. The rest was not detected. It was assumed that thallium is at least partially contained in leaching solutions which were discharged periodically when the concentrations of certain heavy metals became too high [219, 364].

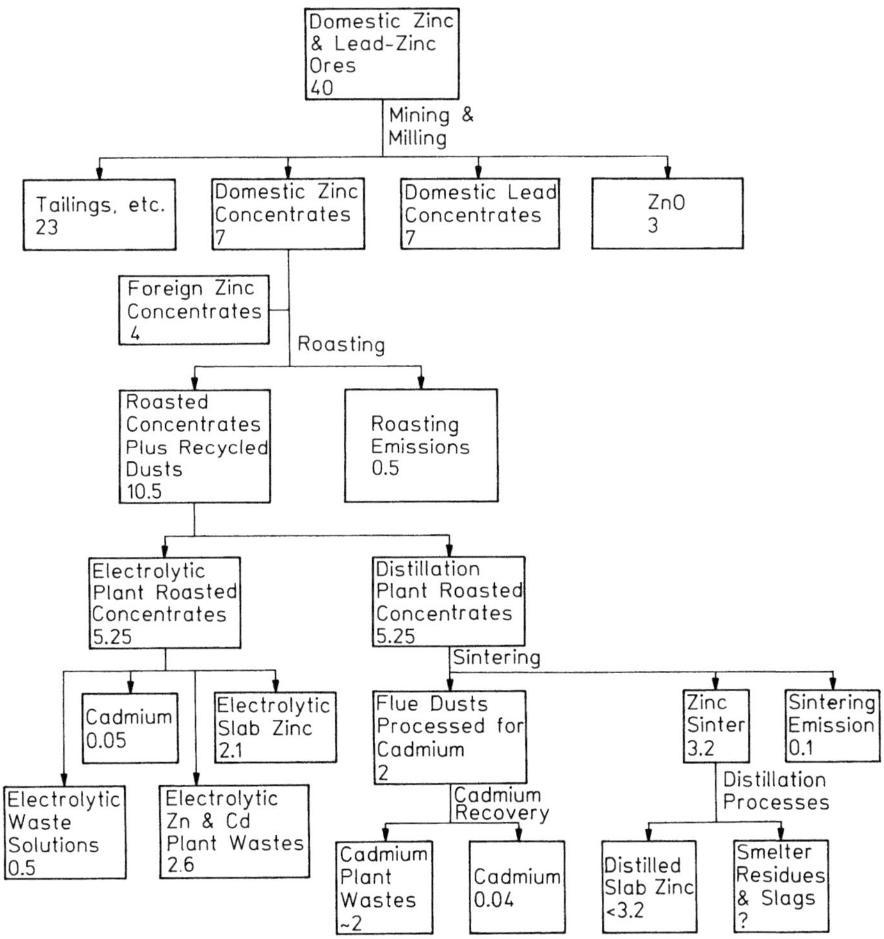

Fig. 6. Simplified flow of thallium in the U.S. zinc and cadmium industry [364] (numbers in short tons)

The flow of thallium in the production process of zinc and cadmium is schematically illustrated in Fig. 6. The distribution of thallium in the different production stages based on US concentrations and Soviet distribution data is given in short tons.

From the 11 short tons (st) of the thallium in concentrates 0.6 st are released into the atmosphere, 0.5 st into rivers, 4.6 st are contained in solid wastes, and 5.4 st remain in the final products. These values strongly depend on the procedure applied, the thallium content of the raw materials, the type of filtration as well as the influence of the techniques employed in the various stages of the process.

The processing of lead concentrates generally involves sintering, smelting, drossing, and refining. During sintering, temperatures are reached that allow various thallium compounds to evaporate. Chizhikov [68] found half of the thallium in lead concentrates to be enriched in sintering dusts in concentrations as high as 3,000 μg/g. Other authors report similar distributions: 30%–50% of the thallium in dusts, 15% in the slag and the rest in the final lead product [299].

The sintering dusts are partially recycled and partially deposited. Especially if recycling is used, large amounts of thallium are emitted even at a sufficiently high rate of filtration.

Copper processing involves other methods. Roasting and smelting processes are followed by conversion and refining. Data on the fate of thallium during copper production vary greatly. Some of them are summarized in Table 9.

Much more thallium has been found in dusts and gases using oxygen enrichment of shaft furnace smelting. Smith and Carson [364] postulate that with 2 mill. t Cu produced yearly about 500 t thallium enter the process.

Table 9. Percentage of thallium in different production materials

Material	Total Thallium (%)			
	[82]	[82]	[68]	[82, 83, 329]
Matte	46	24	31	70
Slag	34	45	62	18
Dusts/gases	20	31	7	3/9

Table 10. Lead, copper, zinc iron, and steel production data of 1979 and 1980. Values are in mill. t

	Lead refinery product (1980)	Copper (1980)		Zinc (1980) smelting	Production (1979)	
		Smelting	Refinery		Raw iron	Steel
EEC	1.19	0.15	0.97	1.33	98.3	140.8
USA	1.15	1.01	1.68	0.37	78.9	127
USSR	0.78	1.15	1.45	1.06	109	149
Japan	0.30	0.89	1.01	0.74	83.8	112
Africa	0.15	1.28	0.91	0.19	8.5	10.9
World	5.43	7.54	9.36	6.15	536	747

Although there is no definite relationship between production and emission values, the reported data nevertheless give an impression of the magnitude of possible thallium pollution caused by nonferrous metal smelting processes. Production and consumption of some base metals necessary for this evaluation are listed in Table 10 together with the relevant statistical data of the iron industry.

Miscellaneous

Thallium is also released into the environment during the production of gold (tailing dumps containing 100–2,000 µg/g thallium have been reported [414]), the production of sulfuric acid (about 4 t thallium/a in the USA [364]) and the use of potash fertilizer in agriculture (5 t thallium/a [364])

Cement Production

Portland cement is produced by burning a mixture of limestone and clay at 1,450 °C. This temperature at which many pollutants evaporate is also attained in many industries other than the cement industry in which similar mass flows are encountered. The study of the behaviour of thallium during cement burning therefore provides information on the nature of volatile pollutants in various thermal production processes and on possible emission rates of substances which are involved in these processes only as trace components of the raw materials used in large quantities.

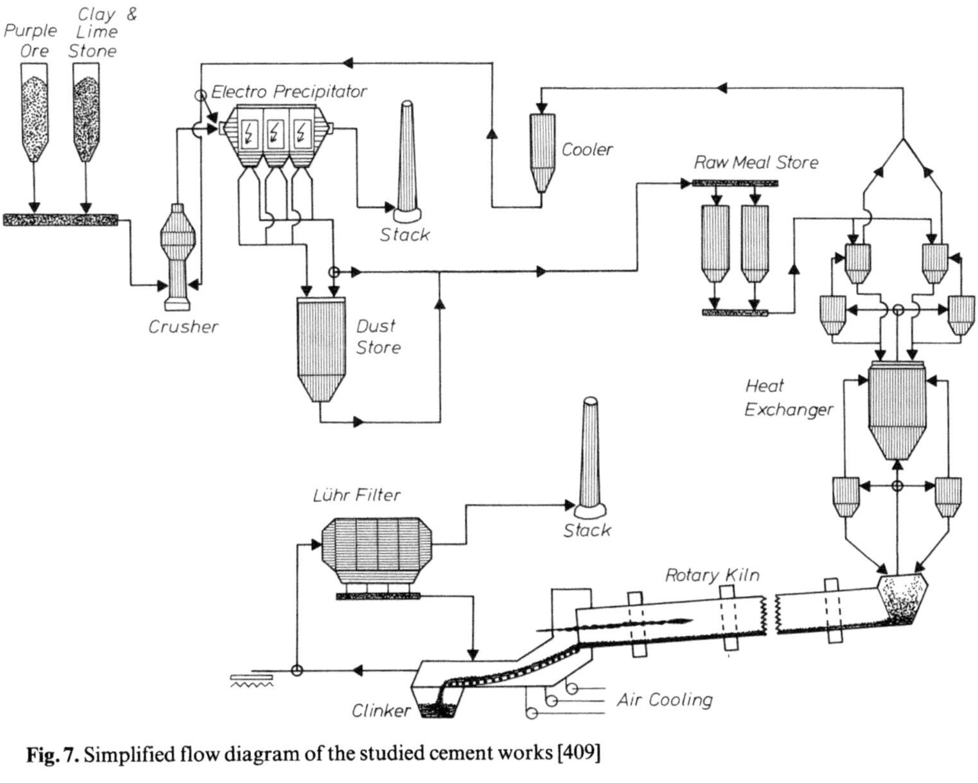

Fig. 7. Simplified flow diagram of the studied cement works [409]

Furthermore, the results obtained are much more reliable than those which have been reported earlier because more recent sampling and analytical techniques are applied today and production conditions have been taken into account [409]. Therefore, thallium emissions from cement plants will be described in more detail.

The investigated plant has a rotary kiln with a cyclon-heat exchanger. During the usual production of 45 t/h, the average solid substance concentration in the flue gas is 72 mg/m^3, which is equivalent to a mass emission of 8.75 kg/h.

The material and gas flow are schematically illustrated in Fig. 7.

Thallium contents of samples taken from different sites in Fig. 7 are compiled in Table 11:

Table 11. Thallium contents of samples taken from different process sites [409]

Site	Tl content (ppm)	Remarks
1. Stone material	0.34	Max. value
2. Purple ore	4.8	
3. Raw meal	297	
4. Cooler dust	3,903	
5. Precipitator dust	3,237	
6. Clinker	0.18	
7. Clinker dust	0.08	Max. value
8. Pulverized coal	2.2–1.8	Max. value
9. Pulverized lignite	0.2	Max. value

From the values in Table 11 the characteristics of the transport and distribution of thallium during cement production processes were derived. Thallium is introduced into the process via stone material, purple ore and fuel. During the period of investigation a thallium input of 35.2 g/h was found. The stone material contributed up to 80% and the other two media 10% each. Most of the thallium introduced is volatilized almost quantitatively in the heat exchanger where a temperature of about 900 °C is reached. The thallium originating from the stone which is bound into silicates and the thallium contained in the fuel evaporate in the rotary kiln.

After condensation onto dust particles in the cooler parts of the plant, thallium is transported to the crusher where the particles are mixed with new raw materials forming again raw meal. The thallium-containing particles are transported with the new material to the heat exchanger where a another cycle of the process starts. The amount of thallium transported in this cycle increases with increasing filter quality.

In the investigated plant, 0.4% of the thallium input is fixed in clinkers and another 0.4% emitted with the flue gas dusts. At a retention rate for thallium of 99.2% the maximum volume in the cycle at an average input of 29.3 g/h is 36,625 g/h. If 0.4% of this is emitted via stacks, approximately 15 g/h will be emitted into the environment.

During the investigation period, the thallium concentration of the raw meal and the thallium emission showed nearly stationary behaviour over a longer period of time. Thus equilibrium conditions were maintained although the theoretical

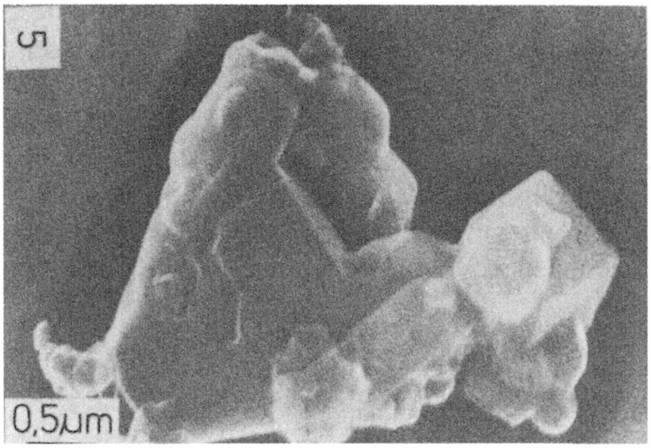

Fig. 8. Scanning electron micrograph of thallium containing pure gas dust [331]

Table 12. Cement production and potassium consumption in different economical units [255]. Figures are in mill. tons

	Cement production (1980)	Potassium consumption (1979/80)
EEC	135.5	4–5
USA	75.2	5.95
USSR	124.8	4.41
Japan	87.8	0.74
Africa	27.3	–
World	866.2	23.5

equilibrium load of 36,625 g/h was never reached. The maximum load actually amounted to 22,849 g/h. It appears therefore that under normal operating conditions the process is not left completely undisturbed. Starting and firing off the plant, CO interuptions of the precipitator, reorganization of production and temperature changes in the kiln during the removal of baked-on crusts may lead to an increased emission of thallium which clearly exceeds the calculated amount of 15 g/h during short periods of time. Under inclement metereological conditions severe pollution of the neighbouring region may result.

Mixed crystals of flue gas dust with the composition Tl(Cl, Br) which are mainly intergrown with arcanite and syngenite and which attain a diameter of < 5 µm were detected in the scanning electron micrographs. The chlorine concentration was much higher than that of bromine. The thallium compounds have been crystallized as spherical particles or idiomorphic cubes with a face edge of 0.5 µm in length as shown in Fig. 8.

These results correspond to the observation that thallium emitted by cement plants into the environment is easily soluble in water and therefore slightly available to the biosphere [340].

In calculations of total Tl emissions originating from the cement industry not only the various Tl concentrations from the raw material must be taken into account but also the various production techniques. In 1972 about 53% of the clinker were produced using the heat exchanger technique. At present old production methods (11%) are being replaced by this process. The other producers use Lepol furnaces. There is no difference between these two techniques with respect to the evaporation of Tl in the rotary kiln. However, because the amount of dust is much lower using the Lepol technique it does not need to be recycled. Therefore, no enrichment of thallium occurs in the furnace. The cement production and potassium consumption are compiled in Table 12.

Behavior and Effects of Thallium in the Environment

Thallium is released into the environment either via aqueous solution or as a component of dust particulates whereby the latter are subject to varying distribution processes according to their grain size, specific weight and the prevailing climatic conditions.

There are no data available on possible chemical reactions of the emitted thallium compounds during the transport in the atmosphere.

Thallium compounds found or assumed to be contained in emitted dust particulates such as Tl(I) chloride, bromide, sulfate, and oxide are stable under normal atmospheric conditions with the exception of Tl(I) oxide which may react with moist air to form Tl(I) hydroxide. Therefore, chemical reactions are not very likely to occur in the atmosphere.

Input and Chemistry in Soils

Heinrichs and Mayer [159] have studied the transfer of atmospheric thallium to soils and plants of two forest ecosystems. The concentrations and annual fluxes of thallium in the precipitation input to the forest canopy, beneath the canopy, and in the seepage water below the humus layer as well as beneath the root zone were determined in a relatively unpolluted central European beech and spruce forest. The two forest systems were so close together that the soils and surroundings could be assumed to be identical.

Compared to the open-field input into the soil, there is a decrease of the thallium flux during passage through the beech forest canopy even if stem flow is taken into account, due to adsorption onto leaf surfaces, bark of twigs and stems, and on to epiphytes (Fig. 9).

Furthermore, input into the humus layer is greater than output. The same is true for the transfer from humus to the root zone.

The situation appears to be different in the spruce forest. Thallium precipitation under the canopy is greater than that in the open field, mainly due to filtering of atmospheric substances by the tree canopy, whereas leaching of metabolic substances, secreted at the surfaces of leaves and needles is of minor importance [247].

The thallium input to the spruce forest floor differs with respect to precipitation under canopy and seepage below the humus layer. This may be attributed mainly to leaching and mineralization of litter.

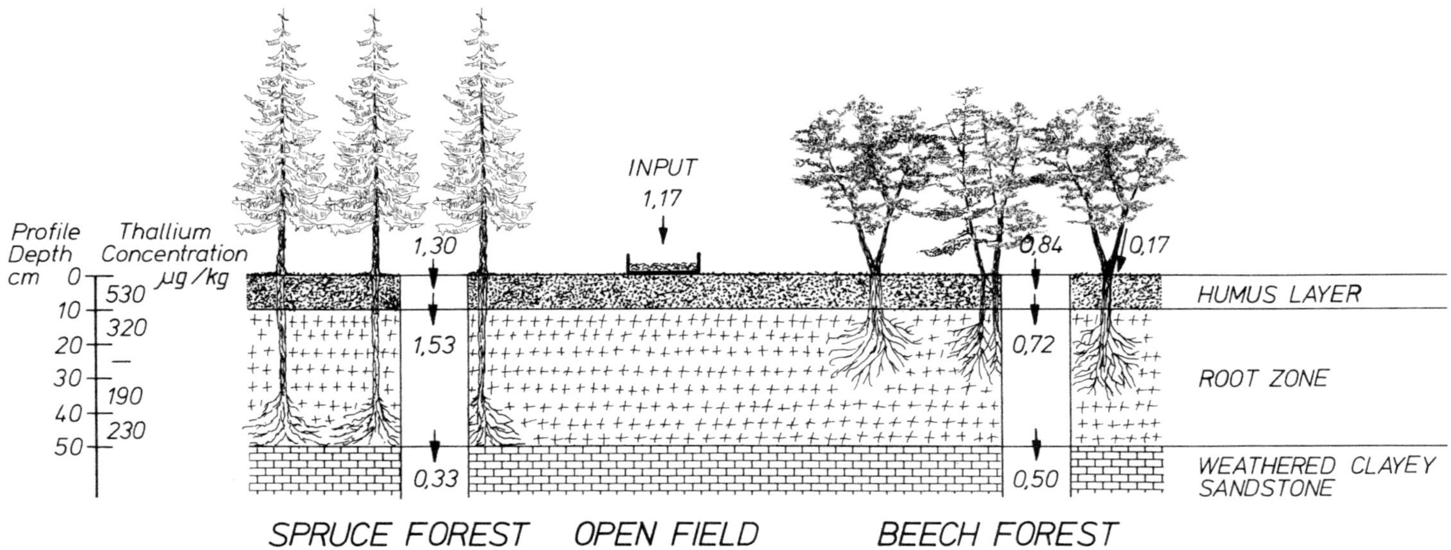

Fig. 9. Thallium fluxes (g/ha · a) after transfer into different Central European forest ecosystems [159]

Comparing the seepage flux of thallium below the humus layer with the one below the root zone of the spruce system, a significant retention of thallium in the root zone is obvious. The total thallium retention in the root zone of beech and spruce forests was determined to be 0.67 and 0.84 g/ha · a, respectively; the total store in the soil layer between the surface and 50 cm depth was 1.89 kg/ha. Most of this thallium is fixed in the uppermost centimeters which indicates its low mobility in the soil.

Various research groups have arrived at similar results for soil profiles taken from the proximity of intensive thallium emitters such as cement plants. Samples from 2 cores taken near a cement plant in Leimen, in southwestern Germany, from the 0–10 cm, 40–50 cm, and 60–70 cm layers were analyzed for thallium. The values obtained for the respective layers of the first core were 4.2 μg/g, 1.3 μg/g, and 0.1 μg/g, and for the second 3.0 μg/g, 0.1 μg/g, and 0.1 μg/g [338].

Correspondingly, the analysis of the uppermost layers of cores of polluted soils showed values of 0.7–12.7 μg/g Tl, whereas at depths below 35 cm only 0.2 μg/g were measured [340].

Thus, thallium in soil appears to occur in a chemical form that does not allow a rapid "wash-out". Nevertheless, thallium introduced from the atmosphere is generally available to the biosphere. In fact, 75%–100% of the thallium contents of flue gas dust from the cement industry is soluble in water [340]. However, this rate rapidly decreases with time after the introduction into the soil. After injection of 2.5 mg/kg Tl into the soil in easily soluble form, only 40% could be determined to be still in this form after 13 days [61].

The extent of the availability of thallium strongly depends on its source. Thallium immissions that originate from the cement industry can be eluted from soils to 40% with an extraction reagent (1 *N* NH_4Ac with 0.01 *M* EDTA). When the same procedure is applied to tailing material, however, only 0.5%–5% Tl is mobilized.

The extractability of soil-associated thallium that originates from several sources varies so greatly that the ammonium acetate extraction has even been applied to distinguish contaminations originating from different emitters at a single location [338].

Findings obtained extraction investigations could be verified during soil-vegetation transfer determinations as demonstrated by Table 13.

Table 13. Yield, Tl content and uptake of *rape* as well as EDTA extractability of thallium from contaminated soils

Soil	Tl content μg/g dry weight	Tl uptake (%) from Tl-supply in soil	
		Plant	EDTA[a]
1	191	22.3	40
2	61	5.1	5
3	25	1.4	0.5
4	39	2.5	2.5

[a] EDTA = extractable amount of Tl with 1/100 m EDTA/1 *N* ammonium acetate solution

Concentrations and Speciation in Natural Waters

It can be derived from the normal potential of the reaction $Tl^{1+} - 2e^- \rightleftharpoons Tl^{3+}$ that in natural aqueous solutions Tl should generally be monovalent and that Tl(III) may occur only under strongly oxidizing conditions [79].

On the other hand, Tl(III) forms compounds with nearly all ligands that are more stable than those of the monovalent form. These compounds exhibit even a higher stability than those of the corresponding Fe(III) and Cr(III) complexes [56]. Therefore, in complexing reactions a preference for the trivalent form is observed.

Considering the widely known complexing and solution constants as well as the composition of natural waters, the redox reaction between Tl(I) and Tl(III) in sea water can be formulated as

$$Tl(OH)_2Cl + 2H^+ + 2e^- \rightleftharpoons TlCl + 2H_2O\ .$$

From this formula it can be calculated that in sea water thallium should occur in the trivalent form. The same result is obtained for fresh water at pH 6.5 [32].

Such formal computations were confirmed by studies on the adsorption behaviour of thallium compounds in natural waters. About 80% of the thallium content was found to be fixed on anionic ion exchangers and the rest was presumed to exist as neutral or cationic complexes [32].

By dosing controlled amounts of monovalent Tl-204 into natural water samples it was indicated that 50% of the Tl was rapidly fixed on an anion exchanger. Within a few days this amount increased markedly.

However, some authors have expressed doubts as to the validity of these results [364]. The contrast between anionic adsorption and the speciation reflected in the redox formula is conspicuous. Apparently, other reactions can play an important part in such a complex environment as a water body, and these were not considered in Bratley and Florence's [32] work.

Table 14. Thallium concentrations in natural and waste waters

Location	Thallium conc. (ng/l)	Ref.
Pacific Ocean surface	13.0	[32]
Fresh water	3.7	[32]
Bay of Biscay	10.1	[245]
Irish Sea	18.7	[245]
Several German rivers	2–40	[154]
Hot springs (New Zealand)	7,000	[407]
"Table" mineral waters (Germany)	– 3,500	[203]
Waste water; cement industry	– 20,000	[16]
Ground water, cement industry	< 1,000	[16]
River water, cement industry	< 1,000	[16]
Mining waste water (Canada)	– 88,300	[418]
Oil drill. waste water (USA)	–672,000	[10]
Ground water of zinc smelter	–810,000	[10]
Mining waste water (Germany)	23,000	[109]
Smelter waste water (Germany)	800,000	[27]

Only few data on the thallium content of natural and waste water have been reported so far. Some of them are compiled in Table 14.

Uptake and Distribution Mechanisms in Plants and Plant Toxicology

Plants take up thallium via roots and foliage. Geilmann et al. [127] have determined the following thallium concentrations (μg/kg dry weight) in plant parts above the ground in a non-polluted environment:

Clover	8–10	Red cabbage	40
Hay	20–25	Kale	125
Turnip leaf	20–25	Leech	75
Carrot	20–25	Endive	80
lettuce	21	Potatoe	5

The proportion between the thallium content in soils and the concentration in plants is relatively constant under otherwise equal conditions. The thallium contents in the leafy material of several food and fodder plant species and in corresponding soils are given in Fig. 10.

Similar results have been obtained in other studies. In soils contaminated with thallium via tailing waste material originating from non-ferrous metal mining activities (Tl concentration: 2.3–19.6 μg/g) transfer factors of great similarly for each plant species were determined: rape 0.26–0.29; parsley 0.15–0.21 and cress 0.45–0.59.

The transfer factors for cellery (leafy part), green cabbage and Brussel sprouts are 0.1. Mushrooms strongly enrich thallium (F = 2.9) [338].

Root uptake and thallium distribution in soybeans were investigated in detail by Cataldo and Wildung [61]. They showed besides soil factors, which determine the metal speciation and the free metal ion concentration, uptake is facilitated by root activities.

There is a pattern of thallium uptake peculiar to higher plant species which is referred to as multiphasic. This means that as the concentration of the thallium

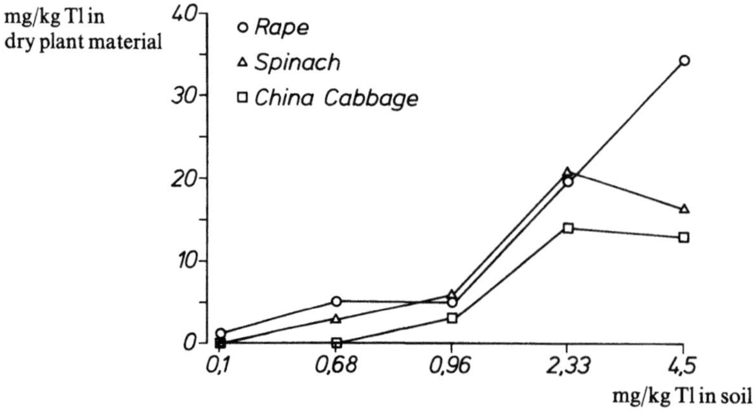

Fig. 10. Tl contents in soils and plants

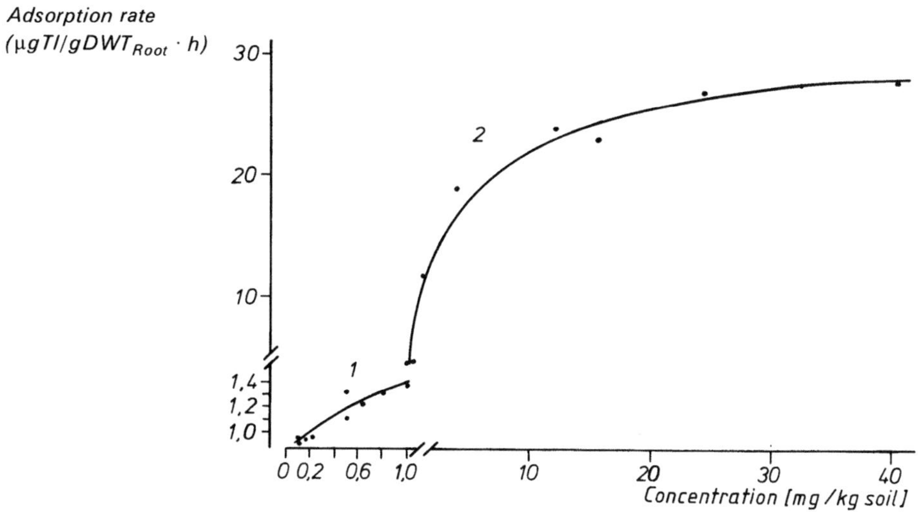

Fig. 11. Relation between thallium uptake and soil concentration in the case of intact soy bean plants [61]

Table 15. Tl contents of various plant parts and rhizoidal soils [177]

Plant	Tl content of rhizoidal soil (mg/kg)	Tl content of various plant parts (mg/kg DW)	
Kohlrabi	3.7	Leaves	35.0
		Tubes	0.10
Turnip	9.4	Leaves	5.90
		Root	0.40
Zuchinni squash	5.2	Leaves	0.90
		Stalk	0.02
Cucumber	5.4	Leaves	0.70
		Fruit	0.10
Beet	5.2	Leaves	2.40
		Root	0.60
Carrot	3.5	Leaves	0.30
		Root	0.10
Onion	0.9	Stalk	0.10
		Tube	0.01

surrounding the roots increases, uptake and absorption by the plants exhibit two distinct isotherms, each of which has different kinetic characteristics. Figure 11 demonstrates that thallium uptake occurs more readily at lower concentrations.

Furthermore, thallium absorption by the root is reduced to 57% in the presence of a 10fold higher concentration of potassium, demonstrating that potassium is a noncompetitive inhibiter of thallium transport [61].

The distribution of thallium in plants is similar to that of nutrient ions, especially potassium. In Table 15 the thallium concentrations of different parts of various edible plants are compiled.

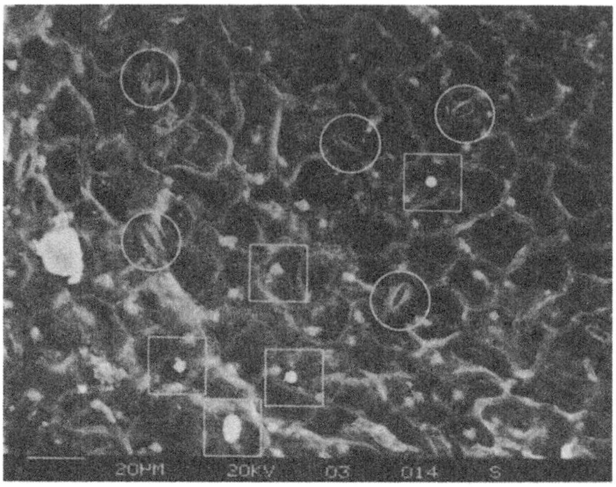

Fig. 12. Scanning electron micrographs of fly ash from rotary kilns used in the Dyckerhoff Cement Plant in Lengerich (West Germany) covering bean leaf. Openings on the leaf surface (circles), overlying thallium-containing particles (squares) [16]

Electron scanning micrographs of thallium-containing particles in fly ash from cement plants reveal that particles with a grain size < 5 μm predominate. However, these particles are too large to enter the plant via leaves as Fig. 12 impressively illustrates.

Detailed information on the bonding of Tl onto particles from other emission sources or from other plant types are not available. It can be assumed, however, that the particles sizes in emissions from coal-burning facilities are of similar magnitude [77].

However, an indirect uptake of thallium compounds after dissolution on the leaf surface is possible. This uptake mechanism causes pale, round chlorotic spots distributed over the entire leaf which can be reproduced in laboratory experiments by spraying thallous solutions or sprinkling thallium sulfate crystals onto leaves. In contrast, thallium uptake via roots shows discoloration, which results from the degradation of chlorophyll, beginning at the base of the leaf vein, progressing along the vein, reaching the tip and spreading over the whole blade until it is totally covered by intercoastal chloroses.

If a more intensive Tl uptake has occurred, effects may include necrotisms and finally litterfall at every season, because scar material forms at the end of the stem which is identical to that developing before normal autumnal litterfall [302].

This phytotoxic effect has also been described by other authors. 2 mg/l thallium in a nutrient solution leads to a reduction in the overall photosynthesis of corn and sunflowers [60]. Further observed effects caused by thallium contamination of plants include impairment of chlorophyl formation and seed sprouting [416], inhibition of transpiration due to the interference of thallium with the stomatal function [34], reduction of growth [34], other growth problems and leaf discolorations, stunting of root systems, and chlorosis for tobacco plants [248]. In the latter case, effects may occur with 0.1 mg Tl per kg soil, depending on the soil type. Thallium has also caused chlorophyll mutations in pot experiments with peas [322].

Fig. 13. Chlorotic and necrotic damage of leaves caused by thallium

The dependence of the yield decreases, and the outright plant damage caused by thallium in breeding soils as well as the thallium uptake of various domesticated plant species were investigated by Hoffmann et al. [177]. Table 16 gives some results of this study.

Plants studied in these experiments showed great differences in their susceptibility to metals and actual metal uptake. A similar statement can be also made in respect to other plant investigations.

Whereas coniferous trees are scarcely affected by thallium, the resistance of deciduous trees is highly dependent on the species. At pratically the same locality summer lindens are affected as described above, but winter lindens are not affected at all. Ash, chestnut and walnut trees are affected similarly, being much more susceptible to thallium than poplar or oak trees.

Table 16a. Yields (dry matter) of agricultural crops from the same soil after application of $TlNO_3$ [177]

	Soil of Dautmergen (with 3 mg Tl/kg)				
	0 Tl	+10 mg Tl/kg	+50 mg Tl/kg	+200 mg Tl/kg	+500 mg Tl/kg
Row I					
Kohlrabi (1979)					
Older leaves	4.3±0.4	4.0±0.1	4.0±0.3	3.9±0.5	3.4±0.5
Younger leaves	7.1±0.3	8.8±0.9	8.8±0.8	7.6±0.4	5.3±0.7
Total	11.4	12.8	12.8	11.5	8.7
Row II					
Green rape (1979)	10.2±0.4	9.9±0.4	9.9±0.4	6.3±0.3	0.9±0.4
Rows II and III mixed (1980)					
Radish					
Tubes	8.2±0.8	9.0±0.3	5.0±0.5	6.5±0.4	4.0±0.5[a]
Leaves	11.4±1.2	9.5±0.5	10.1±0.2	10.4±0.7	11.4±0.8
Total	19.6	18.5	15.1	16.9	15.4
Lettuce	19.3±1.2	14.8±1.1	10.9±0.7	7.4±0.9[a]	5.3±0.8

Table 16b. Tl contents of agricultural crops from the same soil after application of $TlNO_3$

	0 Tl	+10 mg Tl/kg	+50 mg Tl/kg	+200 mg Tl/kg	+500 mg Tl/kg
Row I					
Kohlrabi (1979)					
Older leaves	30.0	382	1,011	1,936	2,354[a]
Younger leaves	6.0	116	299	591	1,080[a]
Row II					
Green rape	10.0	180	499	1,656[a]	3,326
Rows II and III mixed (1980)					
Radish					
Tubes	1.1	8.6	18.4	35.1	31.2[a]
Leaves	4.5	31.6	64.4	150	331[a]
Lettuce	2.2	20	28	33[a]	25[a]

[a] Heavy damage

Concerning fruit trees, significant differences in resistance against thallium caused damages can also be determined. Sweet cherry trees are more sensitive to thallium than are sour cherry trees. When apple, mirabels (small yellow plums) and plum trees have lost their leaves, the foliage of pear trees remains healthy.

A similar division can also be noted for garden vegetables. Green beans, cucumbers, squash species, and potatoes are affected, whereas lettuce, radishes, strawberries and asparagus are less susceptible. Cabbage species generally display a strong resistance towards thallium compounds.

Fodder plant species such as rape and carrots or turnips are not very susceptible whereas corn in the same locality is known to show damages in leaves and reduced yields [16].

Generally, plant species with a hard leaf surface are more resistant than those with soft fuzzy ones.

In the same investigation, characteristic differences in the thallium contents for individual species were determined. Green cabbage takes up large quantities of thallium. However, also other vegetable species such as kohlrabi, savoy and white cabbage can contain much thallium with no apparent external damage.

The highest thallium amounts were determined in green rape; ternips and occasionally cereal grains (before and after cover is removed) indicated a high enrichment, demonstrating that thallium is highly mobile in these plant species.

The thallium content of grass species and hay is only rarely above the allowed limits [16]. There was no distinction made between external and internal contamination in this study.

The tolerance of plants to high levels of both nutrient and non-nutrient elements appears to be genetically controlled and therefore governed by selection [61].

Only on the basis of high thallium concentrations in some plant species of a naturally thallium-rich region and without knowledge of prevailing soil contents it cannot be concluded that there is a species-specific uptake relationship. It is possible that within a plant species group a marked variability may occur, some

Table 17. Thallium contents of selected plants of Alsar Ore Mining Region in Macedonia, Greece [421]

Plant species (selected)	Tl in ash[a] (mg/kg)	Tl FS[a] (mg/kg)
Carnation (Dianthus spec., caryophyllaceae)	5,200	39,10
Lavatera spec., Malvaceae (blossoms)	45	0.34
Centaurea spec., compositae (fruit)	17,000	127.82
Campanula spec. campanulaceae (blossoms, leaves stem)	5,900	45.04

[a] Concersion factor Tl in FS to Tl in ash 1:133

varieties being more thallium-tolerant than others. Such cases have been described in Greece. Plants were studied that had a level of thallium contamination which, in other localities, had caused serious damage (see Table 17).

Mushrooms have extensively been studied because of their efficient heavy metal uptake rates. Nevertheless, a study of 421 mushroom species (1,107 samples taken between 1967 and 1980 in northern, central and southern Europe) revealed that 119 contained no Tl at all and that the other species were contaminated at rates between 0.01 and 1.22 mg Tl/kg fresh mushrooms [346].

Thallium uptake kinetics suggest that more than one transport system operates in yeast, possibly reflecting a multiplicity of potassium transport systems [279]. Prolonged incubation of yeast with Tl_2SO_4 results in the formation of insoluble Tl_2O_3 in mitochondria as a process of oxidative detoxification [227].

Concentration factors of 9.3–430 and more have been reported for algae growing in contaminated waters [364]. The growth of different kinds of algae is inhibit in waters containing 0.5 mg/l thallium; severe inhibition occurs in the range of 5–10 mg/l. Thallium uptake by algae often depends on light intensity [364].

The principal effect of Thallium on Chlamidomas reinhardii results from the inhibition of the NADP reduction or dark reactions [288]. Other studies have shown that thallium only poorly replaces potassium in the activation of L-threonine deaminase in planktonic algae but utilizes the potassium transport system during accumulation by Chlorella [366].

Toxicity to Animals

The uptake of thallium by animals normally occurs with food which is either enriched with thallium from environmental pollution or poisoned by rodenticide residues. The sensitivity of different organisms, which varies widely, is discussed in detail in the following.

Mammals

The symptoms of an acute, subacute or chronic thallium poisoning of mammals greatly resembles those for humans; they are described later together with the thallium distribution in body organs.

Table 18. Toxicity of several thallium compounds to mammals [14]

Compound	Animal	Route	Toxicity	Dosage/kg body weight			
				Compound	Metal		
				mg	mg	m*M*	pT
Thallium oxide Tl_2O	Rat	oral	MLD	20.8	20	0.10	4.01
	Rat	oral	LD_{50}	40.6	39	0.19	3.72
	Rat	i.p.	LD_{50}	75	72	0.35	3.45
	Rabbit	oral	MLD	31.2	30	0.15	3.83
	Rabbit	i.p.	MLD	62.4	60	0.29	3.53
	Rabbit	i.v.	MLD	40.6	39	0.19	3.72
	Dog	oral	MLD	31.2	30	0.15	3.83
	Dog	i.p.	MLD	31.2	30	0.15	3.83
Thallic oxide Tl_2O_3	Guinea pig	oral	MLD	5.2	5.0	0.024	4.61
	Guinea pig	i.p.	MLD	34.0	31	0.15	3.82
Thallium chloride TlCl	Mouse	oral	LD_{50}	24.0	20.5	0.10	4.00
	Mouse	i.p.	MLD	5.66	4.8	0.025	4.74
	Mouse	i.p.	LD_{50}	24.0	20.5	0.10	4.00
	Rat	i.p.	MLD	5.3	4.5	0.022	4.66
Thallium bromide TlBr	Mouse	oral	MLD	29.0	20.9	0.102	3.99
Thallium iodide TlJ	Mouse	oral	MLD	28.0	17.3	0.086	4.07
Thallium nitrate $TlNO_3$	Mouse	oral	LD_{50}	32.5	25	0.12	3.91
	Rat	s.c.	LD_{50}	26.0	20	0.095	4.01
	Rat	s.c.	LD_{50}	20.0	15.3	0.074	4.13
	Rabbit	i.v.	LD_{50}	14.0	10.7	0.052	4.28
	Dog	oral	LD_{50}	45.0	34.5	0.17	3.77
	Dog	s.c., i.v.	LD_{50}	15–20	13.5	0.066	4.18
Thallium sulfate Tl_2SO_4	Mouse	oral	LD_{50}	29.0	23.2	0.11	3.94
	Rat	oral	LD_{50}	25.0	20	0.098	4.01
	Rat	oral	LD_{50}	15.8	12.6	0.062	4.21
	Man	oral	MLD	3	2.4	0.012	2.93
Thallium carbonate Tl_2CO_3	Rat	s.c.	MLD	18.0	7.9	0.039	4.41
Thallium selenite Tl_2SeO_3	Rat	oral	MLD	50.0			
Thallium acetate $Tl(C_2H_3O_2)$	Mouse	oral	LD_{50}	35.2	27	0.13	3.87
	Rat	oral	MLD	37.4	29	0.14	3.85
	Rat	oral	LD_{50}	41.3	32	0.16	3.81
	Rat	i.p.	MLD	25.8	20.0	0.098	4.01
	Rat	i.p.	LD_{50}	29.6	23.0	0.11	3.95
	Guinea pig	oral	MLD	15.5	12.0	0.059	4.23
	Guinea pig	i.p.	MLD	9.8	7.6	0.034	4.47
	Rabbit	oral	MLD	24.5	19.0	0.094	4.03
	Rabbit	i.p.	MLD	16.8	13.6	0.063	4.20
	Rabbit	i.v.	MLD	25.8	20.6	0.098	4.01
	Dog	oral	MLD	25.8	20.6	0.098	4.01

Generally, LD_{50} lies between 5 and 70 mg/kg body weight. The toxicity of the listed compounds does not show any great systematic differences. Likewise, mammals do not exhibit marked differences in their susceptibility of Tl poisoning.

The toxicity of Tl(III) compounds is presumed to be higher than that of the Tl(I) analogues [91]. However, this is not confirmed by the data in Table 18.

Effects of subacute and chronic thallium contamination on mammals have been investigated by several authors.

After a dosage of 10 mg Tl/l in drinking water given for 9 and 24 weeks (groups of 4–6 rats), no visible effects of poisoning, except for varying degrees of fur loss, were observed. There was also no difference in thallium concentration between the various tissues of the animals which had been treated with Tl sulfate for 9 or 24 weeks [211].

Herman and Bensch [171] first injected a group of 15 rats with 10–20 mg Tl(I) acetate per kg body weight. Subsequently, the rats were injected subcutaneously thallium doses of 5 mg/kg body weight once per week over a period of 26 weeks.

These doses were reduced to 2.5 mg or no dose was given at all when symptoms of toxicity occurred. The rats were killed singularly at regular intervals between four and 26 weeks after initial doses had been injected. Their internal organs were investigated in detail for microscopic or electronmicroscopic alterations. Clinical symptoms included loss of hair, irritability, diarrhea; during the entire experiments the weight of the rats did not increase. With a normal microscope no Tl-induced morphological change of any organ was observed. However, in electron microscopic studies alterations were detected not only in the kidneys, liver and in the intestinal tract but also in the brain, seminal vesicle and pancreas.

A daily dose of Tl(I) carbonate or sulfate either orally or subcutaneously of 0.25 mg/kg body weight (corresponding to ca. 1/100 of LD_{50}) to groups of 4 domestic rabbits over a period of 6 months caused the following effects: behavioural disorders, aggressivity, crippled hind legs, diarrhea, hair loss, decrease in the SAP activity and in the SH-groups in blood serum, histologic alterations of the liver, kidneys and stomach membranes [384].

Kennedy and Cavanagh [196] carried out similar experiments on cats. The chief functional disturbance in the nervous system was hypotonia and ataxia. The pathological changes were confined to the central and peripheral parts of primary sensory neurons; no motor nerve fibre degeneration was found.

While sensory nerves were greatly affected those in the muscle tissue were apparently unaffected (Table 19).

For sheep injected with a daily dose of 0.03–1.0 Tl mg/kg over a period of 4–15 weeks, a dosage- and time-dependent thallium accumulation could be detected in the liver, kidneys, spleen, muscle, and heart muscle tissues. Preference for one certain tissue type could not be ascertained, even though the liver and kidneys consistently showed the highest thallium contents [151].

Concerning the thallium contents found in the liver and kidneys, a dependence from the doses can be discerned. However, this dependence is not linear. Additionally, for thallium accumulation in the skeletal musculature and heart muscles it is only observed in the lower dosage range. Doses above 0.1 mg Tl/kg body weight per day do not lead to a stronger accumulation of thallium in these tissues (Table 20).

Table 19. H: onset of hypotonia: A: onset of ataxia. P: neurological involvement: + mild; + + moderate; + + + severe

Animal	Mean weekly dose mg/kg/wk (total dose mg/kg)	Amount of thallium given in mg/kg body wt. by weeks												Degree of neuro-logical disturb-ance
		1	2	3	4	5	6	7	8	9	10	11	12	
C8	3.3 (85.8)	Chronic experiment lasting 26 weeks												+
C12	4.0 (27)	6	3	3	3	4 H	4 A	4 P						+
C13	2.3 (14)	6	0	4	4	0 HA	0	0 P						+ +
C2	2.4 (12)	6	0	0	6	0 HA	0 P							+ +
C5	3.0 (12)	6	6 H	0	0 AP									+ +
C16	3.8 (24)	6	0	6	4	4	4 H	0 AP						+ +
C6	2.6 (18)	6	6	0	0	0	0	3 H	3 AP					+ + +
C9	2.7 (16)	6	0	3	3	4 A	0 HP							+ + +
C14	3.7 (20)	6	0	4	4	4 AH	2 P							+ + +
C15	4.5 (50)	6	0	4	4	4	4	4	4	6 H	7	5	2 AP	+ + +

Table 20. Thallium content of tissue material (μg/g FW) after 4–15 weeks of exposure taking into account the concentration in the feed (μg/g DW) [151]

Dosis mg thallium/ kg body weight	(n)	Tl conc. in feed (mg/kg)	Liver	Kidneys	Spleen	Muscles	Heart	a
0.00 (Contr.)	(3)	–	0.05	0.05–0.06	0.05–0.08	0.05–0.06	0.05–0.11	
0.03	(2)	1.5	0.16–0.28	0.32–0.81	0.10–0.13	0.05–0.25	0.06–0.36	15
0.1	(3)	5	8.7	5.1	4.8	6.2	7.1	
			1.0–13.4	1.3–9.5	1.1– 8.9	0.5–12.6	0.6–13.9	11
0.3	(3)	15	10.0	10.6	8.2	5.0	7.6	
			6.9–15.4	3.6–17.2^{++}	4.8–11.6	1.5–7.2	6.7– 8.3^{++}	4
1.0	(3)	50	16.7	16.4	10.2	5.1	9.0	
			7.8–22.7	6.2–34.4	5.5–14.6	1.2–11.6	2.8–15.1	6

FW = fresh weight
DW = dry weight
a = exposure time (weeks)
$^{++}$ n = 2

These results indicate that daily applications of 0.03 mg Tl/kg body weight lead to a measurable thallium accumulation in body tissues which corresponds to a thallium concentration in feed products of 1.5 mg/kg DW. A concentration causing no effect was not found in these experiments.

A pathological study has been performed on dogs [419] and cats [420] accidentally poisoned by rodenticides or by eating vermin poisoned by thallium. The more severely poisoned animals developed a hemorrhagic gastroenteritis with hepatic or renal damage. Animals surviving longer developed gross skin lesions with inflammation, bleeding, crusting and shedding of fur. Lack of coordination, trembling and paralysis developed at this stage. The most important findings include: in the kidneys, tubular necrosis with glomerular hyalinization; in the brain, focal edema and chromatolysis; degenerative changes in peripheral nerves; necrosis of skeletal and cardiac muscle fibers; edema and bronchopneumonia in the lungs; hemorrhage and ulceration in the gut. In contrast to the severe damage seen in other organs, liver damage in the dog was mild, and only moderate centilobular necrosis was seen in the cat.

Clausen and Karlog [71] examined 60 wild martens and badgers in Denmark over a 2-year period and found a thallium load in 48% of these with evidence of poisoning in 22% of all studied cases. Acute hemorrhagic intestinal inflammation was seen in most of the cases of poisoning. In the 13 poisoned animals, thallium concentrations in the liver ranged from 4.7 to 57 mg/kg, in the kidney from 1.8 to 92 mg/kg and in the intestine from 0.2 to 42 mg/kg. Thallium concentration was estimated in 34 red foxes too, which had been found sick or dead in 1971 [270]. In 4 out of 7 foxes, where the cause of death could not immediately be established, a thallium content was found in excess of 1 mg/kg in at least one organ. Of the 27 foxes where an alternative diagnosis had been made, thallium concentrations were higher than 1 mg/kg in 5, between 0.5 and 1 mg/kg in 8 and 0.5 mg/kg or less in the remaining 14 animals.

The time-dependent distribution of thallium in mouse body organs was determined after feeding with 80, 130, and 160 mg/kg WT [3]. Heart, liver, kidney, and stomach showed an organ-specific initial uptake during the first 2–3 h. This was followed by a period of wash-out to result in relatively low Tl levels.

Thallium uptake by the brain was comparatively low and constant during the first 12 h but increased later. Brain levels after a thallium dose of 130 mg were approximately four times higher than those measured after a 80-mg dose, suggesting that the brain has a barrier to thallium which is overcompensated only by high amounts of the metal.

Birds

The minimum lethal dose (MLD) intraperitoneally induced in female mallard ducks was determined to be about 25 mg/kg body weight [403]. Wild geese had died after eating barley contaminated with thallium. The oral MLD from this source was about 50 mg/kg. Feathers on the back just above the uropyial gland were lost after the ingestion of sublethal doses. Liver and kidney contained 6%–11% of the administered amount following death which occurred within 1.25–13 days. Of 37 bald eagles found sick or dying in 1971 to 1972 in the United States, 9 had been poisoned by thallium [73].

Fish

It was reported that 10–15 mg thallium nitrate/l is lethal within three days to rainbow trout, 60 mg Tl/l to perch, and 40–60 mg Tl/l to roach [277].

If longer periods of time are considered, MLD values fall in a corresponding order of magnitude. Zitko [416] has reported the death of juvenile Atlantic salmom after some months in water with 30 μg thallium/l. There was no incipient lethal level visible as a sharp break-off point on the lethality curve (time to 50% mortality versus thallium concentration). The thallium toxicity proved to be independent of water hardness and was not affected when copper and zinc were added [419].

The relative rate of uptake of monovalent cations by goldfish intestinal mucosa decreased in the order: thallium > potassium > rubidium > cesium > sodium > lithium [95].

Reproduction, Teratogenicity and Carcinogenicity

In the literature there are numerous references to the teratogenic effects of thallium as well as the influence of thallium on sexual behaviour and reproduction.

According to Smith and Carson [364], thallium poisoning deleterously affects sexual behaviour, reproductive organs and egg and fetal development and survival. For example, chicks fathered by cocks chronically poisoned by thallium suffered higher mortality during the first three weeks after hatching than did chicks fathered by normal cocks [213].

An injection of 0.6 mg thallium dissolved in 0.5 ml saline onto the chorioallantoic membrane of chicken eggs at the embryonic osteogenesis caused a reduction of long bone growth and areas of necrotic cartilage [148].

Fig. 14. Malformations in mice after application of Tl

Other authors described growth retardation, parrot-beak deformity, microphthalmia, microcephaly, reductive size of all skeletal elements, micromelia and achondroplasia [364, 365]. Dose levels of 25.5 mg thallium(I) sulfate/kg in the initial stages of gestation prevented the phalanges from ossifying [130].

It was suggested that bone deficiencies were caused by insufficient primary growth of the nervous trunks which determine the length of the bone by the space available along them [326].

In SWS mice, a dose of 1 mg/kg Tl on the 9th day after fertilization caused achondroplasia on 14% of the fetuses. The teratogenic dosage of 50% was near 8 mg/kg with respect to pure thallium. This rate represents 1/30 of LD_{50}. An example of malformations is shown in Fig. 14.

Furthermore, at 1/1,000 of LD_{50}, a malformation frequency of 12.5% has been determined [2]. With the same dosage abortus occurred after application before the 9th day; later, malformations were observed.

These results could not be reproduced using NMRI mice which showed, under similar conditions with 8 mg/kg, just a slight increase in the rate of malformation. Statistically, this was not significant [384].

With oral or cutaneous thallium administration, degeneration of the female genital tract occurred; papillomas, precancerous lesions, and cancers were found in the genital tract of mice that had never developed spontaneous cancers [67].

Bacteria

Thallium is toxic to bacteria but higher levels are generally required to obtain measurable effects. It inhibits the nitrification by Nitrobacter agilis at concentrations between 0.8 and 8 ppm [380], and the growth of Pseudomonoas aeruginose is prevented in 120 mg/l thallium acetate solution while Staphylococcus aureus and Streptococcus faecalis were not affected at a concentration of 8,000 mh thallium/l [208]

Effects of Thallium on Humans

Thallium is incorporated via ingestion, inhalation or skin contact. Resorption into the gastro-intestinal tract is more than 80% in the jejunum part of the intestine and

60% of the remainder in the ileum [221] and therefore nearly complete. Thallium uptake with food in unaffected areas is estimated to be less than 2 µg/day [149]. In similar areas, respiratory uptake may lie under 5 ng/day, this calculation being based on the thallium content of atmospheric particles, a daily inhalation air volume of 15 m^3 and a retention of 50% [14]. This indicates that in uncontaminated areas ingestion is clearly responsible for thallium uptake. Adsorption via the skin, which can reach 2.5 $\mu g/cm^2$ min in the case of 2% Tl_2O_3 paste [330], occurs almost exclusively during production and refining of thallium compounds and can therefore be ignored here.

Investigations in the proximity of a thallium-emitting cement plant have shown a clear relation between immissions and the body burden of persons living in the area. Moreover, it was also apparent that increased thallium contamination affected persons who had consumed fruit and vegetables from gardens near the plant.

Since other epidemiologic investigations have shown that inhalation of thallous fine dusts is relatively insignificant it can be assumed that even for contaminated areas ingestion is the predominant uptake form of thallium [16].

Uptake, Circulation, and Elimination

Uptake and distribution within the body occurs rapidly and independently of how the thallium is dispensed. In experiments on rats which were injected Tl-204 as thallium(I) nitrate solution and by oral, intratracheal, subcutaneous, interperitoneal, intramuscular or intravenous application the body burden as a function of the administered dose was similar [223]. From these data it was concluded that complete absorption had occurred from the gastro-intestinal tract following ingestion and from the respiratory tract following intratracheal injection.

After being resorbed into blood, thallium ions disappear quickly from the blood flow. Apparently, they are transported like potassium ions, accumulating in

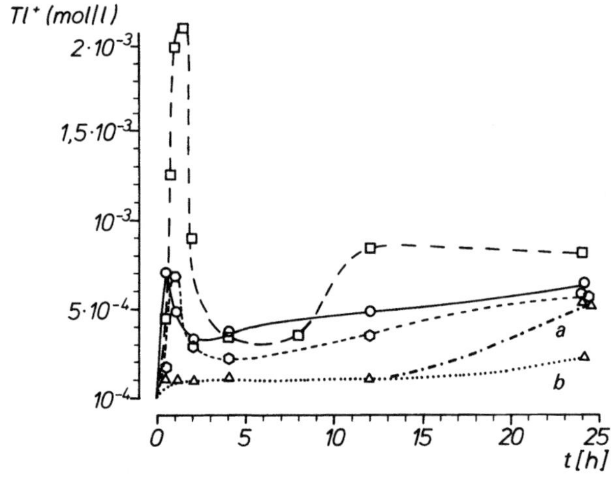

Fig. 15. Time-dependence of the thallium distribution in: heart ○, liver △, kidneys □, and brain △ of mice after a dosage of 160 mg Tl per kg body weight [2]; a = 130 mg, b = 80 mg

the cells of *all* tissue types. This implies that the thallium content of blood does not reflect the general body burden.

The time-dependence of thallium distribution in different body organs is shown in Fig. 15.

In autopsies physiological thallium has been shown to be present in all organs and is generally evenly distributed. In Tables 21 and 22 data from Wenig and Zink [405] are presented.

Table 21. Thallium concentration in several representative organs of the human body; values in nanograms/gram tissue weight

Autopsies on	1	2	3	4	5	6
Fat	0.15	–	–	–	0.32	–
Skeletal muscle	1.02	2.01	0.52	–	0.95	7.05
Cardial muscle	1.95	2.11	–	0.96	–	–
Lungs	1.80	0.72	0.36	1.44	20.2	29.5
Liver	0.76	1.42	0.56	0.55	0.66	2.85
Kidneys	2.69	3.86	2.14	1.80	1.44	4.10
Colon	4.04	5.40	0.85	0.56	1.07	3.08
Rectum	–	1.38	–	–	4.30	0.92
Compacta	2.12	2.51	–	–	2.18	2.88
Nails	1.89	2.51	0.72	4.93	2.52	2.89
Hair	8.70[a]	15.80	4.80	7.54	14.80	11.0
Urine	–	0.26	0.07	0.38	0.35	1.15
Carcinoma	–	–	–	–	0.75	–

1 = 73-year-old woman, pensioner from Erlangen, Germany. Cause of death: heart failure with shrunken kidneys; 2 = 16-year-old man, construction laborer from Erlangen. Cause of death: destruction of the liver by blunt force; 3 = 64-year-old man, pensioner from Erlangen. Cause of death: heart attack, multiple thrombosis; 4 = 25-year-old man from Marktredwitz, Germany, worker in a porcelain factory. Cause of death: subural hematoma; 5 = 44-year-old man, miner (17 years underground) from the Ruhr industrial region. Cause of death: bronchial-CA and miner's lung; 6 = 65-year-old man, miner (30 years underground) from the Ruhr region. Cause of death: pneumonia (miner's lung)

Table 22. Thallium concentration in various organs of the human body. Values in nanograms per gram tissue wet weight

Organ	73-year-old female (death by heart failure)	16-year-old male
Fat	0.15	–
Skeletal muscle	1.02	2.01
Heart muscle	1.95	2.11
Atrio-ventricular node	–	2.66
Achilles tendon	0.25	0.51
Ligamentum nuchae	0.54	–
Meniscus (intraarticular fibrocartilage)	–	0.74
Rib cartilage	0.49	1.05
Epiglottis	–	1.72
Skin	0.46	0.69
Gingiva	0.91 (atrophied)	2.25
Gall Bladder Wall	0.40	0.79

Table 22 (continued)

Organ	73-year-old female (death by heart failure)	16-year-old male
Bladder (urinary)	0.61	1.16
Aorta (wall)	0.64	–
Cerebrum (gray)	0.56	1.25
Cerebrum (white)	0.97	1.24
Cerebellum	0.65	–
Hypophysis with dural sac	0.49	–
Hypophysis without dural sac	–	1.26
Medulla oblongata	0.55	–
Pineal gland	–	1.51
Corpus striatum (of brain)	–	1.27
Internal capaule (of brain)	0.60	–
Nucleus ruber (of brain)	–	1.23
Nucleus niger (of brain)	0.62	–
Choroid plexus (of brain)	0.42	0.76
Liver	0.76	1.42
Pancreas	0.81	1.54
Spleen	1.19	1.54
Kidney, medulla and cortex	2.69	–
Kidney, medulla	–	4.05
Kidney, cortex	–	3.68
Adrenals	0.92	1.32
Thyroid gland	–	1.37
Parotid gland	0.86	1.87
Lymph node	0.91	1.32
Breast (fatty degeneration)	0.32	–
Lung	1.80	0.72
Ovary (fatty degeneration)	0.47	–
Uterus, degenerated	0.61	–
Testicles	–	2.08
Epididymis	–	1.61
Prostate	–	1.37
Stomach wall	0.84	1.56
Duodenal wall	1.56	–
Middle small intestinal wall	0.89	1.54
Colon wall	4.04	5.40
Rectal wall	–	1.38
Spongiosa of the sternum	0.84	1.57
Cranial roof	1.44	–
Rib next to cartilage	1.99	–
Middle rib	2.15	–
Rib next to vertebral column	1.46	–
Thigh (bone) compacta	2.12	2.51
Tooth, whole	–	4.70
Toenail, whole	1.89	2.51
Hair, head	–	15.80
Hair, head, near root	9.69	–
Hair, head, far from root	7.71	–
Serum	0.26	0.33
Blood	0.33	0.59
Gall	0.35	0.53
Urine	–	0.26

Table 23. Concentration of thallium in tissues and other samples in acute poisoning

Tissue	μg/g
Kidney	20.0
Heart	13.3
Brain (gray matter)	10.0
Skin	6.0
Liver	5.0
Bone	5.0
Muscle	5.0
Brain (white matter)	3.0
Lung	1.8
Gall bladder	1.1
Sciatic nerve	1.0

With a few exceptions thallium levels in uncontaminated persons fluctuate within a narrow range. Thallium concentrations in the kidney are regularly above the average value of the various persons. This may be attributed to the excretion of thallium via this organ.

Thallium contents of colon walls fluctuate considerably. Some values here correspond to those of the kidneys, indicating active cell operation of the mucous membrane with respect to resorption or secretion.

In hair, the deposition of physiologic thallium is obvious as it is in finger nails and bone compacta.

The distribution pattern of thallium in human tissue three days after death resulting from acute poisoning is given in Table 23 [64].

The highest concentrations were found in the kidneys. Heart tissue and the gray brain matter showed high enrichment factors whereas concentrations in the white brain matter was conspicuously lower.

Concerning the effects on unborn children, thallium may pass the placenta barrier and can thus enter the fetal blood circulatory system. In experiments using rats, the thallium concentrations in the fetus were of the order of one-fifteenth of the maternal plasma levels [130].

Thallium is a depot poison which remains in the organism for long periods of time. A portion of the adsorbed thallium is secreted via the kidneys and urine; another portion is transported back into the intestinal tract (transepithelial movement) or via the gall bladder (enterohepathic cycle) where it is partially resorbed and partially secreted with the faeces [111].

To a smaller extent, excretion can also occur via skin appendices such as the hair and via mother's milk [174].

The half-life of thallium excretion via urine is ca. 10 days but it may be individually much longer. This half-life measurement should not be confused with the time period necessary for the elimination of half the body burden which has not been sufficiently investigated experimentally so far. In the study on rats described above [233] the excretion of thallium-204 exhibited a similar pattern regardless of the route of administration. About 6% of the administered dose had been excreted in the urine after one day, excretion decreasing to about 0.5% by the 10th day with a half-life of around 3 days. Fecal excretion exceeded that of urine during the 21 days of the study, the fecal to urinary ratio rising from 2 to 5, due to a gradual decrease of urinary thallium.

The amount of thallium eliminated daily by humans after metal contamination is an exponential function of time. If there has been only one single poisoning event, the entire amount eliminated, and therefore under certain conditions the resorbed amount, can be computed if the daily thallium elimination in urine can be determined and if the time of contamination is approximately known. However, it must be taken into account that at an elimination rate of less than 2 mg/d, the excretion seems to slow down [263].

The proportional distribution of human excretion via urine or faeces has been determined in long-range investigations to be 45:55 [23]. This ratio can fluctuate significantly according to distinctness of poisoning symptoms.

Toxicology, Symptoms, and Therapy

Toxicology and Symptoms

Acute toxicity, which is defined either as produced from single or multiple doses administered in short intervals or from a corresponding exposure with always severe symptoms, appears upon doses of more than 100 mg, that is about 1.5 to 2 mg/kg body weight.

The deadly dose of thallium compounds which generally are colorless, odorless, and tasteless differs individually. Values between 0.5 and 3.0 g, corresponding to 6–40 mg/kg body weight, are reported by Müller [263].

Average deadly doses as reported in several publications are 10–15 mg Tl/kg body weight [166, 188, 257]. Accumulation of several lower doses to an acute level is possible.

Lower doses result in subacute toxicity, with the same symptoms on a mild scale.

Chronic poisoning results from multiple doses administered over a long interval and producing severe or mild symptoms, depending on the amounts incorporated.

Acute Poisoning

After relatively large amounts of thallium have been taken internally, there is a period of 1–2 days in which there are nearly no symptoms with the exception of nausea and, in rare cases, vomiting. Other initial symptoms such as persistent obstipation and insomnia as well as an itching and a dull feeling in body extremities do not cause concern to the afflicted person due to their seeming insignificance.

Disorders and symptoms develop slowly. By the end of the first week of affliction, they are recognized as being associated with an illness. Their maximum is reached in the second or third week. Chief symptoms are colic-like stomach pains, which decrease when the stomach is pressed or held firmly, and often pain that is localized behind the sternum (retrosternal pain). At about the same time, the itching feeling increases in intensity to a neuralgiform pain.

The lower body extremities are primarily and most strongly affected, the upper ones irregularly and the cerebral nerves occasionally. High blood pressure and tachycardial effects indicate toxic damage to cardial muscles. Psychic transformations consist mostly of lamentation with a theatralic or hysterical effect which often leads to a misunderstanding of the origin of the illness [263]. These effects culminate in the 3rd or 4th week; peripheric paralysis and chronic disorders can remain.

Frequently loss of head hair occurs on the 13th day after initial poisoning. This long period is explained by the fact that epithelial damage occurs in the hair papilla, the youngest cell layer. Hair development to a phase in which hair loss is possible requires the mentioned period. Medial eyebrows, phylogenetically the younger parts, remain. Secondary, pubic and underarm hair is usually affected but lanuga hair seldom. In borderline cases, loss of hair can be limited to a tonsure-like appearance. Hair growth begins again some time after recovery from the poisoning event, often even ranker than before [112].

The diagnosis of thallium poisoning with evidence from hair samples is possible even before the loss of hair because a dark pigmentation of the hair roots appears 3–5 days after intoxication.

The first regrown hair is often without any pigment. In the case of older people, the loss of hair may remain unnoticed because their hair is no longer alive, remaining attached to the scalp by virtue of the club hair alone. For older women who have been subject to thallium poisoning, the hair may be quite inconspicuous at death even if loss of hair had occurred beforehand.

Symptoms of vegetative disturbances are deficiency or lack of stomach acids, decreasing saliva secretion, unquenchable thirst and trophical disorders of the skin

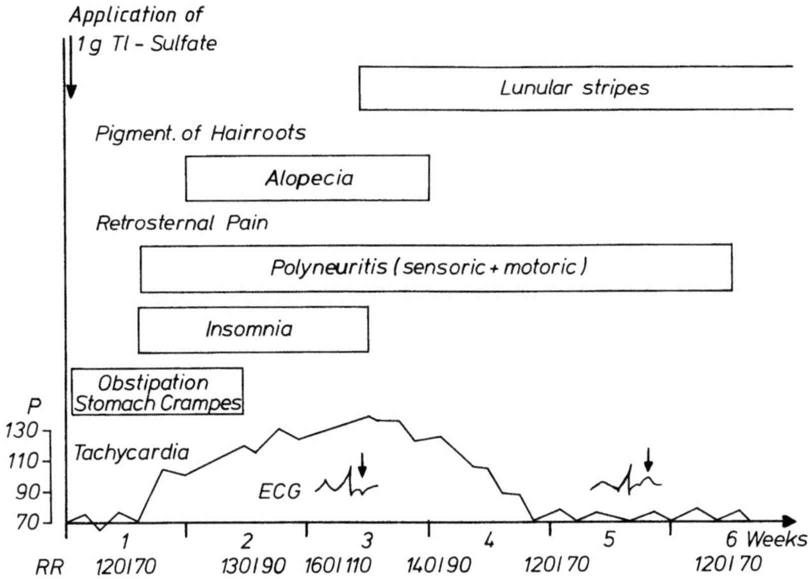

Fig. 16. Time dependencies and symptoms of acute Tl poisoning

with lacking sweat secretion, formation of scurf and secondary infections, inflammation and abcess of mucous membrane of the mouth (stomatitis) and tongue tip (glossitis) as well as conjunctivitis, miction disorders and sphincter debilitation.

Female cycles are arrested. The libido and potency of the male decrease. Loss of weight is a typical symptom. In the third and fourth week after initial thallium poisoning, lunula stripes of the Mees-band type appear on the growing finger or toe nail. These are visible for nearly a year until the nail parts they cover are removed.

Rarely, thallium poisoning can lead to psychoses of various forms, including forms of paranoia. Occasionally, opacity of eye lens forms (cataracts). Upon application of a very large dose of thallium, death can occur even before hair loss begins; in other cases, it occurs in the 3rd or fourth week [263]. Time dependencies and intensities of the described symptoms are compiled in Fig. 16.

Subacute poisoning shows the same symptoms but in weaker form.

Chronic Poisoning

The symptoms of chronic thallium poisoning differ from those of acute or subacute poisoning by the relatively long latent period (several weeks) in which few symptoms are observed: the frequency of eye disorders including less of vision, varying degrees of pain and pain conditions in legs without marked polyneuritis and often only intimated loss of hair. Also observed are stomach anacity, disorders in movement, endocrine stoppages, lack of resistance against infection, decrease in the general health situation, lack of appetite, loss of weight, cerebral disorders with intelligence deficiencies and deviations in behaviour [16].

In general, symptoms are reversible even when complete rehabilitation takes months and is sometimes interrupted by relapses, which are probably caused by the remobilization of thallium from tissue depots. On the other hand, some cases with remaining late infirmities have been reported [312].

The chronological course of a chronic thallium poisoning can be estimated by virtue of the varying thallium contents within a hair.

Symptoms of a large population incurred via industrial emission are discussed later.

Therapy

Therapy for thallium poisoning was generally unsuccessful until the middle 1960's. Attempts to transform thallium chemically after ingestion into forms which are not easily resorbable from the GIT with sodium iodine or reagents containing mercapto groups were not without difficulties. Accelerated excretion of resorbed amounts by various complex-compounds showed no satisfying results. Due to lipophilic and unstable character of some of the compounds the thallium transport into the brain was even increased [111]. Since then, successful therapies have been developed.

Thallium excretion via the kidney can be increased upon dosage of potassium chloride or employment of diuretics [113]. Some authors have recommended hemodialysis and forced diuresis as effective means of decreasing the body burden of thallium [188].

Excretion of thallium via the gastro-intestinal tract can be nearly tripled by the application of Prussian blue [144]. This effect is based on an exchange of potassium ions in the Prussian blue molecule with thallium ions which have an ionic radius similar to that of potassium. Thereby, readsorption from the intestine is prevented and inflow of thallium into the gut is increased by increasing the difference between intra- and extraluminal free concentrations. This therapy requires a constant application of high doses, amounting to 20–30 g per day over a period of some weeks. Negative side effects caused by the release of cyanide have not been observed. Resorbtion of the colloidal Tl-Prussian blue compound from the GIT is negligible. The observation that thallium excretion via faeces continues for up to 20 days, that is up to a point in time where only negligible amounts of unresorbed thallium still remain in the gastro-intestinal tract, suggests that Prussian blue interrupts the enterohepathic cycle or fixes ephitelially transported thallium, whereby the latter process obviously dominates [111].

A success of this method is more likely if treatment is begun as soon as possible. However, positive effects in flushing out thallium may result even if the treatment begins at a later time [307].

Biochemical Mechanisms of Thallium Toxicity

Several symptoms of thallium intoxication, especially peripheric neuropathy, are similar to those of other diseases, which are known to be caused by a diminuation of the sources of nerve-cell energy derived from the activities of the tricarboxylic acid cycle.

The mechanism of cell energy production from pyruvate is shown in Fig. 17. The first step in pyruvate decarboxylation requires thiamin-pyrophosphate, the second step utilizes lipoamide and the third step requires flavine-adenine-dinucleotide to reoxidize lipoamide.

When the pyruvate is decomposed to acetyl-CoA, which then enters the tricarboxylic cycle, FAD also becomes necessary for H^+-transfer to the respiratory chain. Within this chain, in which the protons promote ATP production, the cellular fuel, flavin-mononucleotide (FMN) is required as a redox partner. FAD and FMN are produced from riboflavin. Thallium is known to react with riboflavin [207]. On the other hand, it readily combines with mercapto groups [62]. Both effects may disturb the pyruvate metabolism if thallium enters the nerve cell. Indeed, many of the features of thallium intoxication are similar to those found in riboflavin deficiency [291].

The described cofactors are produced in the rough endoplasmatic reticulum and the Golgi apparatus in the perikaryon of the nerve cell which is schematically shown in Fig. 18. From there they are transported, among others, to axonic mitochondria where energy is produced by the reaction sequence of Fig. 17 to maintain processes such as membrane reactions, ionic pumping, transport mechanisms and syntheses. Under normal circumstances, it is probable that cofactors are present in amounts which are not higher than needed for daily requirements.

If they are blocked, e.g. by thallium infiltration into the perikaryon and the axon, parts of the latter, which are more proximal to the parts where cofactors are synthesized, will fare better than parts that are more distal to the perikaryon.

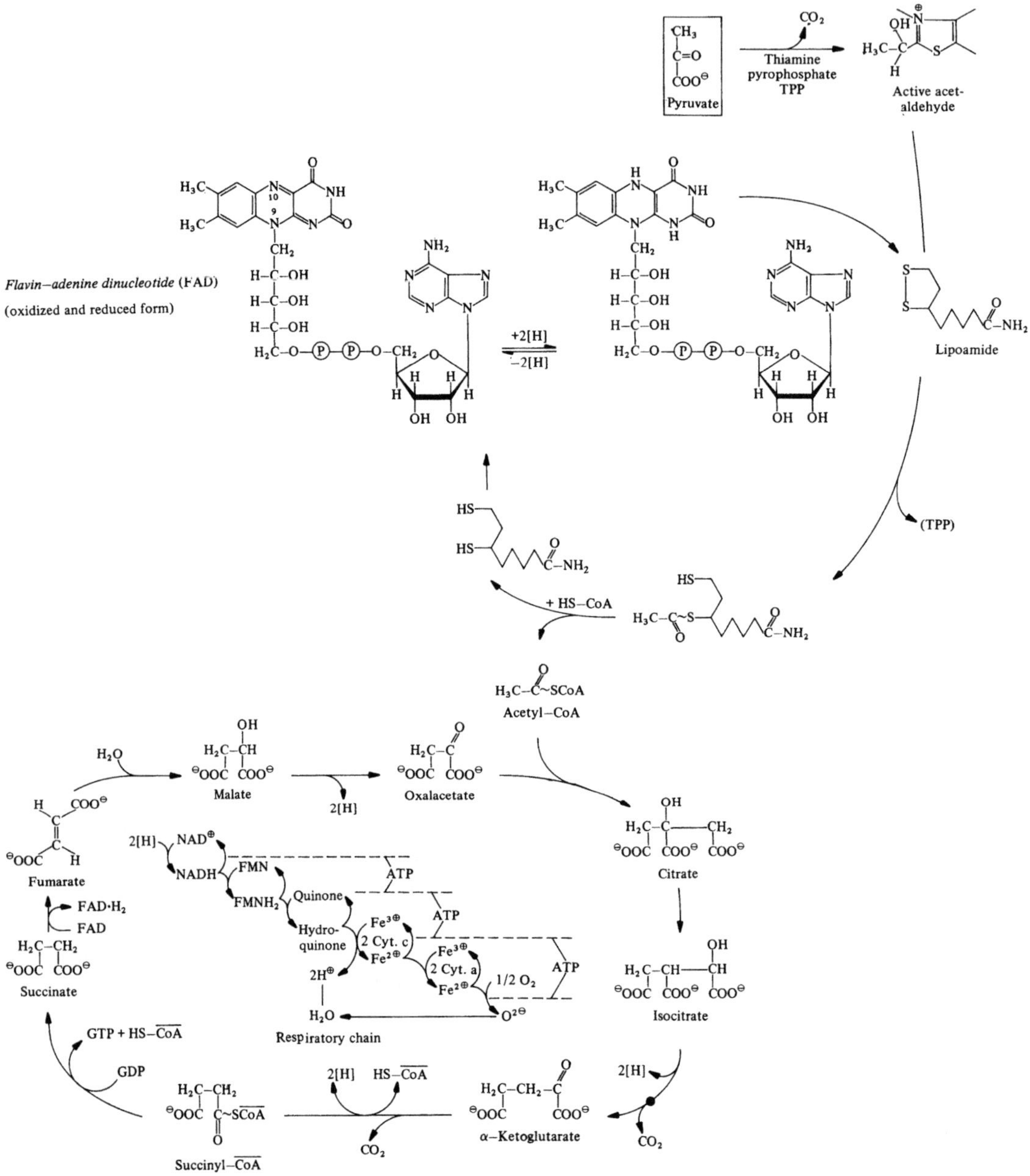

Fig. 17. Mechanism of cell energy production from pyruvate (tricarboxylic acid cycle)

Therefore, distal parts first exhibit deficiency degeneration whereby the known symptoms of peripheric neuropathy occur, because especially the long nerve fibers are affected. This process, which was decribed by Cavanagh [63] as "dying back", is reversible as long as the perikaryon is not destroyed.

Several other biochemical intermediates require FAD in the course of their metabolism and are therefore also affected directly by thallium intoxication, explaining its strong toxicity.

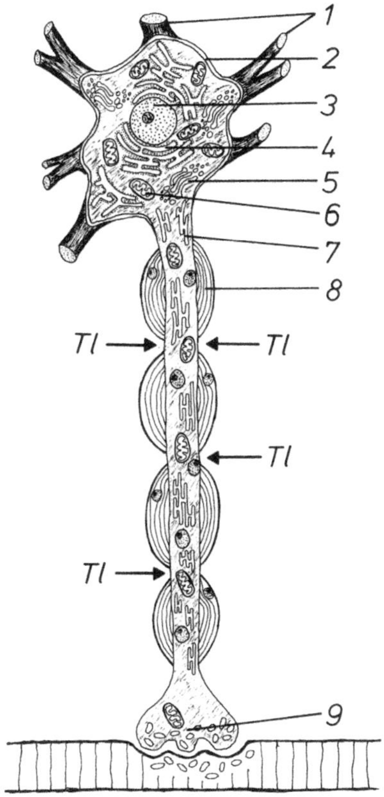

Fig. 18. Nerve cell. *1* dendrite, *2* double membrane, *3* nucleus, *4* rough endoplasmatic reticulum, *5* Golgi apparatus, *6* mitochrondia, *7* smooth endoplasmatic reticulum, *8* neurilemma, *9* synapse

Other effects caused by thallium include:

- influence on enzyme production and amino acid synthesis
- inhibition of the Na^{+}-, K^{+}-dependent ATPase activity
- effects on transport mechanisms
- reduction of mitosis.

The biochemical interactions are reviewed in detail by Carson and Smith [364].

The clinical symptoms of thallium poisoning appear more evident in the light of these observations. Likewise, it is no surprise that subacute thallium poisoning interferes with normal hair growth and skin cell formation. Because of their active proliferation and protein synthetic committments hair and skin mitosis is greatly dependent on the free availability of energy.

It is interesting to consider the mechanism of alopesia in view of the ideas mentioned above. This is not possible without some notes on the structure of hair. The bulbous root rests on the connective tissue of the hair follicle which contains a blood capillary and melanocytes. During growth, cells are pushed upwards from the basis root layer. The cells become callus and are surrounded at the same time by hair cuticle. The hair structure is depicted in Fig. 19.

Hair growth occurs in cycles of which 3 phases can be distinguished: anagen (growth phase), catagen (involution), and telogen (rest phase).

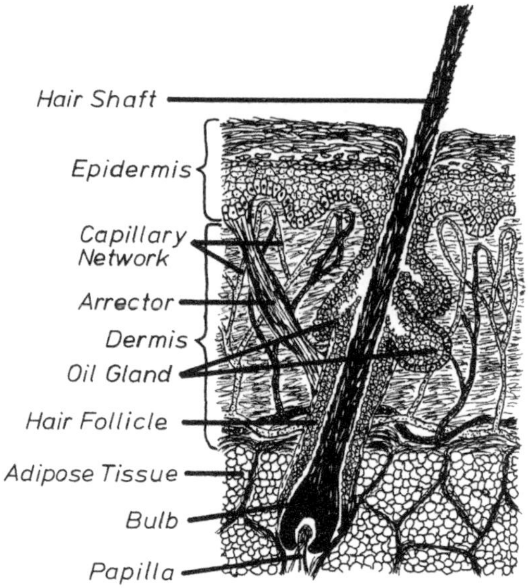

Fig. 19. Schematic structure of a single hair [220]

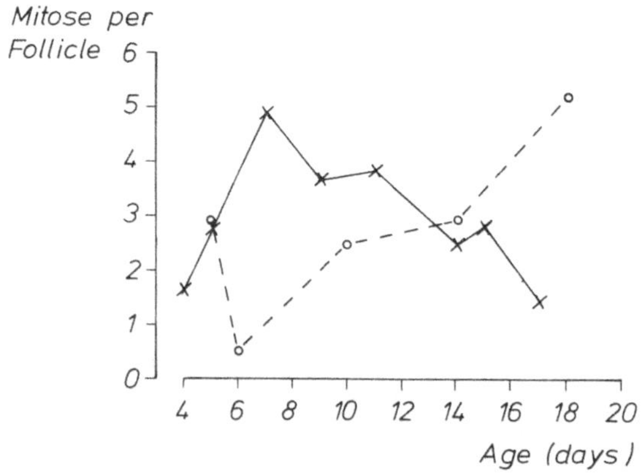

Fig. 20. Mitotic activity in dorsal hair follicles of rat pups given $TlSO_4$ (30 mg/kg) subcutaneously on day 4 (o . . . o). Mitotic activity of control animals (× . . . ×)

After this cycle has been completed, the hair is lost and a new cycle can begin.

Approximately 80% of the terminal hair is in the growth stage. In the rest stage a hair is present as a club hair which falls out later. On the papilla surface new hair grows out of epithelical remains.

Cavanagh and Gregson [65] have studied the effect of thallium salts on the proliferation of hair follicle cells. They showed that no visible effects occur during the first 24 h after subcutaneous injection of thallium sulfate into 4-day-old rat pups. The mitotic rate of the hair follicle cells was normal (Fig. 20). The cells showed no effects.

After 48 h there was a marked drop in the number of cell divisions in the follicles and many pyknotic nuclei were present, almost exclusively in the matrix region of the follicle.

At the end of the catagen, club formation did not occur and the hair shaft did not exhibit the normal firm attachment to the involuted follicle, resulting in hair loss.

Afterwards, the follicles recovered and showed increased mitotic activity with recovery from the insult.

No specific mitotic lesions were found on change in cell cycle indices. Cavanagh and Gregson [65] therefore suggested a failure of energy metabolism as a possible source of the inhibition of hair follicle mitosis by thallium. This argument is supported by the similarity of the experimental results with thallium to those with other reagents such as monoiodo-acetate which inhibits the follicular growth, due undoubtedly to the lack of energy production to drive mitosis. The biochemical mechanism might be the same as in neurons.

Case-Study: Environmental Pollution by Thallium from a Cement Plant in Lengerich/Westfalia

In August 1979 it was discovered that thallium had been emitted in considerable amounts by a cement plant in north-west Germany. Damage to plants and domestic animals had been observed in the vicinity of the plant for a number of years but it was not realized until 1979 that this damage was caused by thallium-containing atmospheric dust fall-out. An inspection of the production process proved pyrite roastings, which were used as iron(III)oxide additives for the production of special cement sorts, as the main thallium source. Chemical analysis revealed the thallium content of the additive to be about 400 µg/g. Between 4% and 5% were added to the pulverized raw meal, depending on the sort of cement to be produced.

To get information about soil contamination around the cement plant, five trajectories in northern, north-western, eastern, south-eastern, and southern direction

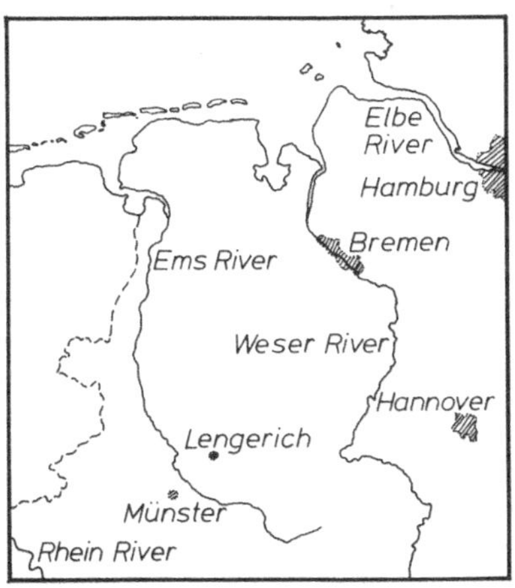

Fig. 21. Geographical site of the cement plant Lengerich

were drawn to a distance of 5 km with sampling points each 500 m. The samples consisted of mixed material from the top 25 cm soil layer.

High thallium concentrations were found at a distance of 500 m from the emitting stack in the eastern and south-eastern direction, namely 6.9 and 6.3 μg/g DWT, respectively. With increasing distance from the emitter thallium concentrations of the soil decreased rapidly, reaching the detection limit of 0.1 μg/g at 2,000 m. The only exception was at a declivity about 3,500 m from the emission centre.

The percentage of distribution of the investigated soil samples was given as:
- 77% less than 0.5 μg/g
- 8% from 0.5– 1.0 μg/g
- 15% from 1.0–10.0 μg/g [16].

The Lengerich cement plant employed a closed cooling and waste water cycle. The water was cleaned mechanically and pumped into ponds where the solid particles settled. After a certain time the water was reused in the production process.

This cycle was connected with the communal waste water plant via a rain overflow which allowed the plant's waste water to reach a small river. The thallium concentration in the cooling water was about 20 μg/l, and the outlet of the municipal waste water plant contained less than 1 μg/l. The pond sludges, which were deposited in an old quarry, showed Tl concentrations of 13 μg/g whereas the sewage sludges of the communal water plant reached 16 μg/g. Neither sludge was used in agriculture.

To investigate possible earlier Tl contaminations, sediments, algae, and fish were analysed in addition to the flowing water.

The maximum thallium concentration in the water was 1 μg/l and the sediments contained up to 1.8 μg/g with an average of less than 1.0 μg/g Tl. Sediments from uncontaminated local rivers had Tl concentrations between 30 and 100 ng/g, demonstrating a more than tenfold increase of Tl by the waste water of the cement plant [16].

Algae and other water plants contained thallium too. Sample concentrations were between 20 and 665 ng/g FW and one sample even reached 6.4 μg/g FW which is of the same order of magnitude as reported from effluents of base-metal mining [416].

None of the samples from the local water authorities and private wells contained more than 1 μg/l.

In 1979, a total of 771 samples of edible plants from a narrow range around the facility were investigated. Within an area with a diameter of 4 km, the center being about 500 m east of the emitting stack, 14% of all plants exceeded the declared limit of 0.5 μg/g FWT. Depending on site and species, characteristic differences in the thallium content of plants were found. The concentration in kale was extraordinarily high, in one case 45 μg/g FWT. Other vegetables such as kohlrabi, savoy and white cabbage contained concentrations of 1–10 μg/g FWT without any external sign of damage.

It was recommended that consumption of vegetables from the described area be discontinued. The concentrations in fruits were, with very few exceptions, less than 0.5 μg/g, making such recommendations unnecessary.

The most important indication of thallium as the origin of the plant damages in Lengerich has been delivered by wild rabbits. After fur loss was observed, studies

Table 24. Fodder amounts and Tl concentration in the feed of investigated bulls [151]

Fodder	Daily consumption (kg) per 400 kg animal	Tl concentration (mg/kd DWT) in fodder
Silomaize	12	0.45
Crushed barley	3	0.10
Crushed soya	0.45	0.28
Minerals	0.1	0.05
Hay	3	–0.44

Table 25. Tl concentrations in organs of investigated bulls

Organ	Mean Tl concentration (mg/kg FWT)
Liver	0.52
Kidneys	1.66
Muscle	0.10
Rumen content	0.30
Testicle	0.40

revealed that the fur had a thallium content of 12.8 μg/g DWT. Hair which was not affected showed 8.7 μg/g whereas regrown hair still contained 4.7 μg/g DWT. "Normal" values were determined to lie near 1.5 μg Tl/g DWT for wild rabbit in an industrial centre [16].

During field research with eight bulls, thallium concentrations in different organs were investigated in reference to uptake from feed to animal. The bulls were given the feed products listed in Table 24 which were harvested from fields near the cement plant.

From the amount of the feed consumed and the general Tl contents of the various feed components it can be calculated that the animals had a daily consumption of 10 mg Tl. This corresponds to a daily dosis of 0.025 mg/kg body weight. The bulls were given the feed listed in Table 24 for at least 6 months. However, they had been supplied with Tl contaminated fodder for more than a year whereby the feed could not be analysed for thallium during the first few months.

Thallium contents in organs of bulls fed in this manner are compiled in Table 25.

Measurements indicated preferred enrichment in the kidneys; liver and testicles contained less thallium and the muscle tissue was relatively uncontaminated [151].

The results were confirmed by the analysis of more than 1,000 samples in the region under discussion (Fig. 22).

As a consequence, it was either recommended that entrails from animals supplied with fodder from the proximity of the cement plant are not consumed or total meat products were deemed unsuitable for humans and destroyed.

Evident increases in thallium excretions were recorded in the local population in the proximity of Lengerich Cement Plant.

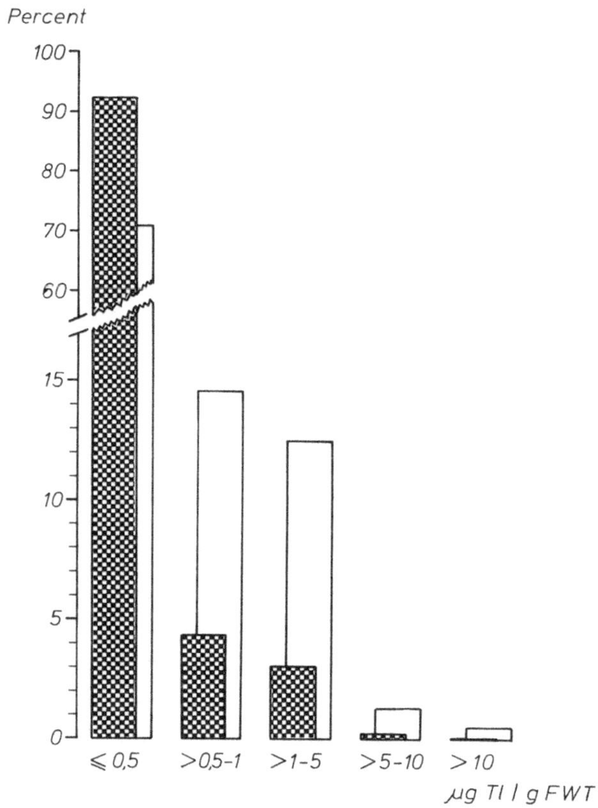

Fig. 22. Frequency distribution of Tl concentrations in meat samples. White: entrails, black: without entrails [16]

Table 26. Normal values of thallium in urine and hair of unaffected persons [101]

Urine	Mean conc. (µg/l)	Upper normal value (µg/l)	Ref.
0.05– 0.1 µg/l			[127]
0.1 – 0.5 µg/kg			[134]
0.13– 1.69 µg/l	0.3	1.8	[405]
0.1 – 1.0 µg/l			[401]
0.3 – 1.1 µg/g creat.		1.0 creat.	[333]
0.1 – 1.2 µg/g	0.4	0.8	[46]
0.02– 0.7 µg/l	0.5	0.6	[46]
Hair (ng/g)			
9.0 –20.0			[127]
4.8 –15.8			[405]
< 10.0			[46]

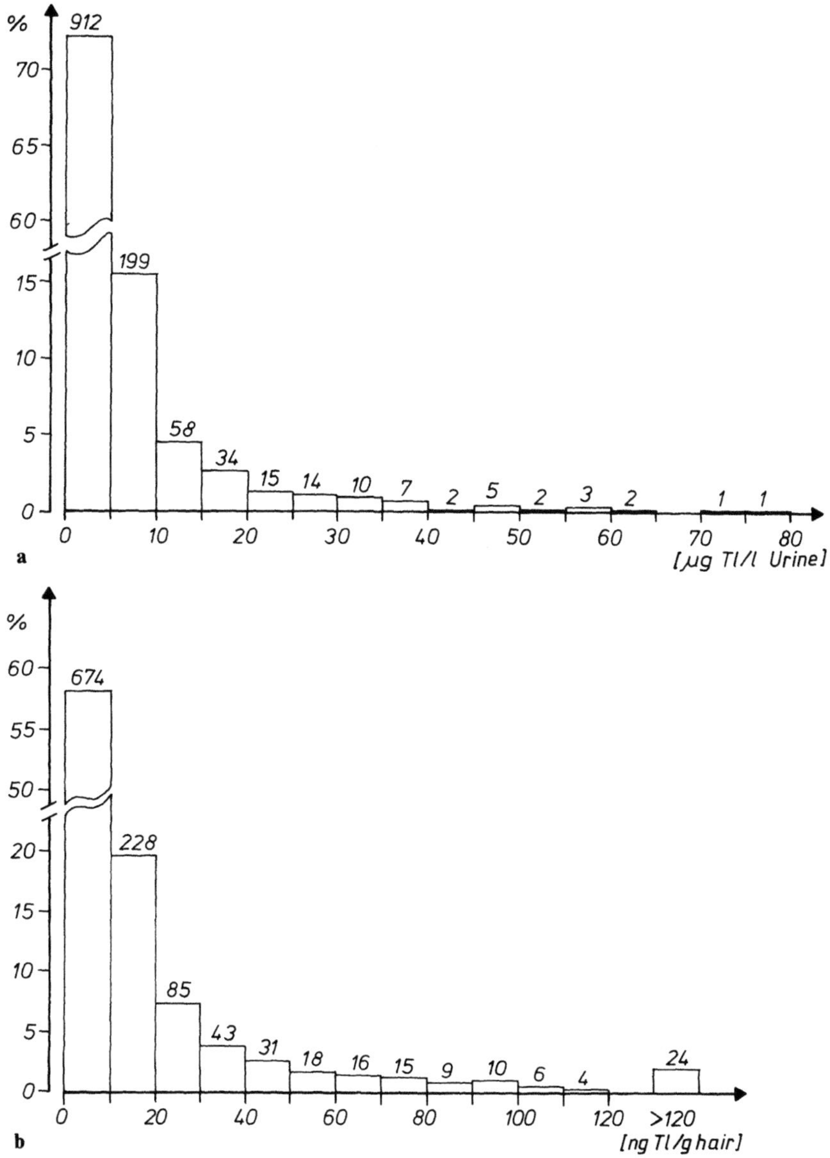

Fig. 23. a Frequency distribution of thallium concentrations in 24-h urine samples of a population (N = 1,265) living near a cement plant emitting thallium-containing dust [46]. **b** Frequency distribution of thallium concentrations in hair samples of a population (N = 1,163) living near a cement plant emitting thallium-containing dust [46]

As a basis of assessment, the normal values of thallium in the urine and hair of unaffected persons are listed in Table 26.

To assess the degree of thallium exposure in the aforementioned cement plant, thallium levels in 24-h urine samples of 1,265 inhabitants and in hair samples of 1,163 inhabitants were determined. The results of the studies are shown in Fig. 23 and Table 27.

Table 27. Urinary thallium levels (TlU) and hair thallium levels (TlH) of different population groups [46]

Population	Parameter	Number of samples	Range	Mean	Geom. mean
Cement plant area	TlU (μg/l)	1,265	<0.1– 76.5	5.2	2.6
	TlH (ng/g)	1,163	0.6–565	20.3	9.5
Rural area	TlU (μg/l)	31	<0.1– 1.2	0.4	0.4
Urban area	TlU (μg/l)	10	<0.1– 0.9	0.3	0.2

As compared to the reference groups it is obvious that the population living in the neighbourhood of the cement plant showed signs of increased thallium intake. On a group basis the investigated individuals had about 10 times more thallium in urine than the reference subjects. A total of 992 (78%) exposed persons had urinary levels >1 μg/l whereas among the reference groups there was only 1 individual (2.5%) with Tl-U>1 μg/l.

Table 26 shows that the normal thallium content in human hair is in the range of 10 ng/g. Again a considerable part of the Lengerich population showed markedly increased Tl-H concentrations reaching up to 565 ng/g.

The distance between the location of residences and cement plant, the intensity of comsumption of home-grown vegetables and fruits, and the site of dwelling relative to the site of the cement plant accounted for about one-third of the Tl-U variation. With respect to the variation of Tl-H these parameters are complemented by age, sex and occupational thallium exposure to reach the same percentages.

A regression line was calculated for the total population examined, yielding a highly significant correlation between log Tl-U and log Tl-H ($r=0.42$; $P<0.001$). The following regression equation was obtained:

$$\log \text{Tl-H} = 0.864 \log \text{Tl-U} - 0.602$$

No positive exposure-response relationship was found between Tl-U or Tl-H as indicators of exposure and the prevalence of certain clear symptoms known to be associated with chronic thallium intoxication. In fact, concerning the prevalence of hair loss, a negative correlation was found.

However, a clear exposure-response relationship was established between Tl-U and a predominance of sleep disorders, tiredness, general bodily weakness, nervousness, headache, and other psychic disorders as well as neurological symptoms such as paraesthesis, muscle and pain in joints.

A similar exposure-response relationship was obtained using Tl-H as an exposure indicator [46].

Altogether, subjective rather than objective symptoms constitute the major health effects associated with the reported Tl-U and Tl-H levels of the environmental intoxication in Lengerich. Such symptoms, however, have to be interpreted carefully. Nevertheless, a slight intoxication could not be excluded [16].

Urinary thallium levels of 300 μg/l have been suggested as the threshold value for chronic poisoning of factory workers [132]. Compared with this value, therapeutic treatment, e.g. oral administration of Prussian blue, was not considered necessary in Lengerich.

Acknowledgements

Thanks are due to all who contributed to this work with valuable discussions and correspondence and especially to Petra and David for their technical assistance. Without their support and that of the citizens of Lengerich and Leimen who first involved me in this subject and advanced my investigations this study would not have been possible.

References

1. Achenbach, C., Hauswirth, O., Heindrichs, C., Ziskoven, R., Köhler, F., Smend, J., Kowalewski, S.: Toxizität und Teratogenität von Thallium (Toxicity and teratogenity of thallium). Dtsch. Ärztebl. *48*, 3189–3192 (1979)
2. Achenbach, C., Ziskoven, R., Köhler, F., Bahr, U., Schulten, H.-R.: Quantitative Spurenanalyse von Thallium in biologischem Material. Angew. Chem *91*, 944–945 (Quantitative trace analysis of thallium in biological material). Angew. Chem. Int. Ed. Engl. *18*, 882–883 (1979)
3. Achenbach, C., Hauswirth, O., Heindrichs, C., Ziskoven, R., Köhler, F., Bahr, U., Heindrichs, A., Schulten, H.R.: Quantitative measurement of time-dependent thallium distribution in organs of mice by field desorption mass spectrometry. J. Toxicol. Environ. Health *6*, 519 (1980)
4. Ahrens, L.H.: The abundance of thallium in the earth's crust. Science *106*, 268 (1947)
5. Ahrens, L.H.: The unique association of thallium and rubidium in minerals. J. Geol. *56*, 578–590 (1948)
6. Ahrens, L.H.: The use of ionization potentials. Part I. Ionic radii of the elements. Geochim. Cosmochim. Acta *2*, 155 (1952)
7. Albert, A.: Quantitative studies of the avidity of naturally occurring substances for trace metals. 3. Phenidine, riboflavin and purines. Biochem. J. *54*, 646–654 (1953)
8. Anon: Thallium. Non-ferrous extractive metallurgy in the U.K. (1968)
9. Anon: Metals: Reference and Encyclopedia (1968)
10. Anon: Institut für Wasser-, Boden- und Lufthygiene des Bundesgesundheitsamtes Berlin: Gutachtliche Stellungnahme zur Frage der Auswirkungen der Ablagerungen arsenhaltiger Schlämme aus der Nievenheimer Zinkhütte auf Grund- und Oberflächenwässer unter besonderer Berücksichtigung hygienischer Gesichtspunkte (Effects of influence of arsenic-containing sludges to ground and surface waters with reference to hygiene). Unpublished (1972)
11. Anon: World Cement Dictionary – 1965–1970, CEM-Bureau, Paris, 1972
12. Anon: Thallium. Lancet *2*, 564–565 (1974)
13. Anon: A.C.G.I.H.: Thallium documentation of the Threshold Limit Values for substances in workroom air. *3*, 256, Cincinatti, 1976
14. Anon: Thallium. Stellungnahme des Bundesgesundheitsamtes (1979)
15. Anon: Wijziging van de toelating van thalliumsulfaat (requirements of permission of thallium sulfate). Rat Muis *28(1)*, 16–17 (1980)
16. Anon: Umweltbelastung durch Thallium. Dokumentation zusammengestellt von der Landesanstalt für Immissionsschutz (LIS) des Landes Nordrhein-Westfalen (eds. E. Koch, G. H. M. Krause) im Auftrag des Ministers für Arbeit, Gesundheit und Soziales und des Ministers für Ernährung, Landwirtschaft und Forsten des Landes Nordrhein-Westfalen. Düsseldorf (1980). (Environmental Pollution by Thallium. Documentation)
17. Anon: LIS-Berichte Nr. 15: Thalliumemission bei der Herstellung von Portlandzement-Klinker (Thallium emission during production of Portland cement clinker), Essen 1981
18. Anon: Thallium: Nippon Kogyo set-up specialist refining plant. Roskill's Letter from Japan *74*, 3 (1982)
19. Anon: Thallium – Ein Überblick (Thallium – An Overview). Schwermetall-Rundbrief *5*, 5–15 (1982)
20. Anon: Rückblick auf die Welt-Zementindustrie 1978–1980 und derzeitige Situation (Retrospective on the cement industry in 1978–1980 and the present situation). Zement-Kalk-Gips *8*, 442–444 (1982)

21. Anon: Yearbook of World Energy Statistics 1980, United Nations, New York, 1982
22. Ariel, M., Bach, D.: The determination of thallium in urine. Analyst *88*, 30–35 (1963)
23. Arnold, W., Herzberg, J.J., Ludwig, E., Sturde, H.: Die Dynamik des Haarausfalles bei Thallium-Vergiftung (The dynamic of hair loss at thallium poisoning). Archiv für klinische und experimentelle Dermatologie *218*, 396–420 (1964)
24. Babitzke, H.R.: Thallium. Minerals, Facts and Problems, 1975 (ed.) Bureau of Mines Bulletin *667*. U.S. Department of the Interior, Washington, D.C., pp. 1109–1114 (1976)
25. Bajpai, K.K., Singh, K., Misra, S.R., Srivastava, T.N., Kumar, V.: Anti-fungal activities of diorgano gallium, indium and thallium diethyl-dithiocarbamates. Indian Phytopathol. *29(3)*, 335–337 (1976)
26. Bakeev, M.I., Buketov, E.A., Andamasov, R.S.: Kompleksnaya Pererabotka Pyli ot Elektroplavki Margantsevykh Rud (Complex Treatment of Dust from the Electric Smelting of Manganese Ores). Tr. Khim.-Met. Inst., Akad. Nauk Kaz. S.S.R. *3*, 60–74 (1967)
27. Balke, K.D., Kussmaul, H., Siebert, G.: Chemische und thermische Kontamination des Grundwassers durch Industriewässer (Chemical and thermal contamination of the ground water via industrial discharges). Z. Dtsch. Geol. Ges. *124*, 447–460 (1973)
28. Bambauer, H.U., Schäfer, H.: Der Mineralbestand eines thalliumhaltigen Reingasstaubes aus der Zementherstellung (Mineral content of thallous fly ash from cement production). Naturwissenschaften *68*, 571 (1981)
29. Bank, W.J., Pleasure, D.E., Suzuki, K., Nigro, M., Katz, R.: Thallium poisoning. Arch. Neurol. *26*, 456–464 (1972)
30. Barckow, J., Jenss, H.: Thalliumvergiftung (Thallium poisoning). Med. Klin. *71*, 1377–1382 (1976)
31. Barclay, R.K., Peacock, W.C., Karnofsky, D.A.: Distribution and excretion of radioactive thallium in the chic embryo, rat and man. J. Pharmacol. Exp. Ther. *107*, 178–187 (1953)
32. Batley, G.E., Florence, T.M.: Determination of thallium in natural waters by anodic stripping voltametry. Electroanal. Chem. Interfac. Electrochem. *61*, 205–211 (1975)
33. Baumgärtner, F.: Table of Neutron Activation Constants, München, Thiemig Verlag 1967
34. Bazzaz, F.A., Carlson, R.W., Rolfe, G.L.: The effect of heavy metals on plants. Part I: Inhibition of gas exchange in sunflower by Pb, Cd, Ni, and Tl. Environ. Pollut. *7*, 241–246 (1974)
35. Bean, J.R., Hudson, R.H.: Acute oral toxicity and tissue residues of thallium sulfate in golden eagles, Aquila chrysaetos. Bull. Environ. Contam. Toxicol. *15(1)*, 118–121 (1976)
36. Bertram, H.P.: Environmental contamination by thallium in the emission area of a cement factory; human exposure in comparison with acute thallium intoxications. 21st Spring Meeting of the Deutsche Pharmakologische Gesellschaft (German Pharmacological Society), Mainz, West Germany. March 18–21, 1980. Naunyn-Schmiedeberg's Arch. Pharmacol. *311*, R 22 (Suppl.) (1980)
37. Bock, K.D., Merguet, P., Schley, G., Schümann, H.-J., Rausch-Stroomann, J.G., Hocevar, V., Schröder, E., Murata, T.: Zur Pathogenese von Hypertonie und Tachycardie bei der akuten Thalliumvergiftung und bei akuter intermittierender Porphyrie (On pathogenesis of hypertonia and tachycardia upon acute thallium poisoning and acute intermittant porphyria). Dtsch. Med. Wochenschr. *93*, 2119–2124 (1968)
38. Bock, R., Monerjan, A.: Abtrennung und gaschromatographische Bestimmung des einwertigen Thalliums als Cyclopentadienyl-Verbindung (Separation and gas chromatographic determination of monovalent thallium as cyclopentadienyl compound). Z. Anal. Chem. *235*, 317–321 (1969)
39. Boenig, G., Heigener, H.: Papierchromatographische Analyse von Thallium in biologischem Material (Paper chromatographic analysis of thallium in biologic material). Landwirtsch. Forsch. *26*, 81–83 (1973)
40. Borzini, G.: Influenza degli ioni tallio sulla germinazione di semi diversi e sullo sviluppo inizale della piantine (Influence of thallium ions on the germination of various seeds and on early growth of plants). Boll. Staz. Patol. Veg. *15*, 200–231 (1935)
41. Bowen, H.J.M.: Trace Elements in Biochemistry, London, Academic Press 1966
42. Bowen, H.J.M.: Environmental Chemistry of the Elements, London, Academic Press 1979
43. Brieger, H.: The Use of Chelating Agents in Occupational Medicine. In: Seven, M.J., Johnson, L.H. (eds.) Metal-Binding in Medicine, Lippincott Company, Philadelphia-Montreal, 1960
44. Brockhaus, A., Dolgner, R., Ewers, U., Wiegand, H., Freier, I., Jerman, E., Krämer, U.: Excessive thallium absorption among a population living near a thallium-emitting cement plant. In: Holm-

stedt, B., Lauwerys, R., Mercier, M., Robberfroid, M. (eds.) Mechanisms of Toxicity and Hazard Evaluation (Proc. of the 2nd International Congress on Toxicology, Brussels 1980). Elsevier/North Holland, Amsterdam-New York-Oxford (Develop. Toxicol. Environ. Sci. *8*, 565–568 (1980)
45. Brockhaus, A., Dolgner, R., Ewers, U, Freier, I., Jermann, E., Krämer, U., Wiegand, H.: Thallium concentrations in urine and hair samples of a population living near a thallium-emitting cement factory, 2nd International Congress on Toxicology, Brussels, Belgium, July 6–11, 1980. Toxicol. Lett (AMST) (Spec. Issue 1), 1980
46. Brockhaus, A., Dolgner, R., Ewers, U., Krämer, U., Soddemann, H., Wiegend, H.: Intake and health effects of thallium among a population living in the vicinity of a cement plant emitting thallium-containing dust. Int. Arch. Occup. Environ. Health *48(4)*, 375–390 (1981)
47. Brockhaus, A., Dolgner, R., Ewers, U., Freier, J., Jermann, E., Wiegand, H.: Repeated surveillance of exposure to thallium in a population living around a cement plant emitting thallium-containing dust. Intern. Conf. Paper: Heavy Metals in the Environment, Amsterdam, pp. 482–485 (1981)
48. Brooks, R.R.: The use of ion-exchange enrichment in the determination of trace elements in sea water. Analyst. *85*, 745 (1960)
49. Brooks, R.R.: Apparent geochemical association of bismuth and thallium. Nature *189*, 910–911 (1961)
50. Brooks, R.R., Ahrens, L.H.: Some observations on the distribution of thallium, cadmium, and bismuth in silicate rocks and the significance of covalency to their degree of association with other elements. Geochim. Cosmochim. Acta *23*, 100 (1961)
51. Brooks, S.C.: Thallium poisoning and soil fertility. Science *75*, 105–106 (1932)
52. Brumsack, H.J.: Potentential metal pollution in grass and soil samples around brickworks. Environ. Geol. *2*, 33–41 (1977)
53. Burger, A., Starke, K.: Beeinflussung der ATP-Ase aminspeichernder Granula von Nebennierenmark und Milznerven durch Thallium (Influence of thallium on ATPase of amino-storing granula of peripheral kidney medulla and spleen nerves) Experientia *25*, 578–579 (1969)
54. Burton, M.A.S., Puckett, K.J., Lesueur, P.: Copper, nickel, and thallium uptake by the lichen cladina-rangiferina. Can. J. Bot. *59*, 91–100 (1981)
55. Buscke, A., Palser, B.: Die biologischen Wirkungen und die praktische Bedeutung des Thalliums (The biologic effects and the practical value of thallium). Erg. Allg. Path. Anal. *25*, 1 (1931)
56. Buser, A.I., Tiptsova, V.G.: In: Vinogradov, A.P., Ryabchikov, D.I. (eds.) Detection and Analysis of Rare Elements, Jerusalem, Israel Prog. Sci. Transl. 1962
57. Cammann, K., Andersson, J.T.: Increased sensitivity and reproducibility through signal averaging in ranges near an instrumental limit of detection; thallium and cadmium trace determinations in rock samples. Z. Analyt. Chem. *310*, 45–50 (1982)
58. Canney, F.C.: Some aspects of the geochemistry of K, Rb, Cs, and Tl in sediments. (Abstr.). Geol. Soc. Amer. Bull. *63*, 1238 (1952)
59. Canterford, G.S., Canterford, D.R.: Toxicity of heavy metals to the marine diatom ditylum-brightwelli; correlation between toxicity and metal speciation. J. Mar. Biol. Assoc. U.K. *60(1)*, 227–242 (1980)
60. Carlson, R.W., Bazzaz, F.A., Rolfe, G.L.: The effect of heavy metals on plants. 2. Net photosynthesis and transpiration of whole corn and sunflower plants treated with Pb, Cd, Ni, and Tl. Environ. Res. *10*, 113–120 (1975)
61. Cataldo, D.A., Wildung, R.E.: Soil and plant factors influencing the accumulation of heavy metals by plants. Workshop on Higher Plant Systems as Monitors of Environmental Mutagens, Marineland, Fla., USA, January 15–19, 1978. Environ. Health Perspect. *27*, 149–160 (1978)
62. Cavanagh, J.B.: Metallic toxicity and the nervous system. In: Smith, W.T., Cavanagh, J.B. (eds.) Recent Advances in Neuropathology. Churchill Livingston, Edinburgh-London-New York 1979
63. Cavanagh, J.B.: The dying-back process. Arch. Pathol. Lab. Med. *103*, 659–664 (1979)
64. Cavanagh, J.B., Fuller, N.H., Johnson, H.R.M., Rudge, P.: The effects of thallium salts, with particular reference to the nervous system changes. Quart. J. Med. *43*, 293–319 (1974)
65. Cavanagh, J.B., Gregson, L.: Some effects of a thallium salt on the proliferation of hair follicle cells. J. Pathology *125*, 179–191 (1978)
66. Cernik, A.A.: The determination of urinary thallium at or near unexposed concentrations. Intern. Sym. on Clinical Chemistry and Chemical Toxicology of Metals. Monte Carlo, March 2–5, 1977 (1977)

67. Champy, C., Hatem, S., Tharanne, J.: Végétations du Tractus Génital de la Souris après Intoxication Thallique Ménagée (Pathology of the genital tract of the female mouse after controlled thallium intoxication). Compt. Rend. Soc. Biol. *152*, 906–907 (1958)
68. Chizhikov, D.M.: Cadmium, New York, Pergamon Press 1966
69. Christmann, W., Lukaszewski, Z., Neeb, R.: Bestimmung des Cadmiums, Blei, Thallium, Kupfer neben Indium in tensidhaltigen Grundlösungen von Polyaminocarbonsäuren mit Hilfe der differentiellen Pulse-Polarographie (Determination of cadmium, lead, thallium, copper, and indium in tensidic solutions of poly-amino carboxylic acids with differential pulse polarography). Z. Anal. Chem. *302*, 32–35 (1980)
70. Clarkson, T.W.: Diseases Associated with Exposure to Metals: Thallium. In: Last, J.M. (ed.) Maxcy-Rosenau Public Health and Preventive Medicine, 11th edit., XXV. Appleton-Century-Crofts: New York, pp. 673–674 (1980)
71. Clausen, B., Karlog, O.: Loading with thallium among wild animals of the marten genus and badgers in Denmark. Nord. Veterinarmed. *26*, 339–350 (1974)
72. Crafts, A.S.: The effects of thallium sulphate upon soils. Science *79*, 62 (1934)
73. Cromartie, E., Reichel, W.L., Locke, L.N., Belisle, A.A., Kaiser, T.E., Lamont, T.G., Mulhern, B.M., Prouty R.M., Swineford, D.M.: Residues of organochlorine pesticides and polychlorinated biphenyls and autopsy data for bald eagles 1971–1972. Pesticides Monit. J. *9(1)*, 11–14 (1975)
74. Csicsaky, M., Wiegand, H.: Effects of thallium on the central and peripheral nervous system of the rat. 22nd Spring Meeting of the Deutsche Pharmakologische Gesellschaft, Mainz, West Germany, March 10–13, 1981. Naunyn-Schmiedeberg's Arch. Pharmacol. *316* (Suppl.), R 25 (1981)
75. Czaplinski, F.: Thallium poisoning in small animals. Med. weteryn. *6*, 531–533 (1950)
76. Davis, R.D., Beckett, P.H.T., Wollan, E.: Critical levels of 20 potentially toxic elements in young spring barley. Plant soil *49*, 395–408 (1978)
77. Davison, R.D., Natusch, F.S., Wallace, J.R., Evans, C.E., Jr.: Trace elements in fly ash – dependence of concentration on particle size. Environ. Sci. Tech. *8*, 1107–1113 (1974)
78. Dawson, G.W.: The Chemical Toxicity of Elements. BNWL-1915 UC-70. Battelle Pacific Northwest Labs., Richland, Wash. (1974)
79. De Albuquerque, C.A.R., Shaw, D.M.: Thallium. In: Wedepohl, K.H. (ed.) Handbook of Geochemistry. Springer, Berlin-Heidelberg-New York 1972
80. De Albuquerque, C.A.R., Muysson, J.R., Shaw, D.M.: Thallium in basalts and related rocks. Chem. Geol. *10*, 41–58 (1972)
81. Deangelis, T.P., Bond, R.E., Brooks, E.E., Heineman, W.R.: Thin-layer differential pulse voltametry. Anal. Chem. *49(12)*, 1792–1797 (1977)
82. Deev, V.I., Kochnev, M.I., Okunev, A.I., Sergin, B.I.: Distribution of cadmium, rare, and uncommon elements among products of oxygen-flash smelting copper-zinc concentrates. Tsvetnye Metally *33(4)*, 19–24 (1960)
83. Deev, V.I., Okunev, A.I., Kochnev, M.I., Vermenichev, S.A., Sergin, B.I.: Povedenie redkikh i rasseyannykh elementov pri plavke sul'fidnykh kontsentratov na kislorode (Behavior of rare and trace elements during the smelting of sulfide concentrates with oxygen). Tr. Inst. Met. Akad. Nauk. S.S.R. Ural'sk Filial *8*, 43–50 (1963)
84. DeFilippo, J.R.: Thallium. In: Commodity Data Summaries 1976. Bureau of Mines, US Department of the Interior, p. 172 (1976)
85. Demiral, B.: Zur Therapie der Thalliumvergiftung mit BAL oder stabilisiertem Schwefelwasserstoff. Klinische und experimentelle Untersuchungen (On the therapy of thallium poisoning with BAL or stabilized hydrogen sulfide. Clinical and experimental studies). Diss. (Thesis). Zürich (1952)
86. Deutsche Forschungsgemeinschaft: Maximale Arbeitsplatzkonzentrationen (Maximum occupational concentration). Boppard, Harald Boldt Verlag 1979
87. Dilling, W.J.: Influence of lead and the metallic ions of copper, zinc, thorium, beryllium, and thallium on the germination of seeds. Ann. App. Biol. *13*, 160–167 (1926)
88. Di Nucci, A., Ferrini, B., Gregotti, C., Richelmi, P., Manzo, L.: Placental transfer of thallium and its modification by Prussian blue. IRCS Med. Sci. Libr. Compend. *7(5)*, 272 (1979)
89. Doak, R.L., Schmidtke, R.P., Wallach, J.D., Davis, L.E., Niemayer, K.H.: Thallium toxicosis, Vet. Med. Small Anal. Clin. *60*, 1227 (1965)
90. Doležahl, J., Hrabanková, E.: Cathodic stripping voltametry of thallium. Anal. Lett. *4(9)*, 585 (1971)

91. Downs, W.L., Scott, J.K., Steadman, L.T., Maynard, E.A.: Acute and subacute toxicity studies of thallium compounds. Am. Ind. Hyg. Ass. J. *21*, 399–406 (1960)
92. Dupuy, C., Fratta, M., Shaw, D.M.: Partition coefficient of thallium compared with rubidium. Earth Planet. Sci. Lett. *19*, 209–212 (1973)
93. Earl, F.L., Vish, T.J.: Teratogenicity of heavy metals. In: Oehme, F.W. (ed.) Hazardous and Toxic Substances, Vol. 2, pp. 617–640. New York, Marcel Dekker 1979
94. Eder, M., Gedigk, P.: Lehrbuch der allgemeinen Pathologie und der Pathologischen Anatomie (General Pathology and Pathological Anatomy), Berlin-Heidelberg-New York, Springer 1977
95. Ellory, J.C., Nibelle, J., Smith, M.W.: The effect of salt adaption on the permeability and cation selectivity of the goldfish intestinal epithelium. J. Physiol. *231*, 105–115 (1973)
96. Emsley, J.: The trouble with thallium. New Scientist *79*, 1114 (1978)
97. Erzinger, J.: Die Bestimmung von Ultraspuren von Cd, Tl und Bi in geologischen Proben (Determination of ultratraces of Cd, Tl, and Bi in geological samples). Zulassungsarbeit Universität Karlsruhe (1977)
98. Erzinger, J.: Zur Geochemie von Se und Tl in magmatischen Gesteinen. Mit einem Beitrag zur Analytik von Se in geochemischen Matrizes (The geochemistry of Se and Tl in magmatic rocks. Section on analytical methods of Se in geochemical matrices). Technische Universität, Diss. (Ph.D. Thesis) Universität Karlsruhe (1981)
99. Eschenauer, H., Neeb, R.: Beiträge zur analytischen Chemie des Weines (Contributions on the analytical chemistry of wine). IX. Mitt. Voltammetrische Bestimmung geringster Thallium-Gehalte im Wein. Z. Lebensm. Unters. *112*, 275 (1960)
100. Escobar, A.: Toxic neuropathies. 1. Anatomical aspects. Bol. Estud. Med. Biol. Univ. Nac. Auton. Mex. *30*, 225–236 (1979)
101. Ewers, U., Brockhaus, A.: Biological monitoring of thallium – Analysis and measured values. Staub-Reinhalt. Luft. *42*, 149–151 (1982)
102. Farber, J.: Thalliumvergiftung bei Haustieren (Thallium poisoning of domestic pets). Tierärztl. Rdsch. *37*, 750 (1931)
103. Feiser, J.: Nebenmetalle. Die Metallischen Rohstoffe (B-Group metals. Metallic raw resources), Vol. 17, Stuttgart, Ferdinand-Enke-Verlag 1966
104. Felix, R.: Die Myokard-Szintigraphie mit Thallium-201 als nichtinvasive Methodik (Myocardial scintigraphy with Tl-201 as non-invasive method). Dtsch. med. Wschr. *100(46)*, 2373 (1975)
105. Filimonova, I.A.: Molds of the genus Aspergilus as analytical indicators. Trudy Khim. Khim. Tekhnol. *3*, 50–51 (1973)
106. Fischer, R.E.: Schriften der Gesellschaft Deutscher Metallhütten- und Bergleute (German Miner's and Metallurgic Plant Scripts). 27: Gewerbliche Vergiftungen durch Thallium und Selen und ihre Verbindungen (Poisoning by Tl and Se and their compounds in industrial establishments). Übersicht über die Veröffentlichungen der Jahre 1945–1973. Clausthal-Zellerfeld, 1974
107. Fitzek, A.: Polarographic determination of thallium in urine. Acta Pharmacol. Toxicol. *36*, 187–189 (1975)
108. Fitzek, A., Henning, A.: Verteilung von Thallium im Organismus einer trächtigen Hauskatze (Distribution of thallium in the organism of a gestating house cat). Dtsch. Tierärztl. Wochenschr. *2*, 66–68 (1976)
109. Förstner, U.: Interner Bericht. Inst. F. Sedimentforschung. Universität Heidelberg (1980)
110. Fogg, A.G., Burgess, C., Thornburn-Burns, D.: A critical study of Brillant Green as a spectrophotometric reagent: Extraction of chloro complexes of antimony, thallium, gallium, and indium, and of tetrabromoindate(III) and improved procedures for the determination of antimony and thallium. Analyst *98*, 347–350 (1973)
111. Forth, W.: Thallium-Vergiftung (thallium poisoning). Münch. med. Wschr. *125*, 65–74 (1983)
112. Forth, W., Henschler, D., Rummel, W.: Allgemeine und spezielle Pharmakologie und Toxikologie (General and special pharmacology and toxicology). Mannheim-Wien-Zürich, Bibliographisches Institut, 1977
113. Forth, W., Henning, C.H.: Thallium-Vergiftungen und ihre Behandlung (Thallium poisoning and its treatment). Dtsch. Ärztebl. *43*, 2803–2807 (1979)
114. Franke, A., Roediek, S., Neu, I.: Die Thalliumvergiftung (The thallium poisoning event). Notfallmed. *5*, 145–151 (1979)
115. Franke, J.P., Coenegracht, P.M.J., de Zeeuw, R.A.: A rapid and specific method for the determination of traces of thallium directly in urine by differential pulse anodic stripping voltametry. Arch. Toxicol. *34*, 137–143 (1975)

116. Fresenius, W., Schneider, W.: Über den Nachweis und die quantitative Bestimmung von Barium und Thallium in Heilwässern (The quantitative determination of barium and thallium in healing waters). Heilbad und Kurort, Z. f. das gesamte Bäderwesen *10*, 378 (1974)
117. Fretwurst, F., Lockmann, F.W.: Akute Thaliumvergiftung (Acute thallium poisoning). Arch. Toxicol. *15*, 327–333 (1954/55)
118. Frey, J., Schlechter, M.: Experimentelle Untersuchungen über die Ausscheidungsgröße des Thalliums in verschiedenen Körperflüssigkeiten (Experimental studies on excretion of thallium in various body fluids). Naunyn-Schmiedeberg's Arch. Pharmacol. *193*, 530–538 (1979)
119. Friberg, J.: Handbook on the Toxicology of Metals. Amsterdam, Elsevier-North Holland 1979
120. Fuller, C.W.: The effect of acids on the determination of thallium by atomic absorption spectrometry with a graphite furnace. Anal. Chem. Acta *81*, 199–202 (1976)
121. Fung, P.C., Shaw, D.M.: Na, Rb, and Tl distributions between phlogopite and sanidine by direct synthesis in a common vapour phase. Geochim. Cosmochim. Acta *42*, 703–708 (1978)
122. Gehring, P.J., Hammond, P.B.: The uptake of thallium by rabbit erythrocytes. J. Pharmacol. Exptl. Therap. *145*, 215 (1964)
123. Gehring, P.J., Hammond, P.B.: Mechanism for the increase in thallium excretion during the administration of potassium. Pharmacologist *8*, 218 (1966)
124. Gehring, P.J., Hammond, P.B.: The interrelationship between thallium und potassium in animals. J. Pharmacol. Exptl. Therap. *155*, 187–201 (1967)
125. Geilmann, W.: Die Verwendung der Verdampfungsanalyse zur Erfassung geringster Stoffmengen (The application of the volatilization method for determining low material contents). Z. Anal. Chem. *160*, 410–426 (1958)
126. Geilmann, W., Neeb, K.H.: Die Verwendung der Verdampfungsanalyse zur Erfassung geringer Stoffmengen (The application of the volatilization method for determining low material contents). Teil 2: Der Nachweis und die Bestimmung kleinster Thalliumgehalte (Thallium). Z. Anal. Chem. *165*, 251–268 (1959)
127. Geilmann, W., Beyermann, K., Neeb, K.-H., Neeb, R.: Thallium, ein regelmäßig vorhandenes Spurenelement im tierischen und pflanzlichen Organismus (Thallium, a regularly occurring trace element in animal and plant organism). Biochem. Z. *333*, 62–70 (1960)
128. Gerdts, E.: Thallium Poisoning. Lancet 1268 (1974)
129. Gibson, J.E., Sigdestad, C.P., Becker, B.A.: Placental transport and distribution of thallium-204 sulfate in newborn rats and mice. Toxicol. Appl. Pharmacol. *10*, 408 (1967)
130. Gibson, J.E., Becker, B.A.: Placental transfer, embryotoxicity, and teratogenicity of thallium sulfate in normal and potassium-deficient rats. Toxicol. Appl. Pharmacol. *16*, 120–132 (1970)
131. Glasby, G.P.: Mechanisms of enrichment of the rarer elements in marine manganese nodules. Marine Chem. *1*, 105–125 (1973)
132. Glomme, J.: Thallium and compounds, in: Encyclopaedia of Occupational Health and Safety, Vol. II, pp. 1402–1403. Geneva, International Labour Office 1972
133. Glomme, J., Sjöström, B.: Yrkesbetingad Talliumförgiftning (Industrial thallium poisoning). Svenska Laekartidningen *52*, 1436–1441 (1955)
134. Goenechea, S., Sellier, K.: Über den natürlichen Thalliumgehalt des menschlichen Körpers (Natural thallium contents of the human body). Dtsch. Z. Gesamte Gerichtl. Med. *60*, 135–141 (1967)
135. Goldschmidt, V.M.: In: Gmelins Handbuch der Anorganischen Chemie (Gmelins Handbook of Anorganic Chemistry). Berlin, Verlag Chemie 1940
136. Gong, H., Rose, A.W., Suhr, N.H.: The geochemistry of cadmium in some sedimentary rocks. Geochim. Cosmochim. Acta *41*; 1687–1692 (1977)
137. Graben, N., Doss, M., Klöppel, H.A.: Störung des Porphyrinstoffwechsels bei der Thalliumvergiftung (Disorder in porphyrin metabolism during thallium poisoning). Med. Klin. *73*, 1114–1116 (1978)
138. Gratzl, E.: Thalliumvergiftung bei kleinen Haustieren (Thallium poisoning in small domestic pets). Tierärztl. Mschr. *23*, 657–674 (1939)
139. Greenspoon, G.N.: Thallium in minerals: facts and problems. US Bur. Mines Bull. *650*, 749–758 (1970)
140. Greving, R., Gagel, O.: Pathologisch-anatomische Befunde am Nervensystem nach experimenteller Thalliumvergiftung (Pathologic-anatomic results of thallium poisoning experimentally done on the nervous system). Z. Ges. Neurol. Psychiat. *120*, 805–814 (1929)

141. Grisham, Ch.M., Cupta, R.K., Barnett, R.E., Mildvan, A.S.: Thallium-205 nuclear relaxation and kinetic studies of sodium and potassium ion-activated triphosphatase. J. Biol. Chem. *249*, 6738–6744 (1974)
142. Grunfeld, O., Hinostroza, G.: Thallium poisoning. Archives of Internal Medicine *114*, 132–138 (1964)
143. Gubernator, K., Moroni, W., Munder, S., Franke, B., Ruske, B., Teufel, D.: Zur Problematik der Thalliumverseuchung in der Umgebung von Zementwerken (Thallium contamination near cement plants). Tutorium Umweltschutz, Heidelberg, IFEU-Institut für Energie- und Umweltforschung Heidelberg, e.V., ÖKO-Institut, Institut für angewandte Ökologie e.V., Freiburg, Oktober 1979
144. Günther, M.: Einfluß von kolloidalem Ferrihexacyanoferrat(II) auf Retention und Toxizität von Thallium(I) bei der Ratte [Influence of colloidal ferrihexacyaniferrate(II) on retention and toxicity of Thallium(I) in the rat]. KFK-Bericht *1326*, Kernforschungszentrum, Institut für Strahlenbiologie, Karlsruhe 1970
145. Haberer, K., Normann, S.: Metallspuren im Wasser – Ihre Herkunft, Wirkung und Verbreitung (Metal traces in water – their origin, effect, and distribution). Vom Wasser *38*, 157–182 (1971)
146. Haberer, K., Normann, S.: Die Bedeutung der Metallspuren in den Gewässern für die Trinkwasserversorgung (The significance of trace metals in water in respect to the drinking water supply). GWF-Wasser/Abwasser *113*, 382–386 (1972)
147. Hagedorn-Götz, H., Stoeppler, M.: Zum forensischen Nachweis von Thallium in menschlichen Haaren durch flammenlose Atomabsorptionsspektrometrie (Thallium in human hair determined by flameless AAS). Arch. Toxikol. *34*, 17–26 (1975)
148. Hall, B.K.: Thallium-induced achondroplasia in the embryonic chick. Develop. Biol. *28*, 47–60 (1972)
149. Hamilton, E.I., Minski, M.J.: Abundance of the chemical elements in man's diet and possible relations with environmental factors. Sci. Total Environ. *1*, 375–394 (1972/73)
150. Hannah, S.A.: Removal of uncommon trace metals by physical and chemical treatment processes. J. Water Poll. Con. Fed. *49*, 2297–2309 (1977)
151. Hapke, H.-J., Barke, E., Spikermann, A.: Ansammlung von Thallium in verzehrbaren Geweben von Hammeln und Bullen in Abhängigkeit von der Thalliummenge im Futter (Thallium in edible tissue of mutton and beef in dependence on the thallium content of fodder). Dtsch. Tierärztl. Wschr. *87*, 376–378 (1980)
152. Hasan, M., Bajpai, V.K., Shipstone, A.C.: Electron microscopic study of thallium-induced alterations in the oligodendrocytes of the rat area postrema. Exp. Pathol. *13*, 338–345 (1977)
153. Hasan, M., Chandra, S.V., Dua, P.R., Raghubir, R., Ali, S.F.: Biochemical and electrophysiologic effects of thallium poisoning on the rat corpus striatum. Toxicol. Appl. Pharmacol. *41*, 353–359 (1977)
154. Heinrichs, H.: Die Untersuchung von Gesteinen und Gewässern auf Cd, Sb, Hg, Tl, Pb und Bi mit der flammenlosen Atom-Absorptions-Spektralphotometrie (The study of rocks and waters for Cd, Sb, Hg, Tl, Pb, and Bi with the flameless AAS). Ph.D. Thesis, Univ. Göttingen (1975)
155. Heinrichs, H.: Emissions of 22 elements from brown coal combustion. Naturwissenschaften *64*, 479–481 (1977)
156. Heinrichs, H.: Determination of bismuth, cadmium, and thallium in 33 international standard reference rocks by fractional distillation combined with flameless atomic absorption spectrometry. Z. Anal. Chem. *294*, 345–351 (1979)
157. Heinrichs, H.: Metalle in kommunalen Klärschlämmen (Metals in communal sludges). Naturwissenschaften *69*, 88–89 (1982)
158. Heinrichs, H.: Trace element discharge from a brown-coal fired power plant. Environ. Tech. Lett. *3*, 127–136 (1982)
159. Heinrichs, H., Mayer, R.: Distribution and cycling of the major trace elements in two European forest ecosystems. J. Environ. Qual. *6*, 401–407 (1977)
160. Heinrichs, H., Schulz-Dobrick, B., Wedepohl, K.H.: Terrestrial geochemistry of Cd, Bi, Tl, Pb, Zn, and Rb. Geochim. Cosmochim. Acta *44*, 1519–1533 (1980)
161. Heinrichs, H., Keltsch, H.: Determination of arsenic, bismuth, cadmium, selenium, and thallium by atomic adsorption spectrometry with a volatilization technique. Anal. Chem. *54*, 1211–1214 (1982)

162. Heit, M.: Variability of the concentrations of 17 trace elements in the muscle and liver of a single-striped bass morone-saxatilis. Bull. Environ. Contam. Toxicol. *23*, 1–5 (1979)
163. Henderson, C.M.B.: Substitution of Rb, Tl, and Cs in potassium feldspars. Progr. Exp. Petrol. N.E.R.C. *1*, 53–57 (1969)
164. Henke, G., Fitzek, A.: Nachweis von Thalliumvergiftungen durch Neutronenaktivierungsanalyse (Thallium poisoning determined by neutron activation analysis). Arch. Toxicol. *27*, 266–272 (1971)
165. Henning, C.H.: Zum Mechanismus der Dekretion von Thallium-Ionen im Gastrointestinal-Trakt. Untersuchungen an Ratten (The mechanism of secretion of thallium ions in the gastrointestinal tract of rats). Ph.D. Thesis Univ. Ruhr-Universität, Bochum (1979)
166. Henning, C.H.: Über die Vergiftung mit Thallium und ihre Therapie (Thallium poisoning and its therapy). Dtsch. Apoth. Z. *119*, 1782–1784 (1979)
167. Henning, C.H., Forth, W.: Adsorption and secretion of thallous ions on isolated intestinal segments in vitro of rats. Naunyn-Schmiedeberg's Arch. Pharmacol. Suppl. *11*, 297 R 2,6 (1977)
168. Henning, C.H., Forth, W.: Secretion of thallous ions from the blood into the gastrointestinal tract of rats. Naunyn-Schmiedeberg's Arch. Pharmacol. Suppl, *302*, R 1,3 (1978)
169. Henry, F.T., Kirch, T.O., Thorpe, T.M.: Determination of trace level arsenic(III), arsenic(V) and total inorganic arsenic by differential pulse polarography. Anal. Chem. *51*, 215–218 (1979)
170. Henschler, D. (ed.): Analysen in biologischem Material (Analyses in biological material), Vol. 2, Weinheim, Verlag Chemie 1978
171. Herman, M.M., Bensch, K.G.: Light and electron microscopic studies of acute and chronic thallium intoxication in rats. Toxicol. Appl. Pharmacol. *10*, 199–222 (1967)
172. Heumann, K.G.: Mass spectrometric isotope dilution analysis for accurate determination of elements in environmental samples. Toxicol. Environ. Chem. Rev. *3(2)*, 111–130 (1980)
173. Heydlauf, H.: Ferric-cyanoferrate(II): an effective antidote in thallium poisoning. Eur. J. Pharmacol. *6*, 340–344 (1969)
174. Heyroth, F.F.: Thallium – A Review and Summary of the Medical Literature (Public Health Reports Supplement No. 197), US Public Health Service (1947)
175. Hillmer, E.: Geochemische Untersuchungen im Bereich der Lagerstätte Meggen, Rheinisches Schiefergebirge (Geochemical studies near the Meggen deposit, Rheinisches Schiefergebirge). Ph.D. Thesis, TH Aachen, pp. 162 (1972)
176. Hirt, H.H., Blaszkiewitz, K.: Giftanalyse im Krankenhaus. Beispiel: Thallium (Poison analysis in the hospital – example: thallium). Krankenhaus-Apotheke *29*, 84 (1980)
177. Hoffmann, G., Schweiger, P., Scholl, W.: Aufnahme von Thallium durch landwirtschaftliche und gärtnerische Nutzpflanzen (Uptake of thallium by agricultural and garden plants). Landwirtsch. Forschung *35*, 45–54 (1982)
178. Ishimori, T., Takashima, Y.: Thallium content of Japanese igneous rocks. Mem. Fac. Sci. Kyushu. Univ., Ser. C. (Chem.) *2(2)*, 65 (1955)
179. Jax, W., Grabensee, B., Schröder, E.: Die Therapie der Thalliumvergiftung (Therapy of thallium poisoning). Med. Welt *24*, 691–693 (1973)
180. Jeantet, A.Y.: Insects resistance to mineral pollution. Bull. Soc. Zool. Fr. *102(3)*, 319–320 (1977)
181. Jeffery, P.G.: Thalliumbestimmung in silikatischen Gesteinen mit Brillantgrün (Thallium determination in silicate rocks with brilliant green). Chemical Methods of Rock Analysis, 2nd edit., p. 436, Braunschweig, Pergamon-Verlag 1970
182. Jeffery, P.G.: Chemical Methods of Rock Analysis, Oxford, Pergamon Press 1978
183. Kaiser, E.P., Herring, B.F., Rabbitt, J.C.: Minor Elements in some Rocks, Ores, and Mill and Smelter Products, Trace Elements Investigations Report TEI-415, Oak Ridge, Tennessee US, Atomic Energy Commission 1954
184. Kamerbeek, H.H.: Therapeutic Problems in Thallium Poisoning. Ph.D. Thesis Utrecht (1971)
185. Kamerbeek, H.H., Rauws, A.G., ten Ham, M., van Heijst, A.N.P.: Prussian blue in the therapy of thallotoxicosis. An experimental and clinical investigation. Acta Med. Scand. *189*, 321–324 (1971)
186. Kamerbeek, H.H., Rauws, A.G., ten Ham, M., von Heijst, A.N.P.: Dangerous redistribution of thallium by treatment with sodium diethyldithiocarbamate. Acta Med. Scand. *189*, 149–154 (1971)
187. Kautz, K., Kirsch, H., Laufhütte, D.W.: Über Spurenelementgehalte in Steinkohlen und den daraus entstehenden Reingasstäuben (Trace element contents in coal and fly ash). VGB-Kraftwerkstechnik *10*, 672–676 (1975)

188. Kazantzis, G.: Thallium. In: Friberg, L. et al. (eds.) Handbook of the Toxicology of Metals, pp. 599–612. Amsterdam, Elsevier 1979
189. Keil, R.: Spurenbestimmung von Thallium in Gesteinen durch Flammen- oder flammenlose Atomabsorptions-Spektralphotometrie nach Vorkonzentrierung durch Extraktion (Trace determination of thallium in rocks by flame or flameless atomic absorption spectrophotometry following preconcentration by extraction). Z. Anal. Chem. *309(3)*, 181–185 (1981)
190. Keinhorst, H.: Thalliumemissionen aus Zementdrehofenanlagen. Gedanken zur Festlegung von Emissionsgrenzwerten für Thalliumverbindungen (Thallium emissions from cement rotary kilns. Thoughts on limiting emission values for thallium compounds). Staub-Reinhalt. Luft *40*, 26–29 (1980)
191. Keinhorst, H.: Emissionsverhältnisse beim Einsatz von thalliumhaltigen Rohstoffen in Zementdrehöfen mit Schwebegas-Vorwärmer (Emission conditions while using thallium-containing resources in cement rotary kilns with a suspended gas preheater). Zement-Kalk-Gips *33*, 648–652 (1980)
192. Keller, K., Thimme, W., Dissmann, W., Buschmann, H.J., Dross, K., Daugs, J.: Thalliumvergiftung mit Verbrauchskoagulopathie (Thallium poisoning with consumption coagulopathy). Schweiz. Med. Wochenschr. *101*, 511–515 (1971)
193. Kempe, D.P.: Wat niet moet (What isn't necessary). Rat Muis *23*, 93 (1975)
194. Kemper, F.H.: Thallium-Vergiftungen (Thallium-Poisoning). Münch. Med. Wochenschr. *121*, 1357–1358 (1979)
195. Kennedy, P., Cavanagh, J.B.: Spinal changes in the neuropathy of thallium poisoning. A case with neuropathological studies. J. neurol. Sci. *29*, 295–301 (1976)
196. Kennedy, P., Cavanagh, J.B.: Sensory neuropathy produced in the cat with thallous acetate. Acta Neuropath. *39*, 81–88 (1977)
197. Kinard, J.T.: Diagnosis of metal poisoning and evaluation of chelation therapy by differential pulse anodic stripping voltametry coupled to a novel digestion procedure. Anal. Lett. *10*, 1147–1162 (1977)
198. Klahre, P., Valenta, P., Nürnberg, H.W.: Ein normiertes Pulsinversvoltammetrisches Analysenverfahren zur Prüfung von Trinkwasser auf toxische Metalle. Teil 1: Simultanbestimmung von Kupfer, Cadmium, Blei und Zink und von Blei und Thallium (A standardized pulse inverse voltametric analytical method for testing drinking water for toxic metals. Part 1: Simultaneous determination of copper, cadmium, lead, and zinc and lead and thallium. Vom Wasser *51*, 199–219 (1978)
199. Klöppel, A., Weiler, G.: Beitrag zur Schwermetallvergiftung, insbesondere durch Thallium (Heavy metal poisoning especially by thallium). Dtsch. med. Wschr. *103*, 75–76 (1978)
200. Kloke, A.: Richtwerte '80: Orientierungswerte für tolerierbare Gesamtgehalte einiger Elemente in Kulturböden (Orientation values for tolerable total contents of several elements in specially prepared soils). VDLUFA-Sonderdruck 1980
201. Kogan, B.I., Tallii, O.: Thallium. Redk. Elem. *5*, 27–48 (1970)
202. Kolotova, S.S., Kharitonova, N.I.: Effect of thallium on plants. Chem. Abs. *76*, 261, 1855 a (1970)
203. Korkisch, J., Steffan, I.: Determination of thallium in natural waters. Int. J. Environ. Anal. Chem. *6*, 111–118 (1979)
204. Krasovskii, G.N., Kenesariev, U.I., Vasyukobich, L.Y.A., Bonashevskaya, T.I., Solov'eva, T.G., Nosacheva, L.F.: Experimentally determined hygienic standards for thallium in bodies of water. Gig. Sanit. *0(9)*, 11–13 (1980)
205. Krause, G.H.M., Prinz, B.: Vegetationsschäden durch Thallium in der Umgebung eines Zementwerkes in Nordrhein-Westfalen. Ein neues Problem der Umweltverschmutzung (Vegetation damage by thallium near a cement plant in Nordrhein-Westfalen. A new problem in environmental protection). Symposium on Chemical and Toxicological Aspects of Environmental Quality, September 17–19, Neuherberg 1979
206. Kubasik, N.P., Volosin, M.T.: A simplified determination of urinary cadmium, lead, and thallium with use of carbon-rod atomizer and atomic absorption spectrophotometry. Clin. Chem. *19*, 954–958 (1973)
207. Kuhn, R., Rudy, H., Wagner-Jauregg, T.: Über lactoflavin (Vitamin B_2) (On lactoflavin). Ber. dtsch. Chem. Gesell. *66*, 1950–1956 (1933)
208. Kunze, M.: Der Einfluß von Thallium auf das Wachstum von Acholeplasmatacese, Mycoplasmataceae und einigen Bakterienspezien (Influence of thallium on the growth of acholeplasmatacesis, mycoplasmatacesis, and some bacteria species). Zbl. Bak. 1. Abt. Orig. *222*, 535–539 (1972)

209. Kuznecov, V.L., Mjasojedova, G.V.: Possible losses of thallium in silicate analysis. J. Appl. Chem. USSR *29*, 2017–2018 (1956)
210. Lameijer, W., van Zwieten, P.A.: Acute cardiovascular toxicity of thallium(I) ions. Arch. Toxicol. *35*, 49–61 (1976)
211. Lameijer, W., van Zwieten, P.A.: Kinetic behavior of thallium in the rat. Arch. Toxicol. *37*, 265–273 (1977)
212. Lamy, C.-A.: Sur les effects toxiques du thallium (On the toxic effect of thallium). J. Pharm. Chim. *44*, 285–288 (1863)
213. Landauer, W.: Influence of thallium poisoning of the father on his descendants. Experiments on roosters. Arch. Gewerbepathol. Gewerbehyg. *1*, 791–792 (1931)
214. Landry, A.S.: The polarographic determination of thallium in an atmospheric sample. J. Amer. Ind. Hyg. Assoc. *21*, 407–413 (1960)
215. Lee, A.G.: The Chemistry of Thallium, p. 366. Amsterdam-London-New York, Elsevier 1971
216. Lee, A.G.: The coordination chemistry of thallium(I). Coord. Chem. Rev. *8*, 289–349 (1972)
217. Lehmann, W.D., Schulten, H.-R.: Determination of trace elements by field desorption mass spectrometry. Anal. Chem. *49*, 1744–1746 (1977)
218. Lehmann, W.D., Schulten, H.-R.: Quantitative Spurenanalyse von Metallen mit der Felddesorptions--Massenspektrometrie (Quantitative trace analysis of metals by field desorption mass spectrometry). Angew. Chem. *89*, 890–891 (1977)
219. Leitsin, V.A.: Povedenie talliya v protsessakh gidrometalluricheskogo proizvodstva tsinka (Thallium in hydrometallurgical zinc production). Sb. Nauch.-Tekh. Trudov Nauch.-Issled. Inst. Met. Chelyabinsk. Sovnarkhoza *1*, 120–129 (1960)
220. Leonhardt, H.: Histologie, Zytologie und Mikroanatomie des Menschen (Histology, cytology, microanatomy of Man). Stuttgart, Georg Thieme Verlag 1974
221. Leopold, G., Furukawa, E., Forth, W., Rummel, W.: Vergleichende Untersuchungen der Resorption von Schwermetallen in vivo und in vitro (Comparative studies of the resorption of heavy metals in vivo and in vitro). Naunyn-Schmiedeberg's Arch. Pharmak. exp. Path. *263*, 275 (1969)
222. Leschhorn, V.: Probleme der Thalliumbelastungen (Problems of thallium contamination). Öff. Gesund. Wesen *42*, 473–476 (1980)
223. Lie, R., Thomas, R.G., Scott, J.K.: The distribution and excretion of thallium-204 in the rat, with suggested MPC's and a bioassay procedure. Health Physics *2*, 334–340 (1960)
224. Lillie, W.I., Parker, H.L.: Retrobulbar neuritis due to thallium poisoning. JAMA *98*, 1347–1348 (1932)
225. Lindegren, C.C.: Mitochondria as vehicles of entry and of transport. J. Biol. Psychol. *13(2)*, 3–5 (1971)
226. Lindegren, C.C., Lindegren, G.: Thallium as vital stain for yeast mitochondria. Nature *234*, 297 (1971)
227. Lindegren, C.C., Lindegren, G.: Oxidative detoxication of thallium in the yeast mitochondria. Antonie v. Leeuwenhoek *39*, 351–353 (1973)
228. Ling, G.N.: Thallium and cesium in muscle cells compete for the adsorption sites normally occupied by K^+. Physiol. Chem. Phys. *9*, 217–225 (1977)
229. Linton, R.W., Williams, P., Evans, C.A., Jr., Natush, D.: Determination of the surface predominance of toxic elements in airborne particles by ion microprobe mass spectrometry and Auger electron spectrometry. Anal. Chem. *49*, 1514–1521 (1977)
230. Locher, F.W.: Stoffkreisläufe und Emissionen beim Brennen von Zementklinker (Material flows and emissions in the combustion of cement clinker). Fortschr. Miner. *58*, 81–82 (1980)
231. Locher, F.W., Rechenberg, W.: Ergebnisse einer Ringanalyse zur Bestimmung von Thallium (Results of a ring analysis to determine thallium), in: Atomspektrometrische Spurenanalytik (Atomspectrometric trace analysis) (ed.) Welz, B., Weinheim, Verlag Chemie 1982
232. Löbel, J.: Untersuchungen zur Feinreinigung von Thallium – Ein Beitrag zur Amalgamelektrolyse (Investigations considering the refining of thallium – An Essay at amalgamelectrolysis). Ph.D. Thesis, Freiberg (1968)
233. Ludwig, H., Vahrenholt, F.: Anreicherung von Schadstoffen im Umkreis von Industrieanlagen (Enrichment of contaminants near industrial plants). Staub-Reinh. Luft. *39*, 481–483 (1979)
234. Lund, A.: Distribution of thallium in the organism and its elimination. Acta Pharmacol. Toxicol. *12*, 251–259 (1956)

235. Lund, A.: The effect of various substances on the excretion and the toxicity of thallium in the rat. Acta Pharmacol. Toxicol. *12*, 260–268 (1956)
236. Machata, G.: Über den Thalliumgehalt im menschlichen Knochen (Thallium contents in human bones). Z. ges. gerichtl. Med. *54*, 95 (1963)
237. Manheim, F.: Further observations on the composition of manganese nodules with particular reference to some of the rarer elements. Geochim. Cosmochim. Acta *31*, 2169 (1967)
238. Manninen, V., Malkonen, M., Skulskii, I.A.: Elimination of thallium in rats influenced by Prussian blue and sodium chloride. Acta pharmacol. et toxicol. *39*, 256–261 (1976)
239. Marczenko, Z., Kalowska, H., Mojski, M.: Extraction spectrophotometric determination of thallium in high-purity indium. Talanta *21*, 93–97 (1974)
240. Marowsky, G.: Sensitivity ranges in neutron activation analysis of some rare trace elements (Cd, Hg, Tl, Bi) in silicate samples. Proc. NATO Advanced Study Institute on Activation Analysis in Geochemistry and Cosmochemistry, Kjeller, Norway (1970)
241. Marowsky, G.: Aktivierungsanalytische Bestimmung von Cadmium, Quecksilber, Thallium und Wismut in Gesteinen (Activation analysis of cadmium, mercury, thallium, and bismuth in rocks). Z. Anal. Chem. *253*, 276–271 (1971)
242. Marowsky, G.: Aktivierungsanalytische Bestimmung von Thallium in Gesteinen – ein Vergleich mehrerer Meßmethoden (Activation analysis of thallium in rocks – a comparison of several methods). J. Radioanal. Chem. *8*, 27–32 (1971)
243. Marowsky, G., Wedepohl, K.H.: General trends in the behavior of Cd, Hg, Tl, and Bi in some major rock-forming processes. Geochim. Cosmochim. Acta *35*, 1255–1267 (1971)
244. Mathys, W.: Thallium in river sediments of North-West Germany. Its sources of input and its relation to other heavy metals. Intern. Conference Paper: Heavy Metals in the Environment, pp. 347–350, Amsterdam (1981)
245. Matthews, A.D., Riley, J.P.: The determination of thallium in silicate rocks, marine sediments and sea water. Anal. Chim. Acta *48*, 25–34 (1969)
246. Matthys, R.: Neuves observations d'empoisonnement criminel par le thallium (New observations on criminal thallium poisening). Ann. Méd. lég *35*, 237–275 (1955)
247. Mayer, R., Ulrich, B.: Conclusions on the filtering action of forests from ecosystem analysis. Oecol. Plant *9*, 157–168 (1974)
248. McCool, M.M.: Effect of thallium sulphate on the growth of several plants and on nitrification in soils. Contrib. Boyce Thompson Institute *5*, 289–296 (1933)
249. McCrory, B.R., Ward, J.D.: The toxicty of thallium sulfate for cattle. J. Am. Vet. Med. Ass. *88*, 301 (1936)
250. McGoldrick, P.J., Keays, R.R., Scott, B.B.: Thallium: a sensitive indicator of rock/sea water interaction and of sulfur saturation of silicate melts. Geochim. Cosmochim. Acta *43*, 1303–1311 (1979)
251. McKillop, A., Taylor, E.C.: Thallium in organic synthesis. Chem. Brit. *9(1)*, 4–11 (1973)
252. McMurtrey, J.E., Jr.: Effect of thallium on growth of tobacco plants. Science *76*, 86 (1932)
253. Mendelssohn, H.: Effect of toxic chemicals on bird life. The impact of pesticides on bird life in Israel. Bull. Intern. Counc. Bird Preserv. *11*, 75–104 (1972)
254. Metall-Statistik 1970–1980 (ed. Metallgesellschaft AG) Vol. 68, Frankfurt 1981
255. Michler, G., Paesler, R. (eds.): Der Fischer-Weltalmanach (The Fischer World Almanac) 1983. Frankfurt, Fischer-Taschenbuch-Verlag 1982
256. Mock, B.: Versuche zur kolorimetrischen Bestimmung geringer Thallium-Mengen (Attempts at the colorimetric determination of small amounts of thallium). Inaug. Diss. Erlangen-Nürnberg (1975)
257. Moeschlin, S.: Klinik und Therapie der Vergiftungen (Clinic and therapy of poisonings), 5th edit. Stuttgart, Georg Thieme Verlag 1972
258. Moeschlin, S., Zollinger, H., Lüthy, F.: Beitrag zur Klinik und Pathologie der Thallium-Vergiftung (Report on effects and pathology of thallium-poisoning). Dtsch. Arch. klin. Med. *189*, 181–213 (1942)
259. Moeschlin, S., Condrau, G.: Epileptiforme Anfälle bei Thalliumvergiftungen (Epileptic attacts in thallium poisoning). Schweiz. med. Wschr. *1950*, 519
260. Moeschlin, S., Demiral, B.: Therapie der experimentellen Thalliumvergiftungen mit stabilisiertem Schwefelwasserstoff oder BAL (Therapy of experimental thallium poisoning with stabilized hydrogen sulfide or BAL). Schweiz. med. Wschr. *82*, 57 (1952)

261. Morris, D.F.C., Killick, R.A.: The determination of silver and thallium in rocks by neutron activation analysis. Talanta *4*, 51–60 (1960)
262. Mühlendahl, K.E. v., Etzold, R., Krienke, E.G.: Thalliumvergiftungen im Kindesalter (Thallium poisoning in adolescent children). Dtsch. med. Wschr. *103*, 1116–1117 (1978)
263. Mueller, B.: Gerichtliche Medizin (Forensic Medicine), 2nd edit., Vol. 2. pp. 807–814. Springer-Verlag, Berlin-Heidelberg-New York, 1977
264. Müller, L.: Thalliumvergiftung – "Thallium-poisoning", in: Handbuch der gesamten Arbeitsmedizin, (ed.) Badder, E.W., Vol. II/1, München, Urban & Schwarzenberg 1961
265. Müller, L.: Portlandzement (Portland cement). Chemie in unserer Zeit *7*, 19–24 (1973)
266. Mullins, L.J., Moore, R.D.: The movement of thallium ions in muscles. J. Gen. Phys. *43*, 759–773 (1960)
267. Munch, B., Clausen, B., Karlog, O.: Thallium poisoning in red foxes (Vulpes vulpes) and badgers (Meles meles) in Denmark. Nordisk Veterinarmedicin *26(5)*, 323–338 (1974)
268. Munch, J.C.: Human thallotoxicosis. J. Amer. Med. Assoc. *102*, 1929–1934 (1934)
269. Munch, J.C., Silver, J.: The pharmacology of thallium and its use in rodent control. US Dept. Agricol. Tech. Bull. *238*, 28 (1931)
270. Munch, J.C., Ginsburg, H.M., Nixon, C.E.: The 1932 thallotoxicosis outbreak in California. J. Am. med. Ass. *100*, 1315 (1933)
271. Murakami, B.: Tl in minerals in Japan. Bull. Chem. Soc. Japan *23*, 1–2 (1950)
272. Natush, D., Wallace, J.R., Evans, C.A., Jr.: Toxic trace elements. Preferential concentration in respirable particles. Science *183*, 202–204 (1974)
273. Neeb, K.: Die Abtrennung kleinster Thalliummengen durch Verdampfungsanalyse (Separation of smallest thallium amounts by volatilization analysis). Ph.D. Universität Mainz (1957)
274. Neeb, K., Martin, J., Franke, R.: Aktivierungsanalytische Bestimmung kleiner Gehalte an Tellur, Chlor, Thallium und Nickel in Selen (Activation analytical determination of Te, Cl, Tl, and Ni traces in Se). Z. Anal. Chem. *247*, 27–31 (1969)
275. Neeb, R.: Anodische Amalgamvoltammetrie. I.: Grundlagen des Arbeitsverfahrens (Basic Procedures). Z. Anal. Chem. *171(5)*, 321–329 (1959)
276. Neeb, R.: Anodische Amalgamvoltammetrie. II. Die Bestimmung kleiner Mengen von Pb, Cd, Bi und Tl (Anodic amalgam voltametry, II. The determination of small amounts of Pb, Cd, Bi, and Tl). Z. Anal. Chem. *171*, 329–339 (1959)
277. Nehring, D.: Experiments on the toxicological effect of thallium ions on fish and fish-food organisms. Z. Fisch. *11*, 557–562 (1962)
278. Newsom, I.E., Lollus, J.B., Ward, J.C.: The toxicity of thallium sulphate for sheep. J. Am. Vet. med. Ass. *76*, 828 (1930)
279. Norris, P., Man, W.K., Hughes, M.N., Kelly, D.P.: Toxicity and accumulation of thallium in bacteria and yeast. Archives of Microbiology *110(2–3)*, 279–286 (1976)
280. Novokhatskii, I.P., Kalinin, S.R.: Tl in some ores and products of the lead works of Kasachstan. Tsvetnaya Metal *16*, 17–18 (1941)
281. Novokhatskii, I.P., Kalinin, S.R.: The occurrence of Tl in silicates. Akad. Nauk. Dokl. SSSR *56*, 831–832 (1947)
282. Nürnberg, H.W.: Polarography and voltametry in studies of toxic metals in man and his environment. Sci. Total Environ. *12*, 35–60 (1979)
283. Nuriev, A.I., Efendiev, G.K.: Thallium and gallium contents in stratal waters, crude oils and surrounding rocks. Azerb. Khim. Zh. *2*, 112–116 (1966)
284. Oehme, F.W.: Heavy metal toxicity. Clin. Toxicol. *5*, 151–167 (1972)
285. O'Shea, T.A., Mancy, K.H.: Characterization of trace metal organic interactions by anodic stripping voltametry. Anal. Chem. *48*, 1603–1607 (1976)
286. O'Shea, T.A., Mancy, K.H.: The effect of pH and hardness metal ions on the competitive interaction between trace metal ions and inorganic and organic complexing agents found in natural waters. Water Res. *12(9)*, 703–712 (1978)
287. Ottinger, R.S., Blumenthal, J.L., Dal Porto, D.F., Gruber, G.I., Santy, M.J., Shih, C.C.: Recommended Methods of Reduction, Neutralization, Recovery or Disposal of Hazardous Waste. Vol. XIII. Industrial and Municipal Disposal Candidate Waste Stream Constituent Profile Reports. Inorganic Compounds. TRW Systems Group for the US Environ. Pro. Agency, PB-224 592, National Technical Information Service, US Department of Commerce, Springfield, Va. (1973)

288. Overnell, J.: The effct of some heavy metal ions on photosynthesis in a fresh water alga. Pesticide Biochem. Physiol. *5(1)*, 19–26 (1975)
289. Pallaghy, C.K.: Localisation of thallium in stomata is independent of transpiration. Aust. J. Biol. Sci. *25*, 415–417 (1972)
290. Paulson, G., Vergara, G., Young, J., Bird, M.: Thallium intoxication treated with dithizone and hemodialysis. Arch. Intern. Med. *129*, 100 (1972)
291. Pentschew, A., Garro, F.: Thallium encephalopathy in monkeys. J. Neuropathol. Exp. Neurol. *28*, 163 (1969)
292. Peter, W.: Lengerich: Seit Jahren giftiges Thallium über Land und Leute (Lengerich: years of poisonous thallium on land and humans). Landwirtsch. Wochenbl. Westfalen-Lippe Ausg. A *136*, 20–22 (1979)
293. Petersohn, K.L.: Thalliumvergiftung in der Schwangerschaft (Thallium poisoning during pregnancy). Arch. Toxicol. *18*, 160–164 (1960)
294. Piazolo, P.: Behandlung der Thalliumvergiftung mit der Hämodialyse (Treatment of thallium poisoning with hemodialysis). Dtsch. med. Wschr. *96*, 1217 (1971)
295. Pielow, E.: Umweltschäden durch Thallium in der Umgebung eines Zementwerkes (Environmental damages by thallium near a cement plant). Umwelt *9(5)*, 394–396 (1979)
296. Pile, C.H.: Thallium poisoning in domestic felines. Aust. Vet. J. *32*, 18–19 (1956)
297. Plunkert, P.A.: Thallium. Mineral commodity summaries, USBM, pp. 158–159 (1982)
298. Polyakova, M.M., Sidorova, N.V., Spiridonova, V.S., Shabalina, L.P., Shatalov, N.N.: Health status and working conditions of persons handling thallium and its compounds. Gig. Tr. Prof. Zabol. *2*, 14–16 (1977)
299. Polyvyannyi, I.R., Ovcharenko, V.P., Elyakov, I.I.: Distribution of metals between products of shaft-furnace lead smelting with preheating and oxygen enriching of the blast. T. Inst. Met. Obogashch., Akad. Nauk. Kaz., SSSR *34*, 71–76 (1968)
300. Preuss, E.: Beiträge zur spektralanalytischen Methodik. II. Bestimmung von Zn, Cd, Hg, In, Tl, Ge, Sn, Sb, Pb und Bi durch fraktionierte Destillation (Spectral analytical methodology. II. Determination of Zn, Cd, Hg, In, Tl, Ge, Sn, Sb, Pb, and Bi by fractional destillation). Z. Angew. Min. *3*, 8–20 (1940)
301. Prick, J.J.G., Smith, W.G.S., Miller, L.: Thallium poisoning. Amsterdam, Elsevier 1955
302. Prinz, B., Krause, G.H.M., Stratmann, H.: Thalliumschäden in der Umgebung der Dyckerhoff Zementwerke AG in Lengerich, Westfalen (Thallium damage in the proximity of the Dyckerhoff Cement Works in Lengerich, Westfalia.) Staub-Reinhalt. Luft *39*, 457–462 (1979)
303. Proctor, N.H., Hughes, J.P.: Chemical Hazards of the Workplace, Thallium, pp. 477–478. Lippincott-Company, Philadelphia-Toronto, 1978
304. Puchelt, H., Walk, H.: Umweltrelevante Spurenelemente in Böden eines alten Bergbaugebiets (Environmental trace elements in soils of an ancient mining region). Naturwissenschaften *67*, 190–191 (1980)
305. Puerner, N.J., Siegel, S.M.: The effects of mercury compounds on the growth and orientation of cucumber seedlings. Physiol. Plant. *26*, 310–312 (1972)
306. Rankama, K.: A note on the geochemistry of thallium. J. Geol. *57*, 608–613 (1949)
307. Rauws, G.: Thallium pharmacokinetics and its modification by Prussian blue. Naunyn-Schmiedeberg's Arch. Pharmacol. *284*, 295–306 (1974)
308. Rauws, G., Canton, J.H.: Adsorption of thallium ions by Prussian blue. Bull. Environ. Contam. Toxicol. *15*, 335 (1976)
309. Rechenberg, W.: Die photometrische Bestimmung von Thallium im Zementofenstaub (The photometric determination of thallium in cement furnace dust). Zement-Kalk-Gips *35*, 90–95 (1982)
310. Redhead, T.D.: A relationship between rainfall and rat damage to sugar-cane in north Queensland. Queensl. J. Agr. Animal. Sci. *37*, 181–188 (1980)
311. Redhead, T.D., Saunders, I.W.: Evaluation of thallium sulphate baits against rats in Queensland (Australia) sugar-cane fields adjacent to different vegetable types. Protect. Ecol. *2(1)*, 1–19 (1980)
312. Reed, D., Crawley, J., Faro, S.N., Pieper, S.J., Kurland, L.T.: Thallotoxicosis. Acute mainifestations and sequelae. J. Amer. Med. Assoc. *183*, 516–522 (1963)
313. Remy, H.: Lehrbuch der anorganischen Chemie (Textbook of anorganic chemistry), Vol. I. Leipzig, Akademische Verlagsgesell., 1960

314. Rethfeld, H.: Thalliumbestimmung in Pflanzenmaterial mit Hilfe der Röntgenfluorescenzanalyse (Thallium determination in plant material with X-ray fluorescence analysis). Z. Anal. Chem. *301*, 308 (1980)
315. Rey W. van, Heinecker, R., Kaemmerer, E.: Beitrag zur Klinik der akuten Thallium-Vergiftung (Clinic of acute thallium poisoning). Med. 1814–1825 (1958)
316. Richeson, E.M.: Industrial thallium intoxication. Ind. Med. Surg. *27*, 607 (1958)
317. Ridley, W.P., Dizikes, L.J., Wood, J.M.: Biomethylation of toxic elements in the environment. Science *197*, 329–332 (1977)
318. Righetti, P.C., Moeschlin, S.: The therapeutic effect of dithiocarb (DTC) and potassium chloride on experimental thallium-poisoning in guinea pigs. Clinical Toxicol. *4*, 165–171 (1971)
319. Rinck, P.A., Beckmann, H.O.: Myocardial uptake of thallous ions by facilitated diffusion. Naturwissenschaften *67*, 567–568 (1980)
320. Ritchie, J.L., Hamilton, G.W., Wackers, F.J.T.: Thallium-201 myocardial imaging. New York, Raven 1978
321. Robinson, K.: Thallium in the United States mineral resources, in: Brobst, D.A., Pratt, W.P. (eds.). Geol. Surv. Prof. Paper *820*, 631. US Government Printing Office, Washington, D.C. (1973)
322. Rosen, G. von: Mutations induced by the action of metal ions in Pisum. Hereditas *43*, 644–664 (1957)
323. Roshchina, T.A.: Aerosols of magnesium based alloys and their hygienic evaluation. Gig. Tr. Prof. Zabol. 0(10), 9–13 (1980)
324. Roth, L.: Wassergefährdende Stoffe (Material Endangering Water). Landsberg, Ecomed. Verlagsgesellschaft 1982
325. Roth, L., Dauderer, H.: Giftliste (List of Poisons). 5. Erg. Lief., T *16/17*, Verlag Moderne Industrie 1979
326. Roth, M.: Teratogenic micromelia in the chick embryo: Osteovasculo-neural macrorelations. Folia Morphol. *23(3)*, 236–246 (1975)
327. Rusanov, A.K., Alekseeva, V.M.: Methods for the quantitative spectral analysis of minerals and solutions. VII. Direct determination of Tl, In, Ga in zinc blende. Zavodskaya Lab. *9(1)*, 66–69 (1940)
328. Rusnyak, I., Gyorgi, L., Ormai, S., Mullner, T.: Some potassium-like qualities of the thallium ion. Experimentia *24*, 809–810 (1968)
329. Ryabov, Y.F., Okunev, A.I., Kirr, L.D., Oshman, V.A.: Raspredelenie nekotorykh redkikh i rasseyannykh elementov v produktakh pererabotki mednykh rud i kontsentratov (Distribution of certain rare and disseminated elements in the products obtained from the processing of copper ores and concentrates). Byull. Tsvetnoi Met. *22*, 24–27 (1957)
330. Sanotskii, I.V.: Issledovanie toksichnosti soedinenii talliya (Toxicity of thallium carbonate, iodide, and bromide). Toksikol. Novykh. Prom. Khim. Veshchestv *2*, 94–104 (1961)
331. Schäfer, H., Baumbauer, H.U.: Elektronenmikroskopische Phasenanalyse eines thalliumhaltigen Reingasstaubes (Electron microscopic phase analysis of a thallium-rich fly ash). Beitr. elektronenmikroskop. Direktab. Oberfl. *15*, 325–329 (1982)
332. Schäfer, S.: Intestinal secretion of metals: Zn, Cd, Hg, Tl, Bi, Sr, and Pb. Naunyn-Schmiedeberg's Arch. Pharmacol. Suppl. *307*, R 4, 14 (1979)
333. Schaller, K.H., Manke, G., Raithel, H.J., Buehlmeyer, G., Schmidt, M., Valentin, H.: Investigations of thallium-exposed workers in cement factories. Int. Arch. Occup. Environ. Health. *47(3)*, 223–232 (1980)
334. Scharrer, E.: Histopathologische Befunde im Zentralnervensystem bei Thalliumvergiftung (Histopathologic findings on thallium poisoning in the central nervous system). Z. ges. Neurol. Psychiat. *145*, 454–470 (1953)
335. Schmid, R.: Grenzen der Entsorgung von Siedlungsabfällen auf landwirtschaftlichen Flächen (Limits of depositing of domestic wastes onto agricultural fields). Ber. Landwirtsch., Suppl. *197*, 97–104 (1981)
336. Schmidbauer, H., Klingler, D.: Chronische Thalliumvergiftung (Chronic thallium poisoning). Wien. Med. Wochenschr. *12*, 344–336 (1979)
337. Schoer, J.: Thallium – Gefahren und Toxikologie (Thallium – dangers and toxicology). Sich. Chem. Umwelt *2*, 5–9 (1982)

338. Schoer, J., Nagel, U.: Thallium in Pflanzen und Böden (Thallium in plants and soils). Untersuchungen im südlichen Rhein-Neckar-Kreis. Naturwissenschaften *67*, 261 (1980)
339. Scholl, W.: Bestimmung von Thallium in verschiedenen anorganischen und organischen Matrices – ein einfaches photometrisches Routineverfahren mit Brilliantgrün (Determination of thallium in various inorganic and organic matrices – a simple photometric routine method using Brilliant green). Landwirtsch. Forsch. *34*, SH 37, 275–286 (1981)
340. Scholl, G., Metzger, F.: Erhebungen über die Thalliumbelastung von Nutzpflanzen auf kontaminierten Böden im Raum Lengerich (Data on thallium contamination of agricultural products grown in contaminated soils near Lengerich). Landwirtsch. Forsch. Sonderh. *38*, Kongreßband, Trier 1981
341. Schulten, H.-R., Ziskoven, R.: Determination of thallium in the brain tissue of the mouse by stable isotope dilution and field desorption mass spectroscopy. J. Physiol. *284*, 170 (1978)
342. Schulten, H.-R., Lehmann, W.K., Ziskoven, R.: Determination of thallium in brain by stable isotope dilution and field desorption mass spectroscopy. Z. Naturforsch. (C) *33*, 484–487 (1978)
343. Schulz-Dobrick, B.: Verteilung von Cd, Tl und Bi in Sedimentgesteinen (Distribution of Cd, Tl, and Bi in sediment rocks). Fortschr. Mineral. *57(1)*, 222–223 (1979)
344. Schwedt, G., Rüssel, H.A.: Gaschromatographie von Tl, Se, Te, Hg, As, Sb, Bi und Sn als Phenylverbindungen (Gas-chromatographic investigations of Tl, Se, Te, Hg, As, Sb, Bi, and Sn as phenyl compounds) Z. Anal. Chem. *264*, 301–303 (1973)
345. Seeger, R., Gross, M.: Uptake and toxicity of thallium in higher fungi. 22nd Spring Meeting of Deutsche Pharmakologische Gesellschaft, Mainz, West Germany, March 10–13, 1981. Naunyn-Schmiedeberg's Arch. Pharmacol. (Suppl.) *316*, R 25 (1981)
346. Seeger, R., Gross, M.: Thallium in higher fungi. Z. Lebensm. Unters. Forsch. *173(1)*, 9–15 (1981)
347. Seiler, H.: Thallium, in: Analysen im biologischen Material. Analytische Methoden zur Prüfung gesundheitsschädlicher Arbeitsstoffe (Analyses in biological material. Analytical methods for testing matrials that affect health). Henschler, D. (ed.), Vol. 2, Weinheim, Verlag Chemie 1980
348. Shabalina, L.P., Spiridonova, V.S.: Toxicity of thallium formate malonate. Gig. Tr. Prof. Zabol. *8*, 35–37 (1976)
349. Shabalina, L.P., Spiridonova, V.S.: Thallium as an industrial poison. Review of literature. J. Hyg. Epidemiol. Microbiol. Immunol. *23(3)*, 247–255 (1979)
350. Shabalina, L.P., Spiridonova, V.S., Pavlenko, G.I., Fomenko, V.N.: Skin resorptive local and mutagenic activity of thallium carbonate. Gig. Tr. Prof. Zabol. *0(4)*, 54–55 (1980)
351. Shannon, R.D., Prewitt, C.T.: Effective ionic radii in oxides and fluorides. Acta Cryst. *25*, 925–946 (1969)
352. Sharma, R.P., Obersteiner, E.J.: Metals and neurotoxic effects. Cytotoxicity of selected metallic compounds on chick ganglia cultures. J. Comp. Pathol. *91(2)*, 235–244 (1981)
353. Shaw, D.M.: The geochemistry of thallium. Geochimica et Cosmochimica Acta *2*, 118–154 (1952)
354. Shaw, D.M.: The geochemistry of gallium, indium, thallium – a review, in: Ahrens, L.H. et al. (eds.) Physics and Chemistry of the Earth (Progress Series), Pergamon Press, London-New York-Paris, 1957
355. Shevchenko, V.P.: Bestimmung von Thallium in Naturwässern mittels inverser Voltammetrie mit vorausgehender Anreicherung durch Mitfällung (Determination of thallium in natural waters by means of inverse voltametry with preceding enrichment via coprecipitation). Z. Anal. Chem. *32*, 1448–1450 (1977)
356. Shkolnik, G.M., Bevill, F.B.: The determination of thallium in urine and plasma by delves cup atomic absorption. At. Absorpt. News *12*, 112–114 (1973)
357. Siegel, B.Z., Siegel, S.M., Speitel, T.: Selectivity in mercury copper and mercury iron accumulation in plants. Water Air Soil Pollut. *8(3)*, 285–292 (1977)
358. Siegel, S.M.: The cyto toxic response of nicotiana-tabacum protoplasts to metal ions: A survey of the chemical elements. Water Air Soil Pollut. *8*, 293–304 (1977)
359. Siegert, H.F.: Über das Vorkommen von Ga, In, Tl, Ge in Zinkerzen (On the occurrence of Ga, In, Tl, Ge in zinc ores). Ph.D. Thesis, Aachen (1959)
360. Sighinolfi, G.P.: Determination of thallium in geochemical reference samples by flameless atomic absorption spectroscopy. Atom. Absorpt. Newsl. *12(6)*, 136–138 (1973)
361. Sighinolfi, G.P., Santos, A.M.: Thallium in deep-seated crustal rocks. Geochim. Cosmochim. Acta *38*, 641–646 (1974)

362. Singh, N.P., Joselow, M.M.: Determination of thallium in whole blood by delves cup atomic absorption spectrophotometry. At. Absorpt. Newsl. *14*, 42–43 (1975)
363. Slepnev, Y.S.: The ratio of Tl to Rb, Cs, and K in the shists, granites and rare metal pegmatites of the Sayan Mountains. Geochem. Intern. *382*, 382–385 (1961)
364. Smith, I.O., Carson, B.L.: Trace Metals in the Environment, Vol. 1, Thallium, Ann Arbor, Ann Arbor Science Publ. Inc. 1977
365. Snell, F.D.: Photometric and fluorometric methods of analysis. Metals, Part 1. New York-Chichester-Brisbane-Toronto, John Wiley & Sons 1978
366. Solt, J., Paschinger, H., Broda, E.: Die energieabhängige Aufnahme von Thallium durch Chlorella (Energy-depending incorporation of thallium by Chlorella). Planta *101*, 242–250 (1971)
367. Specht, W., Rohner, D.: Bestimmung von Thalliumspuren in menschlichen Haaren mittels Neutronenaktivierungsanalyse (Analysing thalliumtraces in human hair by NAA). Arch. Toxikol. *18*, 359 (1960)
368. Specht, W., Rohner, D.: Über den Nachweis von Metallgiften (Thallium) durch Neutronenaktivierungsanalyse und Elektronenbeugung (Detection of metal poisons (thallium) by neutron activation analysis and electron diffraction). Acta Med. Leg. Soc. *17*, 27 (1964)
369. Spencer, E.L.: Frenching of tobacco and thallium toxicity. Amer. J. Botany *24*, 16–24 (1937)
370. Spencer, P.S., Peterson, E.R., Madrid, R.A., Raine, C.S.: Effects of thallium salts on neuronal mitochondria in organotypic cord-ganglion-muscle combination cultures. J. Cell Biol. *59*, 79–95 (1973)
371. Sprung, S.: Die chemische und mineralogische Zusammensetzung von Zementofenstaub (The chemical and mineralogical composition of cement furnace dust). Tonind.-Ztg. *90(10)*, 441–449 (1966)
372. Srivastava, S.K., Bhattacharjee, G., Sharma, A.K., Oberoi, C.K.: Use of inorganic gels in the removal of heavy metals. Water Res. *14(2)*, 113–116 (1980)
373. Steuer, W.: Ergebnisse der Thalliumuntersuchungen bei der Bevölkerung in Baden-Württemberg (Results of thallium studies on the population of Baden-Württemberg). Öff. Gesundh.-Wesen *42*, 477–479 (1980)
374. Stevens, W., van Peteghem, C., Heyndrick, A., Barbier, F.: Eleven cases of thallium intoxication treated with Prussian blue. Int. J. Clin. Pharmacol. Ther. Toxicol. *10*, 1–22 (1974)
375. Stoeppler, M., Bagschik, U., May, K.: Extraktionsstudien mit Tl-204 zur Ermittlung optimaler Arbeitsbedingungen für Tl-Bestimmung in biologischem Material (Extraction studies with Tl-204 for determining the optimal working conditions for the study of Tl in biological material). Z. Anal. Chem. *301*, 106–107 (1980)
376. Strümpler, A.W.: Trace element compositions in atmospheric particulates during 1973 and the summer of 1974 at Chadron, Nebr. Environ. Sci. Technol. *9*, 1164–1168 (1975)
377. Taber, P.: Chronic thallium poisoning, rapid diagnosis and treatment. J. Pediat. *65*, 461 (1964)
378. Taglia, V.: Isotachophoresis on paper. Part II: The separation of Ag(I), Tl(I), Hg2(2^+), and Pb(II). J. Chromatogr. *79*, 380–382 (1973)
379. Takuni, M.: Lösungsmittelextraktion von Silber- und Thalliumpikraten mit Poly- und Bis-Kronenethern (Solvent extractions of silver and thallium picrates with poly-, and biscrown ethers). Z. Anal. Chem. *298*, 363–366 (1979)
380. Tandon, S.P., Mishra, M.M.: Effect of some rare elements on nitrification by nitrate-forming bacteria in soil suspensions. Zentralb. Bakteriol. Parsitenk., Infektionskr. Hyg. Abt. II *122*, 155–157 (1968)
381. Thyresson, N.: Experimental investigation on thallium. Acta Dermatologica et Venereologica Scandinavica *30*, 417–441 (1950)
382. Thyresson, N.: Experimental investigation on thallium poisoning in rats: Distribution of thallium, especially in skin and excretion of thallium under different experimental conditions. Study with use of radioactive isotope Tl-204. Acta Dermatologica et Venereologica Scandinavica *31*,3–27 (1951)
383. Tichowa, T.S.: Thallium und seine Verbindungen (Thallium and its compounds), in: Nowyje dannije po toksikologii redkich metallow i jich sojedinenii, pp. 24–34. Moskow, Red. S.I. Israeljson 1967
384. Török, P., Schmall, W.: Umweltschadstoff Thalliumsulfat (Environmental pollutant thallium sulfate). Dtsch. Ärztebl. *79*, 31–32 (1982)

385. Toots, H., Parker, R.B.: Thallium in salt substitutes. A possible health hazard. Environ. Res. *14(3)*, 327–328 (1977)
386. Tooze, S.: A thallium affair. The Ecologist *10*, 163–164 (1980)
387. Trotman-Dickenson, A.F. (Exec. Ed.): Comprehensive Inorganic Chemistry: Thallium, pp. 1119–1172. Oxford-New York, Pergamon Press 1973
388. Truhaut, R.: Recherches sur la Toxicologie du Thallium (Studies on the toxicology of thallium). Institut National Securité pour la Prevention des Accidents du Travail, Paris 1959
389. Turekian, K.K.: Trace elements geochemistry. Trans. Am. Geophys. Un. *44*, 526–532 (1963)
390. Veenendaal, W.A., Polak, H.L.: Determination of thallium by atom absorption spectrophotometry. Z. Anal. Chem. *223*, 17–23 (1966)
391. Venugopal, B., Luckey, T.D.: Metal Toxicity in Mammals. Chapter 3: Thallium, pp. 122–127 (1978)
392. Vogel, H.U. von (ed.): Chemiker Kalender. Berlin-Heidelberg-New York, Springer 1966
393. Vogt, H.-H.: Historisches vom Thallium (Historical aspects of thallium). Chem. Lab. Betrieb *30*, 300–301 (1979)
394. Voskresenskaya, N.T.: Colorimetric methods of determining small amounts of thallium. J. Analyst. Chem. USSR *11*, 623–627 (1956)
395. Voskresenskaya, N.T.: Thallium in coal. Geochem. Intern. 158 (1968)
396. Voskresenskaya, N.T., Timofeyeva, N.V., Topkahana, M.: Thallium in some minerals and ores of sedimentary origin. Geochem. Intern. 851 (1962)
397. Wade, K., Banister, A.J.: Comprehensive Inorganic Chemistry, Vol. 1, Chapter 12. The Chemistry of Al, Ga, In, and Tl. New York, Pergamon Press 1973
398. Waggaman, W.H., Heffner, G.G., Gee, E.A.: Thallium. US Bureau Mines, Inf. Circ. *7553*, 50 (1950)
399. Wahler, W.: Pulse-polarographische Bestimmung der Spurenelemente Zn, Cd, In, Tl, Pb und Bi in 37 geochemischen Referenzproben nach Voranreicherung durch selektive Verdampfung (Pulse polarographic determination of the trace elements Zn, Cd, In, Tl, Pb, and Bi in 37 geochemical reference samples after pre-enrichment by selective volatilization). N. Jahrb. Mineral. Abh. *108*, 36 (1968)
400. Walk, H.: Die Gehalte der Schwermetalle Cadmium, Thallium, Blei und Wismut und weiterer Spurenelemente in natürlichen Böden und ihren Ausgangsgesteinen Südwestdeutschlands (Contents of the heavy metals cadmium, thallium, lead, and bismuth and other trace elements in natural soils and their forming rocks of Southwest Germany). Ph.D. Thesis, TH Karlsruhe (1982)
401. Wall, C.D.: The determination of thallium in urine by absorption spectroscopy and emission spectrography. Clin. Chim. Acta *76*, 259–265 (1977)
402. Ward, J.C.: Thallium poisoning in migratory birds. J. Am. Pharm. Assoc. *20*, 1272–1276 (1931)
403. Weast, R.C. (ed.): CRC Handbook of Chemistry and Physics, Boca Ratan (Florida), CRC Press 1980
404. Weinig, E., Schmidt, G.: Zur Verteilung des Thalliums im Organismus bei tödlichen Thalliumvergiftungen (On the distribution of thallium in the organism in mortal thallium poisoning). Arch. Toxicol. *21*, 199–215 (1966)
405. Weinig, E., Zink, P.: Über die quantitative massenspektrometrische Bestimmung des normalen Thallium-Gehalts im menschlichen Organismus (On the quantitative mass spectrometric determination of normal thallium contents in the human organism). Arch. Toxicol. *22*, 255–274 (1967)
406. Weinig, E., Walz, W.: Die Thalliumverteilung in Niere und Leber bei letalen Tl-Vergiftungen (Thallium distribution in the kidneys and liver in lethal Tl poisoning). Arch. Toxicol. *27*, 217–225 (1971)
407. Weissberg, B.G.: Gold-silver ore-grade precipitates from New Zealand thermal waters. Econ. Geol. *64*, 95 (1969)
408. Welz, B.: Atomic Absorption Spectroscopy. Weinheim, Verlag Chemie 1975
409. Welzel, K., Winkler, H.D.: Thalliumemissionen, Ursachen und interner Kreislauf (Thallium emissions, causes, and internal flow). Zement-Kalk-Gips *34*, 530–534 (1981)
410. Willson, J.E.: Thallotoxicosis. J. Am. Vet. Med. Ass. *139*, 1116 (1961)
411. Zasukhina, G.D., Krasovskii, G.N., Vasileva, I.M., Sdirkova, N.I., Sokolovskii, V.V., Kenesariev, U.I., Vasyukovich, L.Yu.: Molecular biological effects of thallium carbonate. Bull. Exp. Biol. Med. *90(12)*, 1731–1733 (1980)

412. Zasukhina, G.D., Vasil'eva, I.M., Sdirkova, N.I.: Determination of the mutagenic potential of environmental pollutants with special reference to the detection of the mutagenic action of thallium carbonate. Dokl. Biol. Sci. *250*, 16–18 (1980)
413. Zaun, H., Neumann, O., Werner, K.G.: Die Feinstruktur dystrophischer Haare bei Vitamin-A und Thallium-Intoxikation (The fine structure of dystrophic hair upon vitamin A and thallium intoxication). Hautarzt *23*, 544–550 (1972)
414. Zimmerley, S.R.: Thallium. Salt Lake City Division, Metallurgical Branch, US Bureau of Mines, Salt Lake City, Utah 1947
415. Ziskoven, R., Achenbach, C., Bahr, U., Schulten, H.-R.: Quantitative investigations of diaplacental transfer of thallium by field desorption mass spectrometry. Z. Naturforsch. *35*, 902–906 (1980)
416. Zitko, V.: Toxicity and pollution potential of thallium. Sci. Tot. Environ. *4*, 185–192 (1975)
417. Zitko, V., Carson, W.V.: Accumulation of thallium in clams and mussels. Bull. Environ. Contam. Toxicol. *14(5)*, 530–533 (1975)
418. Zitko, V., Carson, W.V., Carson, W.G.: Thallium occurrence in the environment and toxicity to fish. Bull. Environ. Contam. Toxicol. *13*, 23–30 (1975)
419. Zook, B.C., Gilmore, C.E.: Thallium poisoning in dogs. J. Am. Vet. Med. Ass. *151*, 206–217 (1967)
420. Zook, B.C., Holdsworth, J., Thornton, G.W.: Thallium poisoning in cats. J. Am. Vet. Med. Ass. *153*, 285–299 (1968)
421. Zyka, V.: Thallium in plants from Alsar. Sbornik Geologickych Ved. *10*, 91–96 (1972)

Subject Index

The Handbook of Environmental Chemistry

Editor: O. Hutzinger

Volume 1
Part A

The Natural Environment and the Biogeochemical Cycles

With contributions by *P. J. Craig, J. Emsley, D. J. Faulkner, P. M. Huang, E. A. Paul, M. Schidlowski, W. Stumm, J. C. G. Walker, P. J. Wangersky, J. Westall, A. J. B. Zehnder, S. H. Zinder*

1980. 54 figures, 59 tables. XII, 258 pages. ISBN 3-540-09688-4

Contents: The Atmosphere. – The Hydrosphere. – Chemical Oceanography. – Chemical Aspects of Soil. – The Oxygen Cycle. – The Sulfur Cycle. – The Phosphorus Cycle. – Metal Cycles and Biological Methylation. – Natural Organohalogen Compounds. – Subject Index.

Part B

The Natural Environment and the Biogeochemical Cycles

With contributions by *H.-J. Bolle, R. Fukai, J. W. de Leeuw, S. W. F. van der Ploeg, T. Rosswall, P. A. Schenck, R. Söderlund, Y. Yokoyama, A. J. B. Zehnder*

1982. 84 figures. XV, 317 pages. ISBN 3-540-11106-9

Contents: Basic Concepts of Ecology. – Natural Radionuclides in the Environment. – The Nitrogen Cycles. – The Carbon Cycle. – Molecular Organic Geochemistry. – Radiation and Energy Transport in the Earth Atmosphere System. – Subject Index.

Volume 2
Part A

Reactions and Processes

With contributions by *G. L. Baughman, L. A. Burns, A. S. Cohen, D. T. Gibson, T. E. Graedel, P. M. Huang, A. R. Isensee, K. Kranck, D. Mackay, C. M. Menzie, T. Mill, I. N. Morrison, H. Parlar, J. W. Winchester, V. Zitko*

1980. 66 figures, 27 tables. XVIII, 307 pages. ISBN 3-540-09689-2

Contents: Transport and Transformation of Chemicals: A Perspective. – Transport Processes in Air. – Solubility, Partition Coefficients, Volatility, and Evaporation Rates. – Adsorption Processes in Soil. – Sedimentation Processes in the Sea. – Chemical and Photo Oxidation. – Atmospheric Photochemistry. – Photochemistry at Surfaces and Interphases. – Microbial Metabolism. – Plant Uptake, Transport and Metabolism. – Metabolism and Distribution by Aquatic Animals. – Laboratory Microecosystems. – Reaction Types in the Environment. – Subject Index.

Part B

Reactions and Processes

With contributions by *W. A. Bruggeman, L. A. Burns, G. G. Choudhry, K. A. Daum, P. J. Gehring, L. W. Newland, D. F. Paris, D. V. Parke, R. H. Reitz, A. A. M. Roof, W. C. Steen, R. G. Zepp*

1982. 63 figures. XV, 205 pages. ISBN 3-540-11107-7

Contents: Basic Principles of Environmental Photochemistry. – Experimental Approaches to Environmental Photochemistry. – Aquatic Photochemistry. – Microbial Transformation Kinetics of Organic Compounds. – Hydrophobic Interactions in the Aquatic Environment. – Interactions of Humic Substances with Environmental Chemicals. – Complexing Effects on Behavior of Some Metals. – The Disposition and Metabolism of Environmental Chemicals by Mammalia. – Pharmacokinetic Models. – Subject Index.

Springer-Verlag
Berlin
Heidelberg
New York
Tokyo

The Handbook of Environmental Chemistry

Editor: **O. Hutzinger**

Volume 3
Part A

Anthropogenic Compounds

With contributions by *R. Anliker, G. C. Butler, E. A. Clarke, U. Förstner, W. Funke, C. Hyslop, G. Kaiser, C. Rappe, J. Russow, G. Tölg, M. Zander, V. Zitko*

1980. 61 figures, 73 tables. XIII, 274 pages.
ISBN 3-540-09690-6

Contents: Mercury. - Cadmium. - Polycyclic Aromatic and Heteroaromatic Hydrocarbons. - Fluorocarbons. - Chlorinated Paraffins. - Chloroaromatic Compounds Containing Oxygen. - Organic Dyes and Pigments. - Inorganic Pigments. - Radioactive Substances. - Subject Index.

Part B

Anthropogenic Compounds

With contributions by *K. J. Bock, K. A. Daum, E. Merian, L. W. Newland, C. R. Pearson, H. Stache, M. Zander*

1982. 38 figures. XVII, 210 pages. ISBN 3-540-11108-5

Contents: Lead. - Arsenic, Beryllium, Selenium and Vanadium. - C_1 and C_2 Halocarbons. - Halogenated Aromatics. - Volatile Aromatics. - Surfactants: Chemistry. Environment. - Subject Index.

This handbook is the first work that covers the chemical and physical behavior of compounds in the evironment.
Under the editorship of Prof. O. Hutzinger, formerly director of the Laboratory of Environmental and Toxicological Chemistry at the University of Amsterdam, now Ecological Chemistry and Geochemistry, University of Bayreuth, 65 international specialists have contributed to Parts A and B of the first three volumes:

- Volume 1: **The Natural Environment and the Biogeochemical Cycles**
- Volume 2: **Reactions and Processes**
- Volume 3: **Anthropogenic Compounds**

For a rapid publication of the material each volume was divided into several parts. Part A of the first three volumes appeared in 1980 and Part B in 1982. Each volume of Part B contains a cumulative subject index. More than 5064 literature references are cited. Future volumes are planned and will cover analytical chemistry, environmental engineering and toxicology.
The Handbook of Environmental Chemistry is a critical and complete outline of our present knowledge and will prove invaluable to environmental scientists, biologists, chemists (biochemists, agricultural and analytical chemists), medical scientists, occupational and environmental hygienists, research geologists and meteorologists as well as to industry and administrative bodies.

Springer-Verlag
Berlin
Heidelberg
New York
Tokyo

GPSR Compliance

The European Union's (EU) General Product Safety Regulation (GPSR) is a set of rules that requires consumer products to be safe and our obligations to ensure this.

If you have any concerns about our products, you can contact us on ProductSafety@springernature.com

In case Publisher is established outside the EU, the EU authorized representative is:

Springer Nature Customer Service Center GmbH
Europaplatz 3
69115 Heidelberg, Germany

Batch number: 10620881

Printed by Printforce, the Netherlands